Probability and Mathematical Statistics (Continued)

WILLIAMS • Diffusions, Markov Processes, and Martingales, Volume I: Foundations
ZACKS • Theory of Statistical Inference

Applied Probability and Statistics

ANDERSON, AUQUIER, HAUCK, OAKES, VANDAELE, and WEISBERG • Statistical Methods for Comparative Studies
ARTHANARI and DODGE • Mathematical Programming in Statistics
BAILEY • The Elements of Stochastic Processes with Applications to the Natural Sciences
BAILEY • Mathematics, Statistics and Systems for Health
BARNETT and LEWIS • Outliers in Statistical Data
BARTHOLEMEW • Stochastic Models for Social Processes, *Second Edition*
BARTHOLOMEW and FORBES • Statistical Techniques for Manpower Planning
BECK and ARNOLD • Parameter Estimation in Engineering and Science
BELSLEY, KUH, and WELSCH • Regression Diagnostics: Identifying Influential Data and Sources of Collinearity
BENNETT and FRANKLIN • Statistical Analysis in Chemistry and the Chemical Industry
BHAT • Elements of Applied Stochastic Processes
BLOOMFIELD • Fourier Analysis of Time Series: An Introduction
BOX • R. A. Fisher, The Life of a Scientist
BOX and DRAPER • Evolutionary Operation: A Statistical Method for Process Improvement
BOX, HUNTER, and HUNTER • Statistics for Experimenters: An Introduction to Design, Data Analysis, and Model Building
BROWN and HOLLANDER • Statistics: A Biomedical Introduction
BROWNLEE • Statistical Theory and Methodology in Science and Engineering, *Second Edition*
BURY • Statistical Models in Applied Science
CHAMBERS • Computational Methods for Data Analysis
CHATTERJEE and PRICE • Regression Analysis by Example
CHERNOFF and MOSES • Elementary Decision Theory
CHOW • Analysis and Control of Dynamic Economic Systems
CLELLAND, BROWN, and deCANI • Basic Statistics with Business Applications, *Second Edition*
COCHRAN • Sampling Techniques, *Third Edition*
COCHRAN and COX • Experimental Designs, *Second Edition*
CONOVER • Practical Nonparametric Statistics, *Second Edition*
CORNELL • Experiments with Mixtures: Designs, Models and The Analysis of Mixture Data
COX • Planning of Experiments
DANIEL • Biostatistics: A Foundation for Analysis in the Health Sciences, *Second Edition*
DANIEL • Applications of Statistics to Industrial Experimentation
DANIEL and WOOD • Fitting Equations to Data: Computer Analysis of Multifactor Data, *Second Edition*
DAVID • Order Statistics, *Second Editiion*
DEMING • Sample Design in Business Research
DODGE and ROMIG • Sampling Inspection Tables, *Second Edition*
DRAPER and SMITH • Applied Regression Analysis
DUNN • Basic Statistics: A Primer for the Biomedical Sciences, *Second Edition*
DUNN and CLARK • Applied Statistics: Analysis of Variance and Regression
ELANDT-JOHNSON • Probability Models and Statistical Methods in Genetics
ELANDT-JOHNSON and JOHNSON • Survival Models and Data Analysis

continued on back

Approximation Theorems of Mathematical Statistics

Approximation Theorems of Mathematical Statistics

ROBERT J. SERFLING

Professor of Mathematical Sciences
The Johns Hopkins University

John Wiley & Sons

New York • Chichester • Brisbane • Toronto

Library of Congress Cataloging in Publication Data:

Serfling, R. J.

Approximation theorems of mathematical statistics.

(Wiley series in probability and mathematical statistics)
Includes bibliographical references and indexes.
1. Mathematical statistics. 2. Limit theorems (Probability theory) I. Title.

QA276.S45 519.5 80-13493
ISBN 0-471-02403-1

Printed in the United States of America

10 9 8 7 6 5 4 3 2 1

To my parents and to
the memory of my wife's parents

Preface

This book covers a broad range of limit theorems useful in mathematical statistics, along with methods of proof and techniques of application. The manipulation of "probability" theorems to obtain "statistical" theorems is emphasized. It is hoped that, besides a knowledge of these basic statistical theorems, an appreciation on the instrumental role of probability theory and a perspective on practical needs for its further development may be gained.

A one-semester course each on probability theory and mathematical statistics at the beginning graduate level is presupposed. However, highly polished expertise is not necessary, the treatment here being self-contained at an elementary level. The content is readily accessible to students in statistics, general mathematics, operations research, and selected engineering fields.

Chapter 1 lays out a variety of tools and foundations basic to asymptotic theory in statistics as treated in this book. Foremost are: modes of convergence of a sequence of random variables (convergence in distribution, convergence in probability, convergence almost surely, and convergence in the rth mean); probability limit laws (the law of large numbers, the central limit theorem, and related results).

Chapter 2 deals systematically with the usual statistics computed from a sample: the sample distribution function, the sample moments, the sample quantiles, the order statistics, and cell frequency vectors. Properties such as asymptotic normality and almost sure convergence are derived. Also, deeper insights are pursued, including R. R. Bahadur's fruitful almost sure representations for sample quantiles and order statistics. Building on the results of Chapter 2, Chapter 3 treats the asymptotics of statistics concocted as transformations of vectors of more basic statistics. Typical examples are the sample coefficient of variation and the chi-squared statistic. Taylor series approximations play a key role in the methodology.

The next six chapters deal with important special classes of statistics. Chapter 4 concerns statistics arising in classical parametric inference and contingency table analysis. These include maximum likelihood estimates, likelihood ratio tests, minimum chi-square methods, and other asymptotically efficient procedures.

Chapter 5 is devoted to the sweeping class of W. Hoeffding's U-statistics, which elegantly and usefully generalize the notion of a sample mean. Basic convergence theorems, probability inequalities, and structural properties are derived. Introduced and applied here is the important "projection" method, for approximation of a statistic of arbitrary form by a simple sum of independent random variables.

Chapter 6 treats the class of R. von Mises' "differentiable statistical functions," statistics that are formulated as functionals of the sample distribution function. By differentiation of such a functional in the sense of the Gâteaux derivative, a reduction to an approximating statistic of simpler structure (essentially a U-statistic) may be developed, leading in a quite mechanical way to the relevant convergence properties of the statistical function. This powerful approach is broadly applicable, as most statistics of interest may be expressed either exactly or approximately as a "statistical function."

Chapters 7, 8, and 9 treat statistics obtained as solutions of equations ("M-estimates"), linear functions of order statistics ("L-estimates"), and rank statistics ("R-estimates"), respectively, three classes important in robust parametric inference and in nonparametric inference. Various methods, including the projection method introduced in Chapter 5 and the differential approach of Chapter 6, are utilized in developing the asymptotic properties of members of these classes.

Chapter 10 presents a survey of approaches toward asymptotic relative efficiency of statistical test procedures, with special emphasis on the contributions of E. J. G. Pitman, H. Chernoff, R. R. Bahadur, and W. Hoeffding.

To get to the end of the book in a one-semester course, some time-consuming material may be skipped without loss of continuity. For example, Sections 1.4, 1.11, 2.8, 3.6, and 4.3, and the proofs of Theorems 2.3.3C and 9.2.6A, B, C, may be so omitted.

This book evolved in conjunction with teaching such a course at The Florida State University in the Department of Statistics, chaired by R. A. Bradley. I am thankful for the stimulating professional environment conducive to this activity. Very special thanks are due D. D. Boos for collaboration on portions of Chapters 6, 7, and 8 and for many useful suggestions overall. I also thank J. Lynch, W. Pirie, R. Randles, I. R. Savage, and J. Sethuraman for many helpful comments. To the students who have taken this course with me, I acknowledge warmly that each has contributed a constructive impact on the development of this book. The support of the Office of Naval Research, which has sponsored part of the research in Chapters 5, 6, 7, 8, and 9 is acknowledged with appreciation. Also, I thank Mrs. Kathy

Strickland for excellent typing of the manuscript. Finally, most important of all, I express deep gratitude to my wife, Jackie, for encouragement without which this book would not have been completed.

ROBERT J. SERFLING

Baltimore, Maryland
September 1980

Contents

Approximation Theorems of Mathematical Statistics

CHAPTER 1

Preliminary Tools and Foundations

This chapter lays out tools and foundations basic to asymptotic theory in statistics as treated in this book. It is intended to reinforce previous knowledge as well as perhaps to fill gaps. As for actual proficiency, that may be gained in later chapters through the process of implementation of the material.

Of particular importance, Sections 1.2–1.7 treat notions of convergence of a sequence of random variables, Sections 1.8–1.11 present key probability limit theorems underlying the statistical limit theorems to be derived, Section 1.12 concerns differentials and Taylor series, and Section 1.15 introduces concepts of asymptotics of interest in the context of statistical inference procedures.

1.1 PRELIMINARY NOTATION AND DEFINITIONS

1.1.1 Greatest Integer Part

For x real, $[x]$ denotes the greatest integer less than or equal to x.

1.1.2 $O(\cdot)$, $o(\cdot)$, and $\sim$

These symbols are called "big oh," "little oh," and "twiddle," respectively. They denote ways of comparing the magnitudes of two functions $u(x)$ and $v(x)$ as the argument x tends to a limit L (not necessarily finite). The notation $u(x) = O(v(x))$, $x \to L$, denotes that $|u(x)/v(x)|$ remains bounded as $x \to L$. The notation $u(x) = o(v(x))$, $x \to L$, stands for

$$\lim_{x \to L} \frac{u(x)}{v(x)} = 0,$$

and the notation $u(x) \sim v(x)$, $x \to L$, stands for

$$\lim_{x \to L} \frac{u(x)}{v(x)} = 1.$$

Probabilistic versions of these "order of magnitude" relations are given in **1.2.6**, after introduction of some convergence notions.

Example. Consider the function

$$f(n) = 1 - \left(1 - \frac{1}{n}\right)\left(1 - \frac{2}{n}\right).$$

Obviously, $f(n) \to 0$ as $n \to \infty$. But we can say more. Check that

$$\begin{aligned} f(n) &= \frac{3}{n} + O(n^{-2}),\ n \to \infty, \\ &= \frac{3}{n} + o(n^{-1}),\ n \to \infty, \\ &\sim \frac{3}{n},\ n \to \infty. \quad ■ \end{aligned}$$

1.1.3 Probability Space, Random Variables, Random Vectors

In our discussions there will usually be (sometimes only implicitly) an underlying *probability space* $(\Omega, \mathscr{A}, P)$, where Ω is a set of points, $\mathscr{A}$ is a σ-field of subsets of Ω, and P is a probability distribution or measure defined on the elements of $\mathscr{A}$. A *random variable* $X(\omega)$ is a transformation of Ω into the real line R such that images $X^{-1}(B)$ of Borel sets B are elements of $\mathscr{A}$. A collection of random variables $X_1(\omega), X_2(\omega), \ldots$ on a given pair $(\Omega, \mathscr{A})$ will typically be denoted simply by $X_1, X_2, \ldots$. A *random vector* is a k-tuple $\mathbf{X} = (X_1, \ldots, X_k)$ of random variables defined on a given pair $(\Omega, \mathscr{A})$.

1.1.4 Distributions, Laws, Expectations, Quantiles

Associated with a random vector $\mathbf{X} = (X_1, \ldots, X_k)$ on $(\Omega. \mathscr{A}, P)$ is a right-continuous *distribution function* defined on R^k by

$$F_{X_1,\ldots,X_k}(t_1, \ldots, t_k) = P(\{\omega: X_1(\omega) \le t_1, \ldots, X_k(\omega) \le t_k\})$$

for all $\mathbf{t} = (t_1, \ldots, t_k) \in R^k$. This is also known as the *probability law* of $\mathbf{X}$. (There is also a left-continuous version.) Two random vectors $\mathbf{X}$ and $\mathbf{Y}$, defined on possibly different probability spaces, "have the same law" if their distribution functions are the same, and this is denoted by $\mathscr{L}(\mathbf{X}) = \mathscr{L}(\mathbf{Y})$, or $F_{\mathbf{X}} = F_{\mathbf{Y}}$.

By *expectation* of a random variable X is meant the Lebesgue–Stieltjes integral of $X(\omega)$ with respect to the measure P. Commonly used notations for this expectation are $E\{X\}$, EX, $\int_\Omega X(\omega)dP(\omega)$, $\int_\Omega X(\omega)P(d\omega)$, $\int X\,dP$, $\int X$, $\int_{-\infty}^{\infty} t\,dF_X(t)$, and $\int t\,dF_X$. All denote the same quantity. Expectation may also be represented as a Riemann–Stieltjes integral (see Cramér (1946), Sections 7.5 and 9.4). The expectation $E\{X\}$ is also called the *mean* of the random variable X. For a random *vector* $\mathbf{X} = (X_1, \ldots, X_k)$, the mean is defined as $E\{\mathbf{X}\} = (E\{X_1\}, \ldots, E\{X_k\})$.

Some important characteristics of random variables may be represented conveniently in terms of expectations, provided that the relevant integrals exist. For example, the *variance* of X is given by $E\{(X - E\{X\})^2\}$, denoted $\mathrm{Var}\{X\}$. More generally, the *covariance* of two random variables X and Y is given by $E\{(X - E\{X\})(Y - E\{Y\})\}$, denoted $\mathrm{Cov}\{X, Y\}$. (Note that $\mathrm{Cov}\{X, X) = \mathrm{Var}\{X\}$.) Of course, such an expectation may also be represented as a Riemann–Stieltjes integral,

$$\mathrm{Cov}\{X, Y\} = \iint (x - E\{X\})(y - E\{Y\})dF_{X,Y}(x, y).$$

For a random vector $\mathbf{X} = (X_1, \ldots, X_k)$, the *covariance matrix* is given by $\mathbf{\Sigma} = (\sigma_{ij})_{k \times k}$, where $\sigma_{ij} = \mathrm{Cov}\{X_i, X_j\}$.

For any univariate distribution function F, and for $0 < p < 1$, the quantity

$$F^{-1}(p) = \inf\{x: F(x) \geq p\}$$

is called the pth *quantile* or *fractile* of F. It is also denoted ξ_p. In particular, $\xi_{1/2} = F^{-1}(\frac{1}{2})$ is called the *median* of F.

The function $F^{-1}(t)$, $0 < t < 1$, is called the *inverse* function of F. The following proposition, giving useful properties of F and F^{-1}, is easily checked (Problem 1.P.1).

Lemma. *Let* F *be a distribution function. The function* $\mathrm{F}^{-1}(\mathrm{t})$, $0 < \mathrm{t} < 1$, *is nondecreasing and left-continuous, and satisfies*

(i) $\mathrm{F}^{-1}(\mathrm{F}(\mathrm{x})) \leq \mathrm{x}$, $-\infty < \mathrm{x} < \infty$,

and

(ii) $\mathrm{F}(\mathrm{F}^{-1}(\mathrm{t})) \geq \mathrm{t}$, $0 < \mathrm{t} < 1$.

Hence

(iii) $\mathrm{F}(\mathrm{x}) \geq \mathrm{t}$ *if and only if* $\mathrm{x} \geq \mathrm{F}^{-1}(\mathrm{t})$.

A further useful lemma, concerning the inverse functions of a weakly convergent sequence of distributions, is given in **1.5.6**.

1.1.5 $N(\mu, \sigma^2)$, $N(\mu, \Sigma)$

The *normal* distribution with mean μ and variance $\sigma^2 > 0$ corresponds to the distribution function

$$F(x) = \frac{1}{(2\pi)^{1/2}\sigma} \int_{-\infty}^{x} \exp\left[-\frac{1}{2}\left(\frac{t-\mu}{\sigma}\right)^2\right] dt, \qquad -\infty < x < \infty.$$

The notation $N(\mu, \sigma^2)$ will be used to denote either this distribution or a random variable having this distribution—whichever is indicated by the context. The special distribution function $N(0, 1)$ is known as the *standard normal* and is often denoted by Φ. In the case $\sigma^2 = 0$, $N(\mu, \sigma^2)$ will denote the distribution *degenerate* at μ, that is, the distribution

$$F(x) = \begin{cases} 0, & x < \mu, \\ 1, & x \geq \mu. \end{cases}$$

A random vector $\mathbf{X} = (X_1, \ldots, X_k)$ has the *k-variate normal* distribution with mean vector $\boldsymbol{\mu} = (\mu_1, \ldots, \mu_k)$ and covariance matrix $\boldsymbol{\Sigma} = (\sigma_{ij})_{k \times k}$ if, for every nonnull vector $\mathbf{a} = (a_1, \ldots, a_k)$, the random variable $\mathbf{aX}'$ is $N(\mathbf{a}\boldsymbol{\mu}', \mathbf{a}\boldsymbol{\Sigma}\mathbf{a}')$, that is, $\mathbf{aX}' = \sum_{i=1}^{k} a_i X_i$ has the normal distribution with mean $\mathbf{a}\boldsymbol{\mu}' = \sum_1^k a_i \mu_i$ and variance $\mathbf{a}\boldsymbol{\Sigma}\mathbf{a}' = \sum_{i=1}^{k} \sum_{j=1}^{k} a_i a_j \sigma_{ij}$. The notation $N(\boldsymbol{\mu}, \boldsymbol{\Sigma})$ will denote either this multivariate distribution or a random vector having this distribution.

The components X_i of a multivariate normal vector are seen to have (univariate) normal distributions. However, the converse does not hold. Random variables $X_1, \ldots, X_k$ may each be normal, yet possess a joint distribution which is *not* multivariate normal. Examples are discussed in Ferguson (1967), Section 3.2.

1.1.6 Chi-squared Distributions

Let $\mathbf{Z}$ be k-variate $N(\boldsymbol{\mu}, \mathbf{I})$, where $\mathbf{I}$ denotes the identity matrix of order k. For the case $\boldsymbol{\mu} = \mathbf{0}$, the distribution of $\mathbf{ZZ}' = \sum_1^k Z_i^2$ is called the *chi-squared with k degrees of freedom.* For the case $\boldsymbol{\mu} \neq \mathbf{0}$, the distribution is called *noncentral chi-squared with k degrees of freedom and noncentrality parameter* $\lambda = \boldsymbol{\mu}\boldsymbol{\mu}'$. The notation $\chi_k^2(\lambda)$ encompasses both cases and may denote either the random variable or the distribution. We also denote $\chi_k^2(0)$ simply by χ_k^2.

1.1.7 Characteristic Functions

The *characteristic function* of a random k-vector $\mathbf{X}$ is defined as

$$\phi_{\mathbf{X}(\mathbf{t})} = E\{e^{i\mathbf{tX}'}\} = \int \cdots \int e^{i\mathbf{tx}'}\, dF_{\mathbf{X}(\mathbf{x})}, \qquad \mathbf{t} \in R^k.$$

In particular, the characteristic function of $N(0, 1)$ is $\exp(-\frac{1}{2}t^2)$. See Lukacs (1970) for a full treatment of characteristic functions.

1.1.8 Absolutely Continuous Distribution Functions

An *absolutely continuous* distribution function F is one which satisfies

$$F(x) = \int_{-\infty}^{x} F'(t)dt, \qquad -\infty < x < \infty.$$

That is, F may be represented as the indefinite integral of its derivative. In this case, any function f such that $F(x) = \int_{-\infty}^{x} f(t)dt$, all x, is called a *density* for F. Any such density must agree with F' except possibly on a Lebesgue-null set. Further, if f is continuous at x_0, then $f(x_0) = F'(x_0)$ must hold. This latter may be seen by elementary arguments. For detailed discussion, see Natanson (1961), Chapter IX.

1.1.9 I.I.D.

With reference to a sequence $\{\mathbf{X}_i\}$ of random vectors, the abbreviation *I.I.D.* will stand for "independent and identically distributed."

1.1.10 Indicator Functions

For any set S, the associated *indicator function* is

$$I_S(x) = \begin{cases} 1, & x \in S, \\ 0, & x \notin S. \end{cases}$$

For convenience, the alternate notation $I(S)$ will sometimes be used for I_S, when the argument x is suppressed.

1.1.11 Binomial (n, p)

The *binomial* distribution with parameters n and p, where n is a positive integer and $0 \leq p \leq 1$, corresponds to the probability mass function

$$p(k) = \binom{n}{k} p^k(1 - p)^{n-k}, \qquad k = 0, 1, \ldots, n.$$

The notation $B(n, p)$ will denote either this distribution or a random variable having this distribution. As is well known, $B(n, p)$ is the distribution of the number of successes in a series of n independent trials each having success probability p.

1.1.12 Uniform (a, b)

The *uniform* distribution on the interval $[a, b]$, denoted $U(a, b)$, corresponds to the density function $f(x) = 1$, $a \leq x \leq b$, and $=0$, otherwise.

1.2 MODES OF CONVERGENCE OF A SEQUENCE OF RANDOM VARIABLES

Two forms of approximation are of central importance in statistical applications. In one form, a given random variable is approximated by another random variable. In the other, a given distribution function is approximated by another distribution function. Concerning the first case, three modes of convergence for a sequence of random variables are introduced in **1.2.1**, **1.2.2**, and **1.2.3**. These modes apply also to the second type of approximation, along with a fourth distinctive mode introduced in **1.2.4**. Using certain of these convergence notions, stochastic versions of the $O(\cdot)$, $o(\cdot)$ relations in **1.1.2** are introduced in **1.2.5**. A brief illustration of ideas is provided in **1.2.6**.

1.2.1 Convergence in Probability

Let $X_1, X_2, \ldots$ and X be random variables on a probability space $(\Omega, \mathscr{A}, P)$. We say that X_n *converges in probability* to X if

$$\lim_{n\to\infty} P(|X_n - X| < \varepsilon) = 1, \qquad \text{every} \quad \varepsilon > 0.$$

This is written $X_n \xrightarrow{p} X$, $n \to \infty$, or $p\text{-}\lim_{n\to\infty} X_n = X$. Examples are in **1.2.6**, Section 1.8, and later chapters. Extension to the case of $X_1, X_2, \ldots$ and X random elements of a metric space is straightforward, by replacing $|X_n - X|$ by the relevant metric (see Billingsley (1968)). In particular, for random k-vectors $\mathbf{X}_1, \mathbf{X}_2, \ldots$ and $\mathbf{X}$, we shall say that $\mathbf{X}_n \xrightarrow{p} \mathbf{X}$ if $\|\mathbf{X}_n - \mathbf{X}\| \xrightarrow{p} 0$ in the above sense, where $\|\mathbf{z}\| = (\sum_{i=1}^k z_i^2)^{1/2}$ for $\mathbf{z} \in R^k$. It then follows (Problem 1.P.2) that $\mathbf{X}_n \xrightarrow{p} \mathbf{X}$ if and only if the corresponding component-wise convergences hold.

1.2.2 Convergence with Probability 1

Consider random variables $X_1, X_2, \ldots$ and X on $(\Omega, \mathscr{A}, P)$. We say that X_n *converges with probability 1* (or *strongly, almost surely, almost everywhere*, etc.) to X if

$$P\left(\lim_{n\to\infty} X_n = X\right) = 1.$$

This is written $X_n \xrightarrow{wp1} X$, $n \to \infty$, or $p1\text{-}\lim_{n\to\infty} X_n = X$. Examples are in **1.2.6**, Section 1.9, and later chapters. Extension to more general random elements is straightforward.

An equivalent condition for convergence $wp1$ is

$$\lim_{n\to\infty} P(|X_m - X| < \varepsilon, \text{ all } m \ge n) = 1, \qquad \text{each} \quad \varepsilon > 0.$$

This facilitates comparison with convergence in probability. The equivalence is proved by simple set-theoretic arguments (Halmos (1950), Section 22), as follows. First check that

$$(*)\ \left\{\omega: \lim_{n\to\infty} X_n(\omega) = X(\omega)\right\} = \bigcap_{\varepsilon>0} \bigcup_{n=1}^{\infty} \{\omega: |X_m(\omega) - X(\omega)| < \varepsilon, \text{ all } m \geq n\},$$

whence

(**)

$$\left\{\omega: \lim_{n\to\infty} X_n(\omega) = X(\omega)\right\} = \lim_{\varepsilon\to 0} \lim_{n\to\infty} \{\omega: |X_m(\omega) - X(\omega)| < \varepsilon, \text{ all } m \geq n\}.$$

By the continuity theorem for probability functions (Appendix), (**) implies

$$P(X_n \to X) = \lim_{\varepsilon\to 0} \lim_{n\to\infty} P(|X_m - X| < \varepsilon, \text{all } m \geq n),$$

which immediately yields one part of the equivalence. Likewise, (*) implies, for any $\varepsilon > 0$,

$$P(X_n \to X) \leq \lim_{n\to\infty} P(|X_m - X| < \varepsilon, \text{ all } m \geq n),$$

yielding the other part.

The relation (*) serves also to establish that the set $\{\omega: X_n(\omega) \to X(\omega)\}$ truly belongs to $\mathscr{A}$, as is necessary for "convergence $wp1$" to be well defined.

A somewhat stronger version of this mode of convergence will be noted in **1.3.4.**

1.2.3 Convergence in *r*th Mean

Consider random variables $X_1, X_2, \ldots$ and X on $(\Omega, \mathscr{A}, P)$. For $r > 0$, we say that X_n *converges in* r*th mean* to X if

$$\lim_{n\to\infty} E|X_n - X|^r = 0.$$

This is written $X_n \xrightarrow{rth} X$ or $L_r\text{-}\lim_{n\to\infty} X_n = X$. The higher the value of r, the more stringent the condition, for an application of Jensen's inequality (Appendix) immediately yields

$$X_n \xrightarrow{rth} X \Rightarrow X_n \xrightarrow{sth} X, 0 < s < r.$$

Given $(\Omega, \mathscr{A}, P)$ and $r > 0$, denote by $L_r(\Omega, \mathscr{A}, P)$ the space of random variables Y such that $E|Y|^r < \infty$. The usual metric in L_r is given by $d(Y, Z) = \|Y - Z\|_r$, where

$$\|Y\|_r = \begin{cases} E|Y|^r, & 0 < r < 1, \\ [E|Y|^r]^{1/r}, & r \geq 1. \end{cases}$$

Thus convergence in the rth mean may be interpreted as convergence in the L_r metric, in the case of random variables $X_1, X_2, \ldots$ and X belonging to L_r.

1.2.4 Convergence in Distribution

Consider distribution functions $F_1(\cdot), F_2(\cdot), \ldots$ and $F(\cdot)$. Let $X_1, X_2, \ldots$ and X denote random variables (not necessarily on a common probability space) having these distributions, respectively. We say that X_n *converges in distribution* (or *in law*) to X if

$$\lim_{n\to\infty} F_n(t) = F(t), \text{ each continuity point } t \text{ of } F.$$

This is written $X_n \xrightarrow{d} X$, or $d\text{-}\lim_{n\to\infty} X_n = X$. A detailed examination of this mode of convergence is provided in Section 1.5. Examples are in **1.2.6**, Section 1.9, and later chapters.

The reader should figure out why this definition would *not* afford a satisfactory notion of approximation of a given distribution function by other ones if the convergence were required to hold for *all* t.

In as much as the definition of $X_n \xrightarrow{d} X$ is formulated wholly in terms of the corresponding distribution functions F_n and F, it is sometimes convenient to use the more direct notation "$F_n \Rightarrow F$" and the alternate terminology "F_n *converges weakly to* F." However, as in this book the discussions will tend to refer directly to various random variables under consideration, the notation $X_n \xrightarrow{d} X$ will be quite useful also.

Remark. The convergences $\xrightarrow{p}$, $\xrightarrow{wp1}$, and $\xrightarrow{rth}$ each represent a sense in which, for n sufficiently large, $X_n(\omega)$ and $X(\omega)$ approximate each other as *functions of* ω, $\omega \in \Omega$. This means that the *distributions* of X_n and X cannot be too dissimilar, whereby approximation in distribution should follow. On the other hand, the convergence $\xrightarrow{d}$ depends *only* on the distribution functions involved and does not necessitate that the relevant X_n and X approximate each other as functions of ω. In fact, X_n and X need not be defined on the same probability space. Section 1.3 deals formally with these interrelationships. ■

1.2.5 Stochastic $O(\cdot)$ and $o(\cdot)$

A sequence of random variables $\{X_n\}$, with respective distribution functions $\{F_n\}$, is said to be *bounded in probability* if for every $\varepsilon > 0$ there exist M_ε and N_ε such that

$$F_n(M_\varepsilon) - F_n(-M_\varepsilon) > 1 - \varepsilon \qquad \text{all} \quad n > N_\varepsilon.$$

The notation $X_n = O_p(1)$ will be used. It is readily seen that $X_n \xrightarrow{d} X \Rightarrow X_n = O_p(1)$ (Problem 1.P.3).

More generally, for two sequences of random variables $\{U_n\}$ and $\{V_n\}$, the notation $U_n = O_p(V_n)$ denotes that the sequence $\{U_n/V_n\}$ is $O_p(1)$. Further, the notation $U_n = o_p(V_n)$ denotes that $U_n/V_n \xrightarrow{p} 0$. Verify (Problem 1.P.4) that $U_n = o_p(V_n) \Rightarrow U_n = O_p(V_n)$.

1.2.6 Example: Proportion of Successes in a Series of Trials

Consider an infinite series of independent trials each having the outcome "success" with probability p. (The underlying probability space would be based on the set Ω of all infinite sequences ω of outcomes of such a series of trials.) Let X_n denote the proportion of successes in the first n trials. Then

(i) $X_n \xrightarrow{p} p$;

(ii) $X_n \xrightarrow{wp1} p$;

(iii) $$\frac{\sqrt{n}(X_n - p)}{[p(1-p)]^{1/2}} \xrightarrow{d} N(0, 1);$$

(iv) $$\frac{\sqrt{n}(X_n - p)}{(\log\log n)^{1/2}} \xrightarrow{p} 0;$$

(v) $$\frac{\sqrt{n}(X_n - p)}{(\log\log n)^{1/2}} \not\xrightarrow{wp1} 0;$$

(vi) $X_n \xrightarrow{2nd} p$.

Is it true that

(vii) $$\frac{\sqrt{n}(X_n - p)}{(\log\log n)^{1/2}} \xrightarrow{2nd} 0?$$

Justification and answers regarding (i)–(v) await material to be covered in Sections 1.8–1.10. Items (vi) and (vii) may be resolved at once, however, simply by computing variances (Problem 1.P.5).

1.3 RELATIONSHIPS AMONG THE MODES OF CONVERGENCE

For the four modes of convergence introduced in Section 1.2, we examine here the key relationships as given by direct implications (**1.3.1–1.3.3**), partial converses (**1.3.4–1.3.7**), and various counter-examples (**1.3.8**). The question of convergence of moments, which is related to the topic of convergence in rth mean, is treated in Section 1.4.

1.3.1 Convergence *wp*1 Implies Convergence in Probability

Theorem. *If* $X_n \xrightarrow{wp1} X$, *then* $X_n \xrightarrow{p} X$.

This is an obvious consequence of the equivalence noted in **1.2.2**. Incidentally, the proposition is not true in general for *all* measures (e.g., see Halmos (1950)).

1.3.2 Convergence in *r*th Mean Implies Convergence in Probability

Theorem. *If* $X_n \xrightarrow{rth} X$, *then* $X_n \xrightarrow{p} X$.

PROOF. Using the indicator function notation of **1.1.10** we have, for any $\varepsilon > 0$,

$$E|X_n - X|^r \geq E\{|X_n - X|^r I(|X_n - X| > \varepsilon)\} \geq \varepsilon^r P(|X_n - X| > \varepsilon)$$

and thus

$$P(|X_n - X| > \varepsilon) \leq \varepsilon^{-r} E|X_n - X|^r \to 0, \; n \to \infty. \quad \blacksquare$$

1.3.3 Convergence in Probability Implies Convergence in Distribution

(This will be proved in Section 1.5, but is stated here for completeness.)

1.3.4 Convergence in Probability Sufficiently Fast Implies Convergence *wp*1

Theorem. *If*

$$(*) \qquad \sum_{n=1}^{\infty} P(|X_n - X| > \varepsilon) < \infty \qquad \textit{for every} \quad \varepsilon > 0,$$

then $X_n \xrightarrow{wp1} X$.

PROOF. Let $\varepsilon > 0$ be given. We have

$$(**) \qquad P(|X_m - X| > \varepsilon \text{ for some } m \geq n) = P\left(\bigcup_{m=n}^{\infty} \{|X_m - X| > \varepsilon\}\right) \leq \sum_{m=n}^{\infty} P(|X_m - X| > \varepsilon).$$

Since the sum in (**) is the tail of a convergent series and hence $\to 0$ as $n \to \infty$, the alternate condition for convergence *wp*1 follows. ■

Note that the condition of the theorem defines a mode of convergence stronger than convergence *wp*1. Following Hsu and Robbins (1947), we say that X_n *converges completely* to X if (*) holds.

1.3.5 Convergence in rth Mean Sufficiently Fast Implies Convergence $wp1$

The preceding result, in conjunction with the proof of Theorem 1.3.2, yields

Theorem. *If* $\sum_{n=1}^{\infty} E|X_n - X|^r > \infty$, *then* $X_n \xrightarrow{wp1} X$.

The hypothesis of the theorem in fact yields the much stronger conclusion that the random series $\sum_{n=1}^{\infty} |X_n - X|^r$ converges $wp1$ (see Lukacs (1975), Section 4.2, for details).

1.3.6 Dominated Convergence in Probability Implies Convergence in Mean

Theorem. *Suppose that* $X_n \xrightarrow{p} X$, $|X_n| \le |Y|$ wp1 (*all* n), *and* $E|Y|^r < \infty$. *Then* $X_n \xrightarrow{rth} X$.

PROOF. First let us check that $|X| \le |Y| wp1$. Given $\delta > 0$, we have $P(|X| > |Y| + \delta) \le P(|X| > |X_n| + \delta) \le P(|X_n - X| > \delta) \to 0$, $n \to \infty$. Hence $|X| \le |Y| + \delta$ $wp1$ for any $\delta > 0$ and so for $\delta = 0$.

Consequently, $|X_n - X| \le |X| + |X_n| \le 2|Y| wp1$.

Now choose and fix $\varepsilon > 0$. Since $E|Y|^r < \infty$, there exists a finite constant $A_\varepsilon > \varepsilon$ such that $E\{|Y|^r I(2|Y| > A_\varepsilon)\} \le \varepsilon$. We thus have

$$\begin{aligned} E|X_n - X|^r &= E\{|X_n - X|^r I(|X_n - X| > A_\varepsilon)\} \\ &\quad + E\{|X_n - X|^r I(|X_n - X| \le \varepsilon)\} \\ &\quad + E\{|X_n - X|^r I(\varepsilon < |X_n - X| \le A_\varepsilon)\} \\ &\le E\{(|2Y|^r I(2|Y| > A_\varepsilon)\} + \varepsilon^r + A_\varepsilon^r P(|X_n - X| > \varepsilon) \\ &\le 2^r\varepsilon + \varepsilon^r + A_\varepsilon^r P(|X_n - X| > \varepsilon). \end{aligned}$$

Since $P(|X_n - X| > \varepsilon) \to 0$, $n \to \infty$, the right-hand side becomes less than $2^r\varepsilon + 2\varepsilon^r$ for all n sufficiently large. ■

More general theorems of this type are discussed in Section 1.4.

1.3.7 Dominated Convergence $wp1$ Implies Convergence in Mean

By **1.3.1** we may replace $\xrightarrow{p}$ by $\xrightarrow{wp1}$ in Theorem 1.3.5, obtaining

Theorem. *Suppose that* $X_n \xrightarrow{wp1} X$, $|X_n| \le |Y|$ wp1 (*all* n), *and* $E|Y|^r < \infty$. *Then* $X_n \xrightarrow{rth} X$.

1.3.8 Some Counterexamples

Sequences $\{X_n\}$ convergent in probability but *not* $wp1$ are provided in Examples A, B and C. The sequence in Example B is also convergent in mean square. A sequence convergent in probability but *not* in rth mean for any $r > 0$ is provided in Example D. Finally, to obtain a sequence convergent *wp*1 but *not* in rth mean for any $r > 0$, take an appropriate subsequence of the sequence in Example D (Problem 1.P.6). For more counterexamples, see Chung (1974), Section 4.1, and Lukacs (1975), Section 2.2, and see Section 2.1.

Example A. The usual textbook examples are versions of the following (Royden (1968), p. 92). Let $(\Omega, \mathscr{A}, P)$ be the probability space corresponding to Ω the interval $[0, 1]$, $\mathscr{A}$ the Borel sets in $[0, 1]$, and P the Lebesgue measure on $\mathscr{A}$. For each $n = 1, 2, \ldots$, let k_n and v_n satisfy $n = k_n + 2^{v_n}$, $0 \le k_n < 2^{v_n}$, and define

$$X_n(\omega) = \begin{cases} 1, & \text{if } \omega \in [k_n 2^{-v_n}, (k_n + 1)2^{-v_n}] \\ 0, & \text{otherwise.} \end{cases}$$

It is easily seen that $X_n \xrightarrow{p} 0$ yet $X_n(\omega) \to 0$ holds *nowhere*, $\omega \in [0, 1]$. ■

Example B. Let $Y_1, Y_2, \ldots$ be I.I.D. random variables with mean 0 and variance 1. Define

$$X_n = \frac{\sum_1^n Y_i}{(n \log \log n)^{1/2}}.$$

By the central limit theorem (Section 1.9) and theorems presented in Section 1.5, it is clear that $X_n \xrightarrow{p} 0$. Also, by direct computation, it is immediate that $X_n \xrightarrow{2\text{nd}} 0$. However, by the law of the iterated logarithm (Section 1.10), it is evident that $X_n(\omega) \to 0$, $n \to \infty$, only for ω in a set of probability 0. ■

Example C (contributed by J. Sethuraman). Let $Y_1, Y_2, \ldots$ be I.I.D. random variables. Define $X_n = Y_n/n$. Then clearly $X_n \xrightarrow{p} 0$. However, $X_n \xrightarrow{wp1} 0$ if and only if $E|Y_1| < \infty$. To verify this claim, we apply

Lemma (Chung (1974), Theorem 3.2.1). *For any positive random variable* Z,

$$\sum_{n=1}^{\infty} P(Z \ge n) \le E\{Z\} \le 1 + \sum_{n=1}^{\infty} P(Z \ge n).$$

Thus, utilizing the identical distributions assumption, we have

$$\sum_{n=1}^{\infty} P(|X_n| \ge \varepsilon) = \sum_{n=1}^{\infty} P(|Y_1| \ge n\varepsilon) \le \frac{1}{\varepsilon} E|Y_1|,$$

$$1 + \sum_{n=1}^{\infty} P(|X_n| \ge \varepsilon) = 1 + \sum_{n=1}^{\infty} P(|Y_1| \ge n\varepsilon) \ge \frac{1}{\varepsilon} E|Y_1|.$$

The result now follows, with the use of the independence assumption, by an application of the Borel–Cantelli lemma (Appendix). ■

Example D. Consider

$$X_n = \begin{cases} n, & \text{with probability } 1/\log n \\ 0, & \text{with probability } 1\text{-}1/\log n. \end{cases}$$

Clearly $X_n \xrightarrow{p} 0$. However, for any $r > 0$,

$$E|X_n|^r = \frac{n^r}{\log n} \to \infty. \quad \blacksquare$$

1.4 CONVERGENCE OF MOMENTS; UNIFORM INTEGRABILITY

Suppose that X_n converges to X in one of the senses $\xrightarrow{d}$, $\xrightarrow{p}$, $\xrightarrow{wp1}$ or $\xrightarrow{rth}$. What is implied regarding convergence of $E\{X_n^s\}$ to $E\{X^s\}$, or $E|X_n|^s$ to $E|X|^s$, $n \to \infty$? The basic answer is provided by Theorem A, in the general context of $\xrightarrow{d}$, which includes the other modes of convergence. Also, however, specialized results are provided for the cases $\xrightarrow{rth}$, $\xrightarrow{p}$, and $\xrightarrow{wp1}$. These are given by Theorems B, C, and D, respectively.

Before proceeding to these results, we introduce three special notions and examine their interrelationships. A sequence of random variables $\{Y_n\}$ is *uniformly integrable* if

$$\lim_{c \to \infty} \sup_n E\{|Y_n| I(|Y_n| > c)\} = 0.$$

A sequence of set functions $\{Q_n\}$ defined on $\mathscr{A}$ is *uniformly absolutely continuous* with respect to a measure P on $\mathscr{A}$ if, given $\varepsilon > 0$, there exists $\delta > 0$ such that

$$P(A) < \delta \Rightarrow \sup_n |Q_n(A)| < \varepsilon.$$

The sequence $\{Q_n\}$ is *equicontinuous at* ϕ if, given $\varepsilon > 0$ and a sequence $\{A_n\}$ in $\mathscr{A}$ decreasing to ϕ, there exists M such that

$$m > M \Rightarrow \sup_n |Q_n(A_m)| < \varepsilon.$$

Lemma A. (i) *Uniform integrability of* $\{Y_n\}$ *on* $(\Omega, \mathscr{A}, P)$ *is equivalent to the pair of conditions*

(a) $\sup_n E|Y_n| < \infty$

and

(b) *the set functions* $\{Q_n\}$ *defined by* $Q_n(A) = \int_A |Y_n| dP$ *are uniformly absolutely continuous with respect to* P.

(ii) *Sufficient for uniform integrability of* $\{Y_n\}$ *is that*

$$\sup_n E|Y_n|^{1+\varepsilon} < \infty$$

for some $\varepsilon > 0$.

(iii) *Sufficient for uniform integrability of* $\{Y_n\}$ *is that there be a random variable* Y *such that* $E|Y| < \infty$ *and*

$$P(|Y_n| \geq y) \leq P(|Y| \geq y), \textit{ all } n \geq 1, \textit{ all } y > 0.$$

(iv) *For set functions* Q_n *each absolutely continuous with respect to a measure* P, *equicontinuity at* ϕ *implies uniform absolute continuity with respect to* P.

PROOF. (i) Chung (1974), p. 96; (ii) note that

$$E\{|Y_n|I(|Y_n| > c)\} \leq c^{-\varepsilon}E|Y_n|^{1+\varepsilon};$$

(iii) Billingsley (1968), p. 32; (iv) Kingman and Taylor (1966), p. 178. ■

Theorem A. *Suppose that* $X_n \xrightarrow{d} X$ *and the sequence* $\{X_n^r\}$ *is uniformly integrable, where* $r > 0$. *Then* $E|X|^r < \infty$, $\lim_n E\{X_n^r\} = E\{X^r\}$, *and* $\lim_n E|X_n|^r = E|X|^r$.

PROOF. Denote the distribution function of X by F. Let $\varepsilon > 0$ be given. Choose c such that $\pm c$ are continuity points of F and, by the uniform integrability, such that

$$\sup_n E\{|X_n|^r I(|X_n| \geq c)\} < \varepsilon.$$

For any $d > c$ such that $\pm d$ are also continuity points of F, we obtain from the second theorem of Helly (Appendix) that

$$\lim_{n\to\infty} E\{|X_n|^r I(c \leq |X_n| \leq d)\} = E\{|X|^r I(c \leq |X| \leq d)\}.$$

It follows that $E\{|X|^r I(c \leq |X| \leq d)\} < \varepsilon$ for all such choices of d. Letting $d \to \infty$, we obtain $E\{|X|^r I(|X| \geq c)\} < \varepsilon$, whence $E|X|^r < \infty$.

Now, for the same c as above, write

$$\begin{aligned}|E\{X_n^r\} - E\{X^r\}| \leq |E\{X_n^r I(|X_n| \leq c)\} - E\{X^r I(|X| \leq c)\}| \\ + E\{|X_n|^r I(|X_n| > c)\} + E\{|X|^r I(|X| > c)\}.\end{aligned}$$

By the Helly theorem again, the first term on the right-hand side tends to 0 as $n \to \infty$. The other two terms on the right are each less than ε. Thus $\lim_n E\{X_n^r\} = E\{X^r\}$. A similar argument yields $\lim_n E|X_n|^r = E|X|^r$. ■

By arguments similar to the preceding, the following partial converse to Theorem A may be obtained (Problem 1.P.7).

Lemma B. *Suppose that* $X_n \xrightarrow{d} X$ *and* $\lim_n E|X_n|^r = E|X|^r < \infty$. *Then the sequence* $\{X_n^r\}$ *is uniformly integrable.*

We now can easily establish a simple theorem apropos to the case $\xrightarrow{rth}$.

Theorem B. *Suppose that* $X_n \xrightarrow{rth} X$ *and* $E|X|^r < \infty$. *Then* $\lim_n E\{X_n^r\} = E\{X^r\}$ *and* $\lim_n E|X_n|^r = E|X|^r$.

PROOF. For $0 < r \leq 1$, apply the inequality $|x + y|^r \leq |x|^r + |y|^r$ to write $||x|^r - |y|^r| \leq |x - y|^r$ and thus

$$|E|X_n|^r - E|X|^r| \leq E|X_n - X|^r.$$

For $r > 1$, apply Minkowski's inequality (Appendix) to obtain

$$|(E|X_n|^r)^{1/r} - (E|X|^r)^{1/r}| \leq (E|X_n - X|^r)^{1/r}.$$

In either case, $\lim_n E|X_n|^r = E|X|^r < \infty$ follows. Therefore, by Lemma B, $\{X_n^r\}$ is uniformly integrable. Hence, by Theorem A, $\lim_n E\{X_n^r\} = E\{X^r\}$ follows. ■

Next we present results oriented to the case $\xrightarrow{p}$.

Lemma C. *Suppose that* $X_n \xrightarrow{p} X$ *and* $E|X_n|^r < \infty$, *all* n. *Then the following statements hold.*

(i) $X_n \xrightarrow{rth} X$ *if and only if the sequence* $\{X_n^r\}$ *is uniformly integrable.*

(ii) *If the set functions* $\{Q_n\}$ *defined by* $Q_n(A) = \int_A |X_n|^r \, dP$ *are equicontinuous at* ϕ, *then* $X_n \xrightarrow{rth} X$ *and* $E|X|^r < \infty$.

PROOF. (i) see Chung (1974), pp. 96–97; (ii) see Kingman and Taylor (1966), pp. 178–180. ■

It is easily checked (Problem 1.P.8) that each of parts (i) and (ii) generalizes Theorem 1.3.6.

Combining Lemma C with Theorem B and Lemma A, we have

Theorem C. *Suppose that* $X_n \xrightarrow{p} X$ *and that either*

(i) $E|X|^r < \infty$ *and* $\{X_n^r\}$ *is uniformly integrable,*

or

(ii) $\sup_n E|X_n|^r < \infty$ *and the set functions* $\{Q_n\}$ *defined by* $Q_n(A) = \int_A |X_n|^r \, dP$ *are equicontinuous at* ϕ.

Then $\lim_n E\{X_n^r\} = E\{X^r\}$ *and* $\lim_n E|X_n|^r = E|X|^r$.

Finally, for the case $\xrightarrow{wp1}$, the preceding result may be used; but also, by a simple application (Problem 1.P.9) of Fatou's lemma (Appendix), the following is easily obtained.

Theorem D. *Suppose that* $X_n \xrightarrow{wp1} X$. *If* $\overline{\lim}_n E|X_n|^r \le E|X|^r < \infty$, *then* $\lim_n E\{X_n^r\} = E\{X^r\}$ *and* $\lim_n E|X_n|^r = E|X|^r$.

As noted at the outset of this section, the fundamental result on convergence of moments is provided by Theorem A, which imposes a uniform integrability condition. For practical implementation of the theorem, Lemma A (i), (ii), (iii) provides various sufficient conditions for uniform integrability. Justification for the trouble of verifying uniform integrability is provided by Lemma B, which shows that the uniform integrability condition is essentially necessary.

1.5 FURTHER DISCUSSION OF CONVERGENCE IN DISTRIBUTION

This mode of convergence has been treated briefly in Sections 1.2–1.4. Here we provide a collection of basic facts about it. Recall that the definition of $X_n \xrightarrow{d} X$ is expressed in terms of the corresponding distribution functions F_n and F, and that the alternate notation $F_n \Rightarrow F$ is often convenient. The reader should formulate "convergence in distribution" for random *vectors.*

1.5.1 Criteria for Convergence in Distribution

The following three theorems provide *methodology* for establishing convergence in distribution.

Theorem A. *Let the distribution functions* $F, F_1, F_2, \ldots$ *possess respective characteristic functions* $\phi, \phi_1, \phi_2, \ldots$. *The following statements are equivalent:*

(i) $F_n \Rightarrow F$;

(ii) $\lim_n \phi_n(t) = \phi(t)$, *each real* t;

(iii) $\lim_n \int g\, dF_n = \int g\, dF$, *each bounded continuous function* g.

PROOF. That (i) implies (iii) is given by the generalized Helly theorem (Appendix). We now show the converse. Let t be a continuity point of F and let $\varepsilon > 0$ be given. Take any continuous function g satisfying $g(x) = 1$ for $x \le t$, $0 \le g(x) \le 1$ for $t < x < t + \varepsilon$, and $g(x) = 0$ for $x \ge t + \varepsilon$. Then, assuming (iii), we obtain (Problem 1.P.10)

$$\overline{\lim_{n\to\infty}} F_n(t) \le F(t + \varepsilon).$$

Similarly, (iii) also gives

$$\varliminf_{n\to\infty} F_n(t) \geq F(t - \varepsilon).$$

Thus (i) follows.

For proof that (i) and (ii) are equivalent, see Gnedenko (1962), p. 285. ■

Example. If the characteristic function of a random variable X_n tends to the function $\exp(-\frac{1}{2}t^2)$ as $n \to \infty$, then $X_n \xrightarrow{d} N(0, 1)$. ■

The multivariate version of Theorem A is easily formulated.

Theorem B (Fréchet and Shohat). *Let the distribution functions* F_n *possess finite moments* $\alpha_k^{(n)} = \int t^k \, dF_n(t)$ *for* $k = 1, 2, \ldots$ *and* $n = 1, 2, \ldots$. *Assume that the limits* $\alpha_k = \lim_n \alpha_k^{(n)}$ *exist (finite), each* k. *Then*

(i) *the limits* $\{\alpha_k\}$ *are the moments of a distribution function* F;
(ii) *if the* F *given by* (i) *is unique, then* $F_n \Rightarrow F$.

For proof, see Fréchet and Shohat (1931), or Loève (1977), Section 11.4. This result provides a convergence of moments criterion for convergence in distribution. In implementing the criterion, one would also utilize Theorem 1.13, which provides conditions under which the moments $\{\alpha_k\}$ determine a unique F.

The following result, due to Scheffé (1947), provides a convergence of densities criterion. (See Problem 1.P.11.)

Theorem C (Scheffé). *Let* $\{f_n\}$ *be a sequence of densities of absolutely continuous distribution functions, with* $\lim_n f_n(x) = f(x)$, *each real* x. *If* f *is a density function, then* $\lim_n \int |f_n(x) - f(x)| dx = 0$.

PROOF. Put $g_n(x) = [f(x) - f_n(x)]I(f(x) \geq f_n(x))$, each x. Using the fact that f is a density, check that

$$\int |f_n(x) - f(x)| dx = 2 \int g_n(x) dx.$$

Now $|g_n(x)| \leq f(x)$, all x, each n. Hence, by dominated convergence (Theorem 1.3.7), $\lim_n \int g_n(x) dx = 0$. ■

1.5.2 Reduction of Multivariate Case to Univariate Case

The following result, due to Cramér and Wold (1936), allows the question of convergence of multivariate distribution functions to be reduced to that of convergence of univariate distribution functions.

Theorem. *In* R^k, *the random vectors* $\mathbf{X}_n$ *converge in distribution to the random vector* $\mathbf{X}$ *if and only if each linear combination of the components of* $\mathbf{X}_n$ *converges in distribution to the same linear combination of the components of* $\mathbf{X}$.

PROOF. Put $\mathbf{X}_n = (X_{n1}, \ldots, X_{nk})$ and $\mathbf{X} = (X_1, \ldots, X_k)$ and denote the corresponding characteristic functions by ϕ_n and ϕ. Assume now that for any real $\lambda_1, \ldots, \lambda_k$,

$$\lambda_1 X_{n1} + \cdots + \lambda_k X_{nk} \xrightarrow{d} \lambda_1 X_1 + \cdots + \lambda_k X_k.$$

Then, by Theorem 1.5.1A,

$$\lim_{n\to\infty} \phi_n(t\lambda_1, \ldots, t\lambda_k) = \phi(t\lambda_1, \ldots, t\lambda_k), \text{ all } t.$$

With $t = 1$, and since $\lambda_1, \ldots, \lambda_k$ are arbitrary, it follows by the multivariate version of Theorem 1.5.1A that $\mathbf{X}_n \xrightarrow{d} \mathbf{X}$.

The converse is proved by a similar argument. ■

Some extensions due to Wald and Wolfowitz (1944) and to Varadarajan (1958) are given in Rao (1973), p. 128. Also, see Billingsley (1968), p. 49, for discussion of this "Cramer–Wold device."

1.5.3 Uniformity of Convergence in Distribution

An important question regarding the weak convergence of F_n to F is whether the pointwise convergences hold uniformly. The following result is quite useful.

Theorem (Pólya). *If* $\mathrm{F}_n \Rightarrow \mathrm{F}$ *and* F *is continuous, then*

$$\lim_{n\to\infty} \sup_t |\mathrm{F}_n(t) - \mathrm{F}(t)| = 0.$$

The proof is left as an exercise (Problem 1.P.12). For generalities, see Ranga Rao (1962).

1.5.4 Convergence in Distribution for Perturbed Random Variables

A common situation in mathematical statistics is that the statistic of interest is a slight modification of a random variable having a known limit distribution. A fundamental role is played by the following theorem, which was developed by Slutsky (1925) and popularized by Cramér (1946). Note that no restrictions are imposed on the possible dependence among the random variables involved.

Theorem (Slutsky). *Let* $X_n \xrightarrow{d} X$ *and* $Y_n \xrightarrow{p} c$, *where* c *is a finite constant. Then*

(i) $X_n + Y_n \xrightarrow{d} X + c$;

(ii) $X_n Y_n \xrightarrow{d} cX$;

(iii) $X_n/Y_n \xrightarrow{d} X/c$ *if* $c \neq 0$.

Corollary A. *Convergence in probability,* $X_n \xrightarrow{p} X$, *implies convergence in distribution,* $X_n \xrightarrow{d} X$.

Corollary B. *Convergence in probability to a constant is equivalent to convergence in distribution to the given constant.*

Note that Corollary A was given previously in **1.3.3**. The method of proof of the theorem is demonstrated sufficiently by proving (i). The proofs of (ii) and (iii) and of the corollaries are left as exercises (see Problems 1.P.13–14).

PROOF OF (i). Choose and fix t such that $t - c$ is a continuity point of F_X. Let $\varepsilon > 0$ be such that $t - c + \varepsilon$ and $t - c - \varepsilon$ are also continuity points of F_X. Then

$$\begin{aligned} F_{X_n+Y_n}(t) &= P(X_n + Y_n \le t) \\ &\le P(X_n + Y_n \le t, |Y_n - c| < \varepsilon) + P(|Y_n - c| \ge \varepsilon) \\ &\le P(X_n \le t - c + \varepsilon) + P(|Y_n - c| \ge \varepsilon). \end{aligned}$$

Hence, by the hypotheses of the theorem, and by the choice of $t - c + \varepsilon$,

$$(*)\qquad \begin{aligned} \overline{\lim_n}\, F_{X_n+Y_n}(t) &\le \overline{\lim_n}\, P(X_n \le t - c + \varepsilon) + \overline{\lim_n}\, P(|Y_n - c| \ge \varepsilon) \\ &= F_X(t - c + \varepsilon). \end{aligned}$$

Similarly,

$$P(X_n \le t - c - \varepsilon) \le P(X_n + Y_n \le t) + P(|Y_n - c| \ge \varepsilon)$$

and thus

$$(**)\qquad F_X(t - c - \varepsilon) \le \underline{\lim_n}\, F_{X_n+Y_n}(t).$$

Since $t - c$ is a continuity point of F_X, and since ε may be taken arbitrarily small, (*) and (**) yield

$$\lim_n F_{X_n+Y_n}(t) = F_X(t - c) = F_{X+c}(t). \quad \blacksquare$$

1.5.5 Asymptotic Normality

The most important special case of convergence in distribution consists of convergence to a normal distribution. A sequence of random variables $\{X_n\}$ converges in distribution to $N(\mu, \sigma^2)$, $\sigma > 0$, if equivalently, the sequence $\{(X_n - \mu)/\sigma\}$ converges in distribution to $N(0, 1)$. (Verify by Slutsky's Theorem.)

More generally, a sequence of random variables $\{X_n\}$ is *asymptotically normal* with "mean" μ_n and "variance" σ_n^2 if $\sigma_n > 0$ for all n sufficiently large and

$$\frac{X_n - \mu_n}{\sigma_n} \xrightarrow{d} N(0, 1).$$

We write "X_n is $AN(\mu_n, \sigma_n^2)$." Here $\{\mu_n\}$ and $\{\sigma_n\}$ are sequences of constants. It is not necessary that μ_n and σ_n^2 be the mean and variance of X_n, nor even that X_n possess such moments. Note that if X_n is $AN(\mu_n, \sigma_n^2)$, it does not necessarily follow that $\{X_n\}$ converges in distribution to anything. Nevertheless in any case we have (show why)

$$\sup_t |P(X_n \le t) - P(N(\mu_n, \sigma_n^2) \le t)| \to 0, \qquad n \to \infty,$$

so that for a range of probability calculations we may treat X_n as a $N(\mu_n, \sigma_n^2)$ random variable.

As exercises (Problems 1.P.15–16), prove the following useful lemmas.

Lemma A. *If* X_n *is* $AN(\mu_n, \sigma_n^2)$, *then also* X_n *is* $AN(\bar{\mu}_n, \bar{\sigma}_n^2)$ *if and only if*

$$\frac{\bar{\sigma}_n}{\sigma_n} \to 1, \quad \frac{\bar{\mu}_n - \mu_n}{\sigma_n} \to 0.$$

Lemma B. *If* X_n *is* $AN(\mu_n, \sigma_n^2)$, *then also* $a_n X_n + b_n$ *is* $AN(\mu_n, \sigma_n^2)$ *if and only if*

$$a_n \to 1, \quad \frac{\mu_n(a_n - 1) + b_n}{\sigma_n} \to 0.$$

Example. If X_n is $AN(n, 2n)$, then so is

$$\frac{n-1}{n} X_n$$

but not

$$\frac{\sqrt{n} - 1}{\sqrt{n}} X_n. \quad \blacksquare$$

We say that a sequence of random *vectors* $\{\mathbf{X}_n\}$ is *asymptotically (multivariate) normal* with "mean vector" $\boldsymbol{\mu}_n$ and "covariance matrix" $\boldsymbol{\Sigma}_n$ if $\boldsymbol{\Sigma}_n$ has nonzero diagonal elements for all n sufficiently large, and for every vector $\boldsymbol{\lambda}$ such that $\boldsymbol{\lambda}\boldsymbol{\Sigma}_n\boldsymbol{\lambda}' > 0$ for all n sufficiently large, the sequence $\boldsymbol{\lambda}\mathbf{X}_n'$ is $AN(\boldsymbol{\lambda}\boldsymbol{\mu}_n', \boldsymbol{\lambda}\boldsymbol{\Sigma}_n\boldsymbol{\lambda}')$. We write "$\mathbf{X}_n$ is $AN(\boldsymbol{\mu}_n, \boldsymbol{\Sigma}_n)$." Here $\{\boldsymbol{\mu}_n\}$ is a sequence of vector constants and $\{\boldsymbol{\Sigma}_n\}$ a sequence of covariance matrix constants. As an exercise (Problem 1.P.17), show that $\mathbf{X}_n$ is $AN(\boldsymbol{\mu}_n, c_n^2\boldsymbol{\Sigma})$ if and only if

$$\frac{\mathbf{X}_n - \boldsymbol{\mu}_n}{c_n} \xrightarrow{d} N(\mathbf{0}, \boldsymbol{\Sigma}).$$

Here $\{c_n\}$ is a sequence of real constants and $\boldsymbol{\Sigma}$ a covariance matrix.

1.5.6 Inverse Functions of Weakly Convergent Distributions

The following result will be utilized in Section 1.6 in proving Theorem 1.6.3.

Lemma. *If* $F_n \Rightarrow F$, *then the set*

$$\{t: 0 < t < 1, F_n^{-1}(t) \nrightarrow F^{-1}(t), n \to \infty\}$$

contains at most countably many elements.

PROOF. Let $0 < t_0 < 1$ be such that $F_n^{-1}(t_0) \nrightarrow F^{-1}(t_0)$, $n \to \infty$. Then there exists an $\varepsilon > 0$ such that $F^{-1}(t_0) \pm \varepsilon$ are continuity points of F and $|F_n^{-1}(t_0) - F^{-1}(t_0)| > \varepsilon$ for infinitely many $n = 1, 2, \ldots$. Suppose that $F_n^{-1}(t_0) < F^{-1}(t_0) - \varepsilon$ for infinitely many n. Then, by Lemma 1.1.4(ii), $t_0 \le F_n(F_n^{-1}(t_0)) \le F_n(F^{-1}(t_0) - \varepsilon)$. Thus the convergence $F_n \Rightarrow F$ yields $t_0 \le F(F^{-1}(t_0) - \varepsilon)$, which in turn yields, by Lemma 1.1.4(i), $F^{-1}(t_0) \le F^{-1}(F(F^{-1}(t_0) - \varepsilon)) \le F^{-1}(t_0) - \varepsilon$, a contradiction. Therefore, we must have

$$F_n^{-1}(t_0) > F^{-1}(t_0) + \varepsilon \qquad \text{for infinitely many} \quad n = 1, 2, \ldots.$$

By Lemma 1.1.4(iii), this is equivalent to

$$F_n(F^{-1}(t_0) + \varepsilon) < t_0 \qquad \text{for infinitely many} \quad n = 1, 2, \ldots,$$

which by the convergence $F_n \Rightarrow F$ yields $F(F^{-1}(t_0) + \varepsilon) \le t_0$. But also $t_0 \le F(F^{-1}(t_0))$, by Lemma 1.1.4(i). It follows that

$$t_0 = F(F^{-1}(t_0))$$

and that

$$F(x) = t_0 \qquad \text{for} \quad x \in [F^{-1}(t_0), F^{-1}(t_0) + \varepsilon],$$

that is, that F is flat in a right neighborhood of $F^{-1}(t_0)$. We have thus shown a one-to-one correspondence between the elements of the set $\{t: 0 < t < 1, F_n^{-1}(t) \nrightarrow F^{-1}(t), n \to \infty\}$ and a subset of the flat portions of F. Since (justify) there are at most countably many flat portions, the proof is complete. ■

1.6 OPERATIONS ON SEQUENCES TO PRODUCE SPECIFIED CONVERGENCE PROPERTIES

Here we consider the following question: given a sequence $\{X_n\}$ which is convergent in some sense *other* than *wp*1, is there a closely related sequence $\{X_n^*\}$ which *retains* the convergence properties of the original sequence but *also* converges *wp*1? The question is answered in three parts, corresponding respectively to postulated convergence in probability, in rth mean, and in distribution.

1.6.1 Conversion of Convergence in Probability to Convergence *wp*1

A standard result of measure theory is the following (see Royden (1968), p. 230).

Theorem. *If* $X_n \xrightarrow{p} X$, *then there exists a subsequence* X_{n_k} *such that* $X_{n_k} \xrightarrow{wp1} X$, $k \to \infty$.

Note that this is merely an *existence* result. For implications of the theorem for statistical purposes, see Simons (1971).

1.6.2 Conversion of Convergence in rth Mean to Convergence *wp*1

Consider the following question: given that $X_n \xrightarrow{rth} 0$, under what circumstances does the "smoothed" sequence

$$X_n^* = \frac{\sum_1^n w_i X_i}{\sum_1^n w_i} \quad \left(w_i \geq 0, \sum_1^\infty w_i = \infty\right)$$

converge *wp*1? (Note that simple averaging is included as the special case $w_i \equiv 1$.) Several results, along with statistical interpretations, are given by Hall, Kielson and Simons (1971). One of their theorems is the following.

Theorem. *A sufficient condition for* $\{X_n^*\}$ *to converge to* 0 *with probability* 1 *is that*

$$\sum_{n=1}^\infty \frac{E|X_n|^r}{n} < \infty.$$

Since convergence in rth mean implies convergence in probability, a competing result in the present context is provided by Theorem 1.6.1, which however gives only an existence result whereas the above theorem is *constructive*.

1.6.3 Conversion of Convergence in Distribution to Convergence *wp*1

Let $\mathscr{B}_{[0,1]}$ denote the Borel sets in [0, 1] and $m_{[0,1]}$ the Lebesgue measure restricted to [0, 1].

Theorem. *In* R^k, *suppose that* $\mathbf{X}_n \xrightarrow{d} \mathbf{X}$. *Then there exist random* k*-vectors* $\mathbf{Y}, \mathbf{Y}_1, \mathbf{Y}_2, \ldots$ *defined on the probability space* $([0, 1], \mathscr{B}_{[0,1]}, m_{[0,1]})$ *such that*

$$\mathscr{L}(\mathbf{Y}) = \mathscr{L}(\mathbf{X}) \quad \text{and} \quad \mathscr{L}(\mathbf{Y}_n) = \mathscr{L}(\mathbf{X}_n), \qquad n = 1, 2, \ldots,$$

and

$$\mathbf{Y}_n \xrightarrow{wp1} \mathbf{Y}, \qquad \text{i.e.,} \quad m_{[0,1]}(\mathbf{Y}_n \to \mathbf{Y}) = 1.$$

We shall prove this result only for the case $k = 1$. The theorem may, in fact, be established in much greater generality. Namely, the mappings $X, X_1, X_2, \ldots$ may be random elements of any separable complete metric space, a generality which is of interest in considerations involving stochastic processes. See Skorokhod (1956) for the general treatment, or Breiman (1968), Section 13.9, for a thorough treatment of the case R^∞.

The device given by the theorem is sometimes called the "Skorokhod construction" and the theorem the "Skorokhod representation theorem."

PROOF (for the case $k = 1$). For $0 < t < 1$, define

$$Y(t) = F_X^{-1}(t) \qquad \text{and} \quad Y_n(t) = F_{X_n}^{-1}(t), \qquad n = 1, 2, \ldots.$$

Then, using Lemma 1.1.4, we have

$$\begin{aligned} F_Y(y) &= m_{[0,1]}(\{t\colon Y(t) \le y\}) = m_{[0,1]}(\{t\colon t \le F_X(y)\}) \\ &= F_X(y), \qquad \text{all } y, \end{aligned}$$

that is, $\mathscr{L}(Y) = \mathscr{L}(X)$. Similarly, $\mathscr{L}(Y_n) = \mathscr{L}(X_n)$, $n = 1, 2, \ldots$. It remains to establish that

$$m_{[0,1]}(\{t\colon Y_n(t) \not\to Y(t)\}) = 0.$$

This follows immediately from Lemma 1.5.6. ■

Remarks. (i) The exceptional set on which Y_n fails to converge to Y is at most countably infinite.

(ii) Similar theorems may be proved in terms of constructions on probability spaces other than $([0, 1], \mathscr{B}_{[0,1]}, m_{[0,1]})$. However, a desirable feature of the present theorem is that it *does permit* the use of this convenient probability space.

(iii) The theorem is "constructive," not existential, as is demonstrated by the proof. ■

1.7 CONVERGENCE PROPERTIES OF TRANSFORMED SEQUENCES

Given that $X_n \to X$ in some sense of convergence, and given a function g, a basic question is whether $g(X_n) \to g(X)$ in the same sense of convergence. We deal with this question here. In Chapter 3 we deal with the related but different question of whether, given that X_n is $AN(a_n, b_n)$, and given a function g, there exist constants c_n, d_n such that $g(X_n)$ is $AN(c_n, d_n)$.

Returning to the first question, the following theorem states that the answer is "yes" if the function g is continuous with P_X-probability 1. A detailed treatment covering a host of similar results may be found in Mann and Wald (1943). However, the methods of proof there are more cumbersome than the modern approaches we take here, utilizing for example the Skorokhod construction.

Theorem. *Let* $\mathbf{X}_1, \mathbf{X}_2, \ldots$ *and* $\mathbf{X}$ *be random* k*-vectors defined on a probability space and let* g *be a vector-valued Borel function defined on* R^k. *Suppose that* g *is continuous with* $P_\mathbf{X}$*-probability* 1. *Then*

(i) $\mathbf{X}_n \xrightarrow{wp1} \mathbf{X} \Rightarrow g(\mathbf{X}_n) \xrightarrow{wp1} g(\mathbf{X})$;

(ii) $\mathbf{X}_n \xrightarrow{p} \mathbf{X} \Rightarrow g(\mathbf{X}_n) \xrightarrow{p} g(\mathbf{X})$;

(iii) $\mathbf{X}_n \xrightarrow{d} \mathbf{X} \Rightarrow g(\mathbf{X}_n) \xrightarrow{d} g(\mathbf{X})$.

PROOF. We restrict to the case that g is real-valued, the extension for vector-valued g being routine. Let $(\Omega, \mathscr{A}, P)$ denote the probability space on which the $\mathbf{X}$'s are defined.

(i) Suppose that $\mathbf{X}_n \xrightarrow{wp1} \mathbf{X}$. For $\omega \in \Omega$ such that $\mathbf{X}_n(\omega) \to \mathbf{X}(\omega)$ and such that g is continuous at $\mathbf{X}(\omega)$, we have $g(\mathbf{X}_n(\omega)) \to g(\mathbf{X}(\omega))$, $n \to \infty$. By our assumptions, the set of such ω has P-probability 1. Thus $g(\mathbf{X}_n) \xrightarrow{wp1} g(\mathbf{X})$.

(ii) Let $\mathbf{X}_n \xrightarrow{p} \mathbf{X}$. Suppose that $g(\mathbf{X}_n) \not\xrightarrow{p} g(\mathbf{X})$. Then, for some $\varepsilon > 0$ and some $\Delta > 0$, there exists a subsequence $\{n_k\}$ for which

$$(*) \qquad P(|g(\mathbf{X}_{n_k}) - g(\mathbf{X})| > \varepsilon) > \Delta, \qquad \text{all } k = 1, 2, \ldots.$$

But $\mathbf{X}_n \xrightarrow{p} \mathbf{X}$ implies that $\mathbf{X}_{n_k} \xrightarrow{p} \mathbf{X}$ and thus, by Theorem 1.6.1, there exists a subsequence $\{n_{k_j}\}$ of $\{n_k\}$ for which

$$\mathbf{X}_{n_{k_j}} \xrightarrow{wp1} \mathbf{X},\ j \to \infty.$$

But then, by (i) just proved, and since $\xrightarrow{wp1} \Rightarrow \xrightarrow{p}$,

$$g(\mathbf{X}_{n_{k_j}}) \xrightarrow{p} g(\mathbf{X}),$$

contradicting (*). Therefore, $g(\mathbf{X}_n) \xrightarrow{p} g(\mathbf{X})$.

(iii) Let $\mathbf{X}_n \xrightarrow{d} \mathbf{X}$. By the Skorokhod construction of **1.6.3**, we may construct on some probability space $(\Omega', \mathscr{A}', P')$ some random vectors $\mathbf{Y}_1, \mathbf{Y}_2, \ldots$ and $\mathbf{Y}$ such that $\mathscr{L}(\mathbf{Y}_1) = \mathscr{L}(\mathbf{X}_1)$, $\mathscr{L}(\mathbf{Y}_2) = \mathscr{L}(\mathbf{X}_2), \ldots$, and $\mathscr{L}(\mathbf{Y}) = \mathscr{L}(\mathbf{X})$, and, moreover, $\mathbf{Y}_n \to \mathbf{Y}$ with P'-probability 1. Let D denote the discontinuity set of the function g. Then

$$\begin{aligned} P'(\{\omega' : g \text{ is discontinuous at } \mathbf{Y}(\omega')\}) &= P'(\mathbf{Y}^{-1}(D)) \\ &= P'_{\mathbf{Y}}(D) = P_{\mathbf{X}}(D) = P(\mathbf{X}^{-1}(D)) \\ &= 0. \end{aligned}$$

Hence, again invoking (i), $g(\mathbf{Y}_n) \to g(\mathbf{Y})$ with P'-probability 1 and thus $g(\mathbf{Y}_n) \xrightarrow{d} g(\mathbf{Y})$. But the latter is the same as $g(\mathbf{X}_n) \xrightarrow{d} g(\mathbf{X})$. ■

Examples. (i) If $X_n \xrightarrow{d} N(0, 1)$, then $X_n^2 \xrightarrow{d} \chi_1^2$.

(ii) If $(X_n, Y_n) \xrightarrow{d} N(\mathbf{0}, \mathbf{I})$, then $X_n/Y_n \xrightarrow{d}$ Cauchy.

(iii) Illustration of g for which $X_n \xrightarrow{p} X$ but $g(X_n) \not\xrightarrow{p} g(X)$. Let

$$g(t) = \begin{cases} t - 1, & t < 0, \\ t + 1, & t \geq 0, \end{cases}$$

$$X_n = -\frac{1}{n} \text{ with probability 1,}$$

and

$$X = 0 \text{ with probability 1.}$$

The function g has a single discontinuity, located at $t = 0$, so that g is discontinuous with P_X-probability 1. And indeed $X_n \xrightarrow{p} X = 0$, whereas $g(X_n) \xrightarrow{p} -1$ but $g(X) = g(0) = 1 \neq -1$.

(iv) In Section 2.2 it will be seen that under typical conditions the sample variance $s^2 = (n - 1)^{-1} \sum_1^n (X_i - \bar{X})^2$ converges $wp1$ to the population variance σ^2. It then follows that the analogue holds for the standard deviation: $s \xrightarrow{wp1} \sigma$.

(v) *Linear and quadratic functions of vectors.* The most commonly considered functions of vectors converging in some stochastic sense are linear transformations and quadratic forms.

Corollary. *Suppose that the* k*-vectors* $\mathbf{X}_n$ *converge to the* k*-vector* $\mathbf{X}$ wp1, *or in probability, or in distribution. Let* $\mathbf{A}_{m \times k}$ *and* $\mathbf{B}_{k \times k}$ *be matrices. Then* $\mathbf{A}\mathbf{X}_n' \to \mathbf{A}\mathbf{X}'$ *and* $\mathbf{X}_n\mathbf{B}\mathbf{X}_n' \to \mathbf{X}\mathbf{B}\mathbf{X}'$ *in the given mode of convergence.*

PROOF. The vector-valued function

$$\mathbf{A}\mathbf{x}' = \left(\sum_{i=1}^{k} a_{1i}x_i, \ldots, \sum_{i=1}^{k} a_{mi}x_i\right)$$

and the real-valued function

$$\mathbf{x}\mathbf{B}\mathbf{x}' = \sum_{i=1}^{k}\sum_{j=1}^{k} b_{ij}x_i x_j$$

are *continuous* functions of $\mathbf{x} = (x_1, \ldots, x_k)$. ■

Some key applications of the corollary are as follows.

Application A. *In* R^k, *let* $\mathbf{X}_n \xrightarrow{d} N(\boldsymbol{\mu}, \boldsymbol{\Sigma})$. *Let* $\mathbf{C}_{m \times k}$ *be a matrix. Then* $\mathbf{C}\mathbf{X}_n' \xrightarrow{d} N(\mathbf{C}\boldsymbol{\mu}', \mathbf{C}\boldsymbol{\Sigma}\mathbf{C}')$.

(This follows simply by noting that if $\mathbf{X}$ is $N(\boldsymbol{\mu}, \boldsymbol{\Sigma})$, then $\mathbf{C}\mathbf{X}'$ is $N(\mathbf{C}\boldsymbol{\mu}', \mathbf{C}\boldsymbol{\Sigma}\mathbf{C}')$.)

Application B. *Let* $\mathbf{X}_n$ *be* $AN(\boldsymbol{\mu}, b_n^2\boldsymbol{\Sigma})$. *Then*

$$\frac{\|\mathbf{X}_n - \boldsymbol{\mu}\|}{b_n} \xrightarrow{d} \textit{a limit random variable.}$$

(Proof left as exercise—Problem 1.P.22) If $b_n \to 0$ (typically, $b_n \sim n^{-1/2}$), then follows $\mathbf{X}_n \xrightarrow{p} \boldsymbol{\mu}$. More generally, however, we can establish (Problem 1.P.23)

Application C. *Let* $\mathbf{X}_n$ *be* $AN(\boldsymbol{\mu}, \boldsymbol{\Sigma}_n)$, *with* $\boldsymbol{\Sigma}_n \to \mathbf{0}$. *Then* $\mathbf{X}_n \xrightarrow{p} \boldsymbol{\mu}$.

Application D. (*Sums and products of random variables converging* wpl *or in probability.*) *If* $X_n \xrightarrow{wp1} X$ *and* $Y_n \xrightarrow{wp1} Y$, *then* $X_n + Y_n \xrightarrow{wp1} X + Y$ *and* $X_nY_n \xrightarrow{wp1} XY$. *If* $X_n \xrightarrow{p} X$ *and* $Y_n \xrightarrow{p} Y$, *then* $X_n + Y_n \xrightarrow{p} X + Y$ *and* $X_nY_n \xrightarrow{p} XY$.

(Proof left as exercise—Problem 1.P.24)

1.8 BASIC PROBABILITY LIMIT THEOREMS: THE WLLN AND SLLN

"Weak laws of large numbers" (WLLN) refer to convergence in probability of averages of random variables, whereas "strong laws of large numbers" (SLLN) refer to convergence *wp*1. The first two theorems below give the WLLN and SLLN for sequences of I.I.D. random variables, the case of central importance in this book.

Theorem A. *Let* $\{X_i\}$ *be I.I.D. with distribution function* F. *The existence of constants* $\{a_n\}$ *for which*

$$\frac{1}{n}\sum_{i=1}^{n} X_i - a_n \xrightarrow{p} 0$$

holds if and only if

$$(*) \qquad t[1 - F(t) + F(-t)] \to 0, \ t \to \infty,$$

in which case we may choose $a_n = \int_{-n}^{n} x \, dF(x)$.

A sufficient condition for (*) is finiteness of $\int_{-\infty}^{\infty} |x| dF(x)$, but in this case the following result asserts a stronger convergence.

Theorem B (Kolmogorov). *Let* $\{X_i\}$ *be I.I.D. The existence of a finite constant* c *for which*

$$\frac{1}{n}\sum_{i=1}^{n} X_i \xrightarrow{wp1} c$$

holds if and only if $E\{X_1\}$ *is finite and equals* c.

The following theorems provide WLLN or SLLN under relaxation of the I.I.D. assumptions, but at the expense of assuming existence of variances and restricting their growth with increasing *n*.

Theorem C (Chebyshev). *Let* $X_1, X_2, \ldots$ *be uncorrelated with means* $\mu_1, \mu_2, \ldots$ *and variances* $\sigma_1^2, \sigma_2^2, \ldots$. *If* $\sum_1^n \sigma_i^2 = o(n^2)$, $n \to \infty$, *then*

$$\frac{1}{n}\sum_{i=1}^{n} X_i - \frac{1}{n}\sum_{i=1}^{n} \mu_i \xrightarrow{p} 0.$$

Theorem D (Kolmogorov). *Let* $X_1, X_2, \ldots$ *be independent with means* $\mu_1, \mu_2, \ldots$ *and variances* $\sigma_1^2, \sigma_2^2, \ldots$. *If the series* $\sum_1^\infty \sigma_i^2/i^2$ *converges, then*

$$(**) \qquad \frac{1}{n}\sum_{i=1}^{n} X_i - \frac{1}{n}\sum_{i=1}^{n} \mu_i \xrightarrow{wp1} 0.$$

Theorem E. *Let* $X_1, X_2, \ldots$ *have means* $\mu_1, \mu_2, \ldots$, *variances* $\sigma_1^2, \sigma_2^2, \ldots$, *and covariances* $\mathrm{Cov}\{X_i, X_j\}$ *satisfying*

$$\mathrm{Cov}\{X_i, X_j\} \le \rho_{j-i}\sigma_i\sigma_j (i \le j),$$

where $0 \le \rho_k \le 1$ *for all* $k = 0, 1, \ldots$. *If the series* $\sum_1^\infty \rho_i$ *and* $\sum_1^\infty \sigma_i^2(\log i)^2/i^2$ *are both convergent, then* (**) holds.

Further reading on Theorem A is found in Feller (1966), p. 232, on Theorems B, C and D in Rao (1973), pp. 112–114, and on Theorem E in Serfling (1970). Other useful material is provided by Gnedenko and Kolmogorov (1954) and Chung (1974).

1.9 BASIC PROBABILITY LIMIT THEOREMS: THE CLT

The central limit theorem (CLT) pertains to the convergence in distribution of (normalized) *sums* of random variables. The case of chief importance, I.I.D. summands, is treated in **1.9.1**. Generalizations allowing non-identical distributions, double arrays, and a random number of summands are presented in **1.9.2**, **1.9.3**, and **1.9.4**, respectively. Finally, error estimates and asymptotic expansions related to the CLT are discussed in **1.9.5**. Also, some further aspects of the CLT are treated in Section 1.11.

1.9.1 The I.I.D. Case

Perhaps the most widely known version of the CLT is

Theorem A (Lindeberg–Lévy). *Let* $\{X_i\}$ *be I.I.D. with mean* μ *and finite variance* σ. *Then*

$$\sqrt{n}\left(\frac{1}{n}\sum_{i=1}^{n} X_i - \mu\right) \xrightarrow{d} N(0, \sigma^2),$$

that is,

$$\frac{1}{n}\sum_{i=1}^{n} X_i \text{ is } AN\left(\mu, \frac{\sigma^2}{n}\right).$$

The multivariate extension of Theorem A may be derived from Theorem A itself with the use of the Cramér–Wold device (Theorem 1.5.2). We obtain

Theorem B. *Let* $\{\mathbf{X}_i\}$ *be I.I.D. random vectors with mean* $\boldsymbol{\mu}$ *and covariance matrix* $\boldsymbol{\Sigma}$. *Then*

$$\sqrt{n}\left(\frac{1}{n}\sum_{i=1}^{n} \mathbf{X}_i - \boldsymbol{\mu}\right) \xrightarrow{d} N(\mathbf{0}, \boldsymbol{\Sigma}),$$

that is (by Problem 1.P.17),

$$\frac{1}{n}\sum_{i=1}^{n} \mathbf{X}_i \text{ is } AN\left(\boldsymbol{\mu}, \frac{1}{n}\boldsymbol{\Sigma}\right).$$

Remark. It is not necessary, however, to assume finite variances. Feller (1966), p. 303, gives

Theorem C. *Let* $\{X_i\}$ *be I.I.D. with distribution function* F. *Then the existence of constants* $\{a_n\}$, $\{b_n\}$ *such that*

$$\frac{1}{n}\sum_{i=1}^{n} X_i \textit{ is } AN(a_n, b_n)$$

holds if and only if

$$(*) \qquad \frac{t^2[1 - F(t) + F(-t)]}{U(t)} \to 0, \qquad t \to \infty,$$

where $U(t) = \int_{-t}^{t} x^2\, dF(x)$.

(Condition (*) is equivalent to the condition that $U(t)$ *vary slowly* at ∞, that is, for every $\alpha > 0$, $U(\alpha t)/U(t) \to 1$, $t \to \infty$.) ■

1.9.2 Generalization: Independent Random Variables Not Necessarily Identically Distributed

The Lindeberg–Lévy Theorem of **1.9.1** is a special case of

Theorem A (Lindeberg–Feller). *Let* $\{X_i\}$ *be independent with means* $\{\mu_i\}$, *finite variances* $\{\sigma_i^2\}$, *and distribution functions* $\{F_i\}$. *Suppose that* $B_n^2 = \sum_1^n \sigma_i^2$ *satisfies*

$$(V) \qquad \frac{\sigma_n^2}{B_n^2} \to 0, \qquad B_n \to \infty, \qquad \textit{as} \quad n \to \infty.$$

Then

$$\frac{1}{n}\sum_{i=1}^{n} X_i \quad \textit{is} \quad AN\left(\frac{1}{n}\sum_{i=1}^{n} \mu_i, \frac{1}{n^2} B_n^2\right)$$

if and only if the Lindeberg condition

$$(L) \qquad \frac{\sum_{i=1}^{n} \int_{|t-\mu_i| > \varepsilon B_n} (t - \mu_i)^2\, dF_i(t)}{B_n^2} \to 0,\ n \to \infty, \qquad \textit{each} \quad \varepsilon > 0,$$

is satisfied.

(See Feller (1966), pp. 256 and 492.) The following corollary provides a practical criterion for establishing conditions (L) and (V). Indeed, as seen in the proof, (V) actually follows from (L), so that the key issue is verification of (L).

Corollary. *Let* $\{X_i\}$ *be independent with means* $\{\mu_i\}$ *and finite variances* $\{\sigma_i^2\}$. *Suppose that, for some* $\nu > 2$,

$$\sum_{i=1}^{n} E|X_i - \mu_i|^\nu = o(B_n^\nu), \qquad n \to \infty.$$

Then

$$\frac{1}{n}\sum_{i=1}^{n} X_i \quad \textit{is} \quad AN\left(\frac{1}{n}\sum_{i=1}^{n}\mu_i, \frac{1}{n^2}B_n^2\right).$$

PROOF. First we establish that condition (L) follows from the given hypothesis. For $\varepsilon > 0$, write

$$\int_{|t-\mu_i|>\varepsilon B_n} (t - \mu_i)^2\, dF_i(t) \le (\varepsilon B_n)^{2-\nu} \int_{|t-\mu_i|>\varepsilon B_n} |t - \mu_i|^\nu\, dF_i(t)$$
$$\le (\varepsilon B_n)^{2-\nu} E|X_i - \mu_i|^\nu.$$

By summing these relations, we readily obtain (L).

Next we show that (L) implies

$$\text{(V*)} \qquad \frac{\max_{j\le n}\sigma_j^2}{B_n^2} \to 0, \qquad n \to \infty.$$

For we have, for $1 \le i \le n$,

$$\sigma_i^2 \le \int_{|t-\mu_i|>\varepsilon B_n} (t - \mu_i)^2\, dF_i(t) + \varepsilon^2 B_n^2.$$

Hence

$$\max_{j\le n}\sigma_j^2 \le \sum_{i=1}^{n}\int_{|t-\mu_i|>\varepsilon B_n} (t - \mu_i)^2\, dF_i(t) + \varepsilon^2 B_n^2.$$

Thus (L) implies (V*).

Finally, check that (V*) implies $B_n \to \infty$, $n \to \infty$. ■

A useful special case consists of independent $\{X_i\}$ with common mean μ, common variance σ^2, and uniformly bounded νth absolute central moments, $E|X_i - \mu|^\nu \le M < \infty$ (all i), where $\nu > 2$.

A convenient multivariate extension of Theorem A is given by Rao (1973), p. 147:

Theorem B. *Let* $\{\mathbf{X}_i\}$ *be independent random vectors with means* $\{\boldsymbol{\mu}_i\}$, *covariance matrices* $\{\boldsymbol{\Sigma}_i\}$ *and distribution functions* $\{F_i\}$. *Suppose that*

$$\frac{\boldsymbol{\Sigma}_1 + \cdots + \boldsymbol{\Sigma}_n}{n} \to \boldsymbol{\Sigma}, \qquad n \to \infty,$$

and that

$$\frac{1}{n}\sum_{i=1}^{n}\int_{\|\mathbf{x}-\boldsymbol{\mu}_i\|>\varepsilon\sqrt{n}}\|\mathbf{x}-\boldsymbol{\mu}_i\|^2\,dF_i(\mathbf{x})\to 0,\qquad n\to\infty,\qquad \textit{each}\quad \varepsilon>0.$$

Then

$$\frac{1}{n}\sum_{i=1}^{n}\mathbf{X}_i \quad \textit{is}\quad \mathrm{AN}\left(\frac{1}{n}\sum_{i=1}^{n}\boldsymbol{\mu}_i, \frac{1}{n}\boldsymbol{\Sigma}\right).$$

1.9.3 Generalization: Double Arrays of Random Variables

In the theorems previously considered, asymptotic normality was asserted for a sequence of sums $\sum_1^n X_i$ generated by a single sequence $X_1, X_2, \ldots$ of random variables. More generally, we may consider a *double array* of random variables:

$$\begin{array}{c} X_{11}, X_{12}, \ldots, X_{1k_1}; \\ X_{21}, X_{22}, \ldots, X_{2k_2}; \\ \vdots \\ X_{n1}, X_{n2}, \ldots, X_{nk_n}; \\ \vdots \end{array}$$

For each $n \geq 1$, there are k_n random variables $\{X_{nj}, 1 \leq j \leq k_n\}$. It is assumed that $k_n \to \infty$. The case $k_n = n$ is called a "triangular" array.

Denote by F_{nj} the distribution function of X_{nj}. Also, put

$$\mu_{nj} = E\{X_{nj}\},$$

$$A_n = E\left\{\sum_{j=1}^{k_n} X_{nj}\right\} = \sum_{j=1}^{k_n}\mu_{nj},$$

$$B_n^2 = \mathrm{Var}\left\{\sum_{j=1}^{k_n} X_{nj}\right\}.$$

The Lindeberg–Feller Theorem of **1.9.2** is a special case of

Theorem. *Let* $\{X_{nj}\colon 1 \leq j \leq k_n;\ n = 1, 2, \ldots\}$ *be a double array with independent random variables within rows. Then the "uniform asymptotic neglibility" condition*

$$\max_{1\leq j\leq k_n} P(|X_{nj}-\mu_{nj}| > \tau B_n)\to 0,\qquad n\to\infty,\qquad \textit{each}\quad \tau>0,$$

and the asymptotic normality condition

$$\sum_{j=1}^{k_n} X_{nj}\quad \textit{is}\quad \mathrm{AN}(A_n, B_n^2)$$

together hold if and only if the Lindeberg condition

$$\frac{\sum_{j=1}^{k_n} \int_{|t-\mu_{nj}|>\varepsilon B_n} (t - \mu_{nj})^2 \, dF_{nj}(t)}{B_n^2} \to 0, \qquad n \to \infty, \qquad \textit{each} \quad \varepsilon > 0,$$

is satisfied.

(See Chung (1974), Section 7.2.) The independence is assumed only *within* rows, which themselves may be arbitrarily dependent.

The analogue of Corollary 1.9.2 is (Problem 1.P.26)

Corollary. *Let* $\{X_{nj}: 1 \le j \le k_n;\ n = 1, 2, \ldots\}$ *be a double array with independent random variables within rows. Suppose that, for some* $\nu > 2$,

$$\sum_{j=1}^{k_n} E|X_{nj} - \mu_{nj}|^\nu = o(B_n^\nu), \qquad n \to \infty.$$

Then

$$\sum_{j=1}^{k_n} X_{nj} \quad \textit{is} \quad AN(A_n, B_n^2).$$

1.9.4 Generalization: A Random Number of Summands

The following is a generalization of the classical Theorem 1.9.1A. See Billingsley (1968), Chung (1974), and Feller (1966) for further details and generalizations.

Theorem. *Let* $\{X_i\}$ *be I.I.D. with mean* μ *and finite variance* σ^2. *Let* $\{\nu_n\}$ *be a sequence of integer-valued random variables and* $\{a_n\}$ *a sequence of positive constants tending to* ∞, *such that*

$$\frac{\nu_n}{a_n} \xrightarrow{p} c$$

for some positive constant c. *Then*

$$\frac{\sum_{i=1}^{\nu_n} (X_i - \mu)}{\sqrt{\nu_n}} \xrightarrow{d} N(0, \sigma^2).$$

1.9.5 Error Bounds and Asymptotic Expansions

It is of both theoretical and practical interest to characterize the error of approximation in the CLT. Denote by

$$G_n(t) = P(S_n^* \le t)$$

the distribution function of the normalized sum

$$S_n^* = \frac{\sum_1^n X_i - E\{\sum_1^n X_i\}}{[\mathrm{Var}\{\sum_1^n X_i\}]^{1/2}}.$$

For the I.I.D. case, an exact bound on the error of approximation is provided by the following theorem due to Berry (1941) and Esséen (1945). (However, the earliest result of this kind was established by Liapounoff (1900, 1901).)

Theorem (Berry–Esséen). *Let* $\{X_i\}$ *be I.I.D. with mean* μ *and variance* $\sigma^2 > 0$. *Then*

$$(*) \qquad \sup_t |G_n(t) - \Phi(t)| \le \frac{33}{4} \frac{E|X_1 - \mu|^3}{\sigma^3 n^{1/2}}, \qquad \textit{all } n.$$

The fact that $\sup_t |G_n(t) - \Phi(t)| \to 0$, $n \to \infty$, is, of course, provided under second-order moment assumptions by the Lindeberg–Lévy Theorem 1.9.1A, in conjunction with Pólya's Theorem 1.5.3. Introducing higher-order moment assumptions, the Berry–Esséen Theorem asserts for this convergence the rate $O(n^{-1/2})$. It is the best possible rate in the sense of not being subject to improvement without narrowing the class of distribution functions considered.

However, various authors have sought to improve the constant 33/4. Introducing new methods, Zolotarev (1967) reduced to 0.91; subsequently, van Beeck (1972) sharpened to 0.7975. On the other hand, Esséen (1956) has determined the following "asymptotically best" constant:

$$\frac{3 + \sqrt{10}}{6\sqrt{2\pi}} = \overline{\lim_{n\to\infty}} \sup_F \left\{ \frac{\sigma^3 \sqrt{n}}{E|X_1 - \mu|^3} \sup_t |G_n(t) - \Phi(t)| \right\}.$$

More generally, independent summands not necessarily identically distributed are also treated in Berry and Esséen's work. For this case the right-hand side of (*) takes the form

$$C \frac{\sum_{i=1}^n E|X_i - \mu_i|^3}{[\mathrm{Var}\{\sum_{i=1}^n X_i\}]^{3/2}};$$

where C is a universal constant. Extension in another direction, to the case of a random number of (I.I.D.) summands, has recently been carried out by Landers and Rogge (1976).

For t sufficiently large, while n remains fixed, the quantities $G_n(t)$ and $\Phi(t)$ each become so close to 1 that the bound given by (*) is too crude. The problem in this case may be characterized as one of approximation of "large deviation" probabilities, with the object of attention becoming the relative error in approximation of $1 - G_n(t)$ by $1 - \Phi(t)$. Cramér (1938) developed a general theorem characterizing the ratio

$$\frac{1 - G_n(t_n)}{1 - \Phi(t_n)}$$

under the restriction $t_n = o(n^{1/2})$, $n \to \infty$, for the case of I.I.D. X_i's having a moment generating function. In particular, for $t_n = o(n^{1/6})$, the ratio tends to 1, whereas for $t_n \to \infty$ at a faster rate the ratio can behave differently. An important special case of $t_n = o(n^{1/6})$, namely $t_n \sim c(\log n)^{1/2}$, has arisen in connection with the asymptotic relative efficiency of certain statistical procedures. For this case, $1 - G_n(t_n)$ has been dubbed a "moderate deviation" probability, and the Cramér result $[1 - G_n(t_n)]/[1 - \Phi(t_n)] \to 1$ has been obtained by Rubin and Sethuraman (1965a) under less restrictive moment assumptions. Another "large deviation" case important in statistical applications is $t_n \sim cn^{1/2}$, a case *not* covered by Cramér's theorem. For this case Chernoff (1952) has characterized the exponential rate of convergence of $[1 - G_n(t_n)]$ to 0. We shall examine this in Chapter 10.

Still another approach to the problem is to refine the Berry–Esséen bound on $|G_n(t) - \Phi(t)|$, to reflect dependence on t as well as n. In this direction, (*) has been replaced by

$$|G_n(t) - \Phi(t)| \leq C \frac{E|X_1 - \mu|^3}{\sigma^3 n^{1/2}} \frac{1}{1 + t^2}, \qquad \text{all} \quad t,$$

where C is a universal constant. For details, see Ibragimov and Linnik (1971). In the same vein, under more restrictive assumptions on the distribution functions involved, an asymptotic expansion of $G_n(t) - \Phi(t)$ in powers of $n^{-1/2}$ may be given, the last term in the expansion playing the role of error bound. For example, a simple result of this form is

$$|G_n(t) - \Phi(t)| \leq \frac{1}{n^{1/2}} \frac{E\{(X_1 - \mu)^3\}}{6\sigma^3(2\pi)^{1/2}} (1 - t^2)e^{-(1/2)t^2} + o(n^{-1/2}),$$

uniformly in t (see Ibragimov and Linnik (1971), p. 97). For further reading, see Cramér (1970), Theorems 25 and 26 and related discussion, Abramowitz and Stegun (1965), pp. 935 and 955, Wilks (1962), Section 9.4, the book by Bhattacharya and Ranga Rao (1976), and the expository survey paper by Bhattacharya (1977).

Alternatively to the measure of discrepancy $\sup_t |G_n(t) - \Phi(t)|$ used in the Berry–Esséen Theorem, one may also consider L_p metrics (see Ibragimov and Linnik (1971)) or weak convergence metrics (see Bhattacharya and Ranga Rao (1976)), and likewise obtain $O(n^{-1/2})$ as a rate of convergence.

The rate of convergence in the CLT is not only an interesting theoretical issue, but also has various applications. For example, Bahadur and Ranga

Rao (1960) make use of such a result in establishing a large deviation theorem for the sample mean, which theorem then plays a role in asymptotic relative efficiency considerations. Rubin and Sethuraman (1965a, b) develop "moderate deviation" results, as discussed above, and make similar applications. Another type of application concerns the law of the iterated logarithm, to be discussed in the next section.

1.10 BASIC PROBABILITY LIMIT THEOREMS: THE LIL

Complementing the SLLN and the CLT, the "law of the iterated logarithm" (LIL) characterises the extreme fluctuations occurring in a sequence of averages, or partial sums. The classical I.I.D. case is covered by

Theorem A (Hartman and Wintner). *Let* $\{X_i\}$ *be I.I.D. with mean* μ *and finite variance* σ^2. *Then*

$$\overline{\lim_{n\to\infty}} \frac{\sum_1^n (X_i - \mu)}{(2\sigma^2 n \log\log n)^{1/2}} = 1 \text{ wp1}.$$

In words: with probability 1, for any $\varepsilon > 0$, only finitely many of the events

$$\frac{\sum_1^n (X_i - \mu)}{(2\sigma^2 n \log\log n)^{1/2}} > 1 + \varepsilon, \qquad n = 1, 2, \ldots,$$

are realized, whereas infinitely many of the events

$$\frac{\sum_1^n (X_i - \mu)}{(2\sigma^2 n \log\log n)^{1/2}} > 1 - \varepsilon, \qquad n = 1, 2, \ldots,$$

occur.

The LIL complements the CLT by describing the precise extremes of the fluctuations of the sequence of random variables

$$\frac{\sum_1^n (X_i - \mu)}{\sigma n^{1/2}}, \qquad n = 1, 2, \ldots.$$

The CLT states that this sequence converges in distribution to $N(0, 1)$, but does not otherwise provide information about the fluctuations of these random variables about the expected value 0. The LIL asserts that the extreme fluctuations of this sequence are essentially of the exact order of magnitude $(2 \log\log n)^{1/2}$. That is, with probability 1, for any $\varepsilon > 0$, all but finitely many of these fluctuations fall within the boundaries $\pm(1 + \varepsilon)(2 \log\log n)^{1/2}$ and, moreover, the boundaries $\pm(1 - \varepsilon)(2 \log\log n)^{1/2}$ are reached infinitely often.

The LIL also complements—indeed, refines—the SLLN (but assumes existence of 2nd moments). In terms of the *averages* dealt with by the SLLN,

$$\frac{1}{n}\sum_{i=1}^{n} X_i - \mu,$$

the LIL asserts that the extreme fluctuations are essentially of the exact order of magnitude

$$\frac{\sigma(2 \log \log n)^{1/2}}{n^{1/2}}.$$

Thus, with probability 1, for any $\varepsilon > 0$, the infinite sequence of "confidence intervals"

$$\left\{\frac{1}{n}\sum_{i=1}^{n} X_i \pm (1+\varepsilon)\frac{\sigma(2 \log \log n)^{1/2}}{n^{1/2}}\right\}$$

contains μ with only finitely many exceptions. In this fashion the LIL provides the basis for concepts of 100% confidence intervals and tests of power 1. For further details on such statistical applications of the LIL, consult Robbins (1970), Robbins and Siegmund (1973, 1974) and Lai (1977).

A version of the LIL for independent X_i's not necessarily identically distributed was given by Kolmogorov (1929):

Theorem B (Kolmogorov). *Let* $\{X_i\}$ *be independent with means* $\{\mu_i\}$ *and finite variances* $\{\sigma_i^2\}$. *Suppose that* $B_n^2 = \sum_1^n \sigma_i^2 \to \infty$ *and that, for some sequence of constants* $\{m_n\}$, *with probability* 1,

$$(*) \qquad |X_n - \mu_n| \le m_n = o\left(\frac{B_n}{(\log \log B_n)^{1/2}}\right), \qquad n \to \infty.$$

Then

$$\overline{\lim_{n\to\infty}} \frac{\sum_1^n (X_i - \mu_i)}{(2B_n^2 \log \log B_n)^{1/2}} = 1 \text{ wp1}.$$

(To facilitate comparison of Theorems A and B, note that $\log \log(ax^b) \sim \log \log x$, $x \to \infty$.)

Extension of Theorems A and B to the case of $\{X_i\}$ a sequence of *martingale differences* has been carried out by Stout (1970a, b).

Another version of the LIL for independent X_i's not necessarily identically distributed has been given by Chung (1974), Theorem 7.5.1:

Theorem C (Chung). *Let* $\{X_i\}$ *be independent with means* $\{\mu_i\}$ *and finite variances* $\{\sigma_i^2\}$. *Suppose that* $B_n^2 = \sum_1^n \sigma_i^2 \to \infty$ *and that, for some* $\varepsilon > 0$,

$$(**) \qquad \frac{\sum_1^n E|X_i - \mu_i|^3}{B_n^3} = O\left(\frac{1}{(\log B_n)^{1+\varepsilon}}\right), \qquad n \to \infty.$$

Then

$$\lim_{n\to\infty} \frac{\sum_1^n (X_i - \mu_i)}{(2B_n^2 \log\log B_n)^{1/2}} = 1 \text{ wp1.}$$

Note that (*) and (**) are overlapping conditions, but very different in nature.

As discussed above, the LIL *augments* the information provided by the CLT. On the other hand, the CLT in conjunction with a suitable rate of convergence *implies* the LIL and thus implicitly contains all the "extra" information stated by the LIL. This was discovered independently by Chung (1950) and Petrov (1966). The following result is given by Petrov (1971). Note the absence of moment assumptions, and the mildness of the rate of convergence assumption.

Theorem D (Petrov). *Let* $\{X_i\}$ *be independent random variables and* $\{B_n\}$ *a sequence of numbers satisfying*

$$B_n \to \infty, \frac{B_{n+1}}{B_n} \to 1, \qquad n \to \infty.$$

Suppose that, for some $\varepsilon > 0$,

$$\sup_t \left| P\left(\frac{\sum_1^n X_i}{B_n} \le t\right) - \Phi(t)\right| = O\left(\frac{1}{(\log B_n)^{1+\varepsilon}}\right), \qquad n \to \infty$$

Then

$$\overline{\lim_{n\to\infty}} \frac{\sum_1^n X_i}{(2B_n^2 \log\log B_n)^{1/2}} = 1 \text{ wp1.}$$

For further discussion and background on the LIL, see Stout (1974), Chapter 5, Chung (1974), Section 7.5, Freedman (1971), Section 1.5, Breiman (1968), pp. 291–292, Lamperti (1966), pp. 41–49, and Feller (1957), pp. 191–198. The latter source provides a simple treatment of the case that $\{X_i\}$ is a sequence of I.I.D. Bernoulli trials and provides discussion of general forms of the LIL.

More broadly, for general reading on the "almost sure behavior" of sequences of random variables, with thorough attention to extensions to *dependent* sequences, see the books by Révész (1968) and Stout (1974).

1.11 STOCHASTIC PROCESS FORMULATION OF THE CLT

Here the CLT is formulated in a *stochastic process* setting, generalizing the formulation considered in **1.9** and **1.10**. A motivating example, which illustrates the need for such greater generality, is considered in **1.11.1**. An appropriate stochastic process defined in terms of the sequence of partial sums, is introduced in **1.11.2.** As a final preparation, the notion of "convergence in distribution" in the general setting of stochastic processes is discussed in **1.11.3**. On this basis, the stochastic process formulation of the CLT is presented in **1.11.4**, with implications regarding the motivating example and the usual CLT. Some complementary remarks are given in **1.11.5**.

1.11.1 A Motivating Example

Let $\{X_i\}$ be I.I.D. with mean μ and finite variance $\sigma^2 > 0$. The Lindeberg–Lévy CLT (1.9.1A) concerns the sequence of random variables

$$S_n^* = \frac{\sum_1^n (X_i - \mu)}{\sigma n^{1/2}}, \qquad n = 1, 2, \ldots,$$

and asserts that $S_n^* \xrightarrow{d} N(0, 1)$. This useful result has broad application concerning approximation of the distribution of the random variable $S_n = \sum_1^n (X_i - \mu)$ for large n. However, suppose that our goal is to approximate the distribution of the random variable

$$\max_{0 \le k \le n} \sum_{i=1}^{k} (X_i - \mu) = \max\{0, S_1, \ldots, S_n\}$$

for large n. In terms of a suitably normalized random variable, the problem may be stated as that of approximating the distribution of

$$M_n = \max\left\{0, \frac{S_1}{\sigma n^{1/2}}, \ldots, \frac{S_n}{\sigma n^{1/2}}\right\}.$$

Here a difficulty emerges. It is seen that M_n is *not* subject to representation as a direct transformation, $g(S_n^*)$, of S_n^* only. Thus it is *not* feasible to solve the problem simply by applying Theorem 1.7 (iii) on transformations in conjunction with the convergence $S_n^* \xrightarrow{d} N(0, 1)$. However, such a scenario *can* be implemented if S_n^* becomes replaced by an appropriate *stochastic process* or *random function*, say $\{Y_n(t), 0 \le t \le 1\}$, and the concept of $\xrightarrow{d}$ is suitably extended.

1.11.2 A Relevant Stochastic Process

Let $\{X_i\}$ and $\{S_i\}$ be as in **1.11.1**. We define an associated random function $Y_n(t)$, $0 \le t \le 1$, by setting

$$Y_n(0) = 0$$

and

$$Y_n\left(\frac{j}{n}\right) = \frac{S_j}{\sigma n^{1/2}}, \qquad 1 \le j \le n,$$

and defining $Y_n(t)$ elsewhere on $0 \le t \le 1$ by linear interpolation. Explicitly, in terms of $X_1, \ldots, X_n$, the stochastic process $Y_n(\cdot)$ is given by

$$Y_n(t) = \frac{\sum_1^{[nt]} (X_i - \mu) + (nt - [nt])(X_{[nt]+1} - \mu)}{\sigma n^{1/2}}, \qquad 0 \le t \le 1.$$

As $n \to \infty$, we have a sequence of such random functions generated by the sequence $\{X_i\}$. The original associated sequence $\{S_n^*\}$ is recovered by taking the sequence of values $\{Y_n(1)\}$.

It is convenient to think of the stochastic process $\{Y_n(t), 0 \le t \le 1\}$ as a random element of a suitable function space. Here the space may be taken to be $C[0, 1]$, the collection of all continuous functions on the unit interval $[0, 1]$.

We now observe that the random variable M_n considered in **1.11.1** may be expressed as a direct function of the process $Y_n(\cdot)$, that is,

$$M_n = \sup_{0 \le t \le 1} Y_n(t) = g(Y_n(\cdot)),$$

where g is the function defined on $C[0, 1]$ by

$$g(x(\cdot)) = \sup_{0 \le t \le 1} x(t), \; x(\cdot) \in C[0, 1].$$

Consequently, a scenario for dealing with the convergence in distribution of M_n consists of

(a) establishing a "convergence in distribution" result for the random function $Y_n(\cdot)$, and
(b) establishing that the transformation g satisfies the hypothesis of an appropriate generalization of Theorem 1.7 (iii).

After laying a general foundation in **1.11.3**, we return to this example in **1.11.4**.

1.11.3 Notions of Convergence in Distribution

Consider a collection of random variables $X_1, X_2, \ldots$ and X having respective distribution functions $F_1, F_2, \ldots$ and F defined on the real line and having respective probability measures $P_1, P_2, \ldots$ and P defined on the Borel sets of the real line. Three equivalent versions of "convergence of X_n to X in distribution" will now be examined. Recall that in **1.2.4** we defined this to mean that

(*) $$\lim_{n \to \infty} F_n(t) = F(t), \text{ each continuity point } t \text{ of } F,$$

and we introduced the notation $X_n \xrightarrow{d} X$ and alternate terminology "weak convergence of distributions" and notation $F_n \Rightarrow F$.

We next consider a condition equivalent to (*) but expressed in terms of $P_1, P_2, \ldots$ and P. First we need further terminology. For any set A, the *boundary* is defined to be the closure minus the interior and is denoted by ∂A. For any measure P, a set A for which $P(\partial A) = 0$ is called a *P-continuity set.* In these terms, a condition equivalent to (*) is

(**) $$\lim_{n\to\infty} P_n(A) = P(A), \text{ each } P\text{-continuity set } A.$$

The equivalence is proved in Billingsley (1968), Chapter 1, and is discussed also in Cramér (1946), Sections 6.7 and 8.5. In connection with (**), the terminology "weak convergence of probability measures" and the notation $P_n \Rightarrow P$ is used.

There is a significant advantage of (**) over (*): it may be formulated in a considerably more general context. Namely, the variables $X_1, X_2, \ldots$ and X may take values in an arbitrary metric space S. In this case $P_1, P_2, \ldots$ and P are defined on the Borel sets in S (i.e., on the σ-field generated by the open sets with respect to the metric associated with S). In particular, if S is a metrizable *function* space, then $P_n \Rightarrow P$ denotes "convergence in distribution" of a sequence of stochastic processes to a limit stochastic process. Thus, for example, for the process $Y_n(\cdot)$ discussed in **1.11.2**, $Y_n(\cdot) \xrightarrow{d} Y(\cdot)$ becomes defined for an appropriate limit process $Y(\cdot)$.

For completeness, we mention a further equivalent version of weak convergence, also meaningful in the more general setting, and indeed often adopted as the primary definition. This is the condition

(***) $$\lim_{n\to\infty} \int_S g\, dP_n = \int_S g\, dP, \text{ each bounded continuous function on } S.$$

The equivalence is proved in Billingsley (1968), Chapter 1. See also the proof of Theorem 1.5.1A.

1.11.4 Donsker's Theorem and Some Implications

Here we treat formally the "partial sum" stochastic process introduced in **1.11.2**. Specifically, for an I.I.D. sequence of random variables $\{X_i\}$ defined on a probability space $(\Omega, \mathscr{A}, P)$ and having mean μ and finite variance σ^2, we consider for each $n(n = 1, 2, \ldots)$ the stochastic process

$$Y_n(t, \omega) = \frac{\sum_{i=1}^{[nt]} (X_i(\omega) - \mu) + (nt - [nt])(X_{[nt]+1}(\omega) - \mu)}{\sigma n^{1/2}}, \qquad 0 \le t \le 1,$$

which is a random element of the space $C[0, 1]$. When convenient, we suppress the ω notation.

The space $C[0, 1]$ may be metrized by

$$\rho(x, y) = \sup_{0 \le t \le 1} |x(t) - y(t)|$$

for $x = x(\cdot)$ and $y = y(\cdot)$ in $C[0, 1]$. Denote by $\mathscr{B}$ the class of Borel sets in $C[0, 1]$ relative to ρ. Denote by Q_n the probability distribution of $Y_n(\cdot)$ in $C[0, 1]$, that is, the probability measure on $(C, \mathscr{B})$ induced by the measure P through the relation

$$Q_n(B) = P(\{\omega: Y_n(\cdot, \omega) \in B\}), \qquad B \in \mathscr{B}.$$

We have this designated a new probability space, $(C, \mathscr{B}, Q_n)$, to serve as a probability model for the partial sum process $Y_n(\cdot)$. In order to be able to associate with the *sequence* of processes $\{Y_n(\cdot)\}$ a limit process $Y(\cdot)$, in the sense of convergence in distribution, we seek a measure Q on $(C, \mathscr{B})$ such that $Q_n \Rightarrow Q$. This will be given by Donsker's Theorem below.

An important probability measure on $(C, \mathscr{B})$ is the *Wiener measure*, that is, the probability distribution of one coordinate of the random path traced by a particle in "Brownian motion," or formally the probability measure defined by the properties:

(a) $W(\{x(\cdot): x(0) = 0\}) = 1$;

(b) for all $0 < t \le 1$ and $-\infty < \alpha < \infty$,

$$W(\{x(\cdot): x(t) \le \alpha\}) = \frac{1}{(2\pi t)^{1/2}} \int_{-\infty}^{\alpha} e^{-u^2/2t}\, du;$$

(c) for $0 \le t_0 \le t_1 \le \cdots \le t_k \le 1$ and $-\infty < \alpha_1, \ldots, \alpha_k < \infty$,

$$W\left(\bigcap_{i=1}^{k} \{x(\cdot): x(t_i) - x(t_{i-1}) \le \alpha_i\}\right)$$
$$= \prod_{i=1}^{k} W(\{x(\cdot): x(t_i) - x(t_{i-1}) \le \alpha_i\}).$$

The existence and uniqueness of such a measure is established, for example, in Billingsley (1968), Section 9.

A random element of $C[0, 1]$ having the distribution W is called a *Wiener process* and is denoted for convenience by $\{W(t), 0 \le t \le 1\}$, or simply by W. Thus, for a Wiener process $W(\cdot)$, properties (a), (b) and (c) tell us that

(a) $W(0) = 0$ with probability 1;

(b) $W(t)$ is $N(0, t)$, each $t \in (0, 1]$;

(c) for $0 \le t_0 \le t_1 \le \cdots \le t_k \le 1$, the increments $W(t_1) - W(t_0), \ldots, W(t_k) - W(t_{k-1})$ are mutually independent.

We are now ready to state the generalization of the Lindeberg–Lévy CLT.

Theorem (Donsker). *Let $\{X_i\}$ be I.I.D. with mean μ and finite variance σ^2. Define $Y_n(\cdot)$ and Q_n as above. Then*

$$Q_n \Rightarrow W.$$

(Alternatively, we may state this convergence as $Y_n(\cdot) \xrightarrow{d} W(\cdot)$ in $C[0, 1]$.) The theorem as stated above is proved in Billingsley (1968), Section 10. However, the theorem was first established, in a different form, by Donsker (1951).

To see that the Donsker theorem contains the Lindeberg–Lévy CLT, consider the set

$$B_\alpha = \{x(\cdot): x(1) \le \alpha\}$$

in $C[0, 1]$. It may be verified that $B_\alpha \in \mathscr{B}$. Since

$$Y_n(1) = \frac{\sum_{i=1}^n (X_i - \mu)}{\sigma n^{1/2}},$$

we have

$$P\left(\frac{\sum_{i=1}^n (X_i - \mu)}{\sigma n^{1/2}} \le \alpha\right) = Q_n(B_\alpha).$$

It may be verified that B_α is a W-continuity set, that is, $W(\partial B_\alpha) = 0$. Hence, by (**) of **1.11.3**, Donsker's Theorem yields

$$\lim_{n\to\infty} Q_n(B_\alpha) = W(B_\alpha).$$

Next one verifies (see **1.11.5**(i) for discussion) that

$$W(B_\alpha) = \Phi(\alpha).$$

Since α is chosen arbitrarily, the Lindeberg–Lévy CLT follows.

Now let us apply the Donsker theorem in connection with the random variable

$$M_n = \max_{0\le k\le n} \frac{\sum_{i=1}^k (X_i - \mu)}{\sigma n^{1/2}} = \sup_{0\le t\le 1} Y_n(t)$$

considered in **1.11.2**. Consider the set

$$B_\alpha^* = \left\{x(\cdot): \sup_{0\le t\le 1} x(t) \le \alpha\right\}.$$

It may be verified that B_α^* belongs to $\mathscr{B}$ and is a W-continuity set, so that

$$\lim_{n\to\infty} P(M_n \le \alpha) = \lim_{n\to\infty} Q_n(B_\alpha^*) = W(B_\alpha^*).$$

By determining (again, see **1.11.5**(i) for discussion) that

$$W(B_\alpha^*) = \left(\frac{2}{\pi}\right)^{1/2} \int_0^\alpha e^{-(1/2)u^2}\, du, \qquad (\alpha > 0),$$

one obtains the limit distribution of M_n.

The fact that the sets $\{B_\alpha^*, \alpha > 0\}$ are W-continuity sets is equivalent to the functional g: $g(x(\cdot)) = \sup_{0 \le t \le 1} x(t)$ being *continuous* (relative to the metric ρ) with W-probability 1. Thus, by an appropriate extension of Theorem 1.7(iii), the preceding argument could be structured as follows:

$$M_n = g(Y_n(\cdot)) \xrightarrow{d} g(W(\cdot)) = \sup_{0 \le t \le 1} W(t).$$

Elaboration of this approach is found in Billingsley (1968).

1.11.5 Complementary Remarks

(i) The application of Donsker's Theorem to obtain the limit distribution of some functional of the partial sum process $Y_n(\cdot)$ requires the evaluation of quantities such as $W(B_\alpha)$ and $W(B_\alpha^*)$. This step may be carried out by a *separate* application of Donsker's Theorem. For example, to evaluate $W(B_\alpha^*)$, the quantity $\lim_n P(M_n \le \alpha)$ is evaluated for a *particular* I.I.D. sequence $\{X_i\}$, one selected to make the computations easy. Then Donsker's Theorem tells us that the limit so obtained is in fact $W(B_\alpha^*)$. Thus $W(B_\alpha^*)$ has been evaluated, so that—again by Donsker's Theorem—the quantity $\lim_n P(M_n \le \alpha)$ is known for the *general* case of I.I.D. X_i's with finite variance. Such a technique for finding $\lim_n P(M_n \le \alpha)$ in the general case represents an application of what is known as the "*invariance principle.*" It is based on the fact that the limit in question is *invariant* over the choice of sequence $\{X_i\}$, within a wide class of sequences.

(ii) Other limit theorems besides the CLT can likewise be reformulated and generalized via the theory of convergence of probability measures on metric spaces. In connection with a given sequence of random variables $\{X_i\}$, we may consider other random functions than $Y_n(\cdot)$, and other function spaces than $C[0, 1]$.

(iii) In later chapters, a number of relevant stochastic processes will be pointed out in connection with various statistics arising for consideration. However, stochastic process aspects will not be stressed in this book. The intention is merely to orient the reader for investigation of these matters elsewhere.

(iv) For detailed treatment of the topic of convergence of probability measures on metric spaces, the reader is referred to Billingsley (1968) and Parthasarathy (1967).

1.12 TAYLOR'S THEOREM; DIFFERENTIALS

1.12.1 Taylor's Theorem

The following theorem is proved in Apostol (1957), p. 96.

Theorem A (Taylor). *Let the function* g *have a finite* n*th derivative* $g^{(n)}$ *everywhere in the open interval* (a, b) *and* (n − 1)*th derivative* $g^{(n-1)}$ *continuous* *in the closed interval* $[a, b]$. *Let* $x \in [a, b]$. *For each point* $y \in [a, b]$, $y \neq x$, *there exists a point* z *interior to the interval joining* x *and* y *such that*

$$g(y) = g(x) + \sum_{k=1}^{n-1} \frac{g^{(k)}(x)}{k!}(y-x)^k + \frac{g^{(n)}(z)}{n!}(y-x)^n.$$

Remarks. (i) For the case $x = a$, we may replace $g^{(k)}(x)$ in the above formula by $g_+^{(k)}(a)$, the kth order right-hand derivative of g at the point a; in place of continuity of $g^{(n-1)}(x)$ at $x = a$, it is assumed that $g_+^{(k)}(x)$ is continuous at $x = a$, for each $k = 1, \ldots, n-1$. Likewise, for $x = b$, $g^{(k)}(x)$ may be replaced by the left-hand derivative $g_-^{(k)}(b)$. These extensions are obtained by minor modification of Apostol's proof of Theorem A.

(ii) For a generalized Taylor formula replacing derivatives by finite differences, see Feller (1966), p. 227. ■

We can readily establish a multivariate version of Theorem A by reduction to the univariate case. (We follow Apostol (1957), p. 124.)

Theorem B (Multivariate Version). *Let the function* g *defined on* R^m *possess continuous partial derivatives of order* n *at each point of an open set* $S \subset R^m$. *Let* $\mathbf{x} \in S$. *For each point* $\mathbf{y}$, $\mathbf{y} \neq \mathbf{x}$, *such that the line segment* $L(\mathbf{x}, \mathbf{y})$ *joining* $\mathbf{x}$ *and* $\mathbf{y}$ *lies in* S, *there exists a point* $\mathbf{z}$ *in the interior of* $L(\mathbf{x}, \mathbf{y})$ *such that*

$$g(\mathbf{y}) = g(\mathbf{x}) + \sum_{k=1}^{n-1} \frac{1}{k!} \sum_{i_1=1}^{m} \cdots \sum_{i_k=1}^{m} \left.\frac{\partial^k g(t_1, \ldots, t_m)}{\partial t_{i_1} \cdots \partial t_{i_m}}\right|_{\mathbf{t}=\mathbf{x}} \cdot \prod_{j=1}^{k} (y_{i_j} - x_{i_j})$$

$$+ \frac{1}{n!} \sum_{i_1=1}^{m} \cdots \sum_{i_n=1}^{m} \left.\frac{\partial^n g(t_1, \ldots, t_m)}{\partial t_{t_1} \cdots \partial t_{i_n}}\right|_{\mathbf{t}=\mathbf{z}} \prod_{j=1}^{n} (y_{i_j} - x_{i_j}).$$

PROOF. Define $H(\alpha) = g(\mathbf{x} + \alpha(\mathbf{y} - \mathbf{x}))$ for real α. By the assumed continuity of the partial derivatives of g, we may apply an extended chain rule for differentiation of H and obtain

$$H'(\alpha) = \sum_{i=1}^{m} \left.\frac{\partial g(t_1, \ldots, t_m)}{\partial t_i}\right|_{\mathbf{t}=\mathbf{x}+\alpha(\mathbf{y}-\mathbf{x})} \cdot (y_i - x_i)$$

and likewise, for $2 \leq k \leq n$,

$$H^{(k)}(\alpha) = \sum_{i_1=1}^{m} \cdots \sum_{i_k=1}^{m} \left.\frac{\partial^k g(t_1, \ldots, t_m)}{\partial t_{i_1} \cdots \partial t_{i_k}}\right|_{\mathbf{t}=\mathbf{x}+\alpha(\mathbf{y}-\mathbf{x})} \cdot \prod_{j=1}^{k} (y_{i_j} - x_{i_j}).$$

Since $L(\mathbf{x}, \mathbf{y}) \subset S$, S open, it follows that the function H satisfies the conditions of Theorem A with respect to the interval $[a, b] = [0, 1]$. Consequently, we have

$$H(1) = H(0) + \sum_{k=1}^{n} \frac{H^{(k)}(0)}{k!} + \frac{H^{(n)}(z)}{n!},$$

where $0 < z < 1$. Now note that $H(1) = g(\mathbf{y})$, $H(0) = g(\mathbf{x})$, etc. ■

A useful alternate form of Taylor's Theorem is the following, which requires the nth order differentiability to hold only at the point x and which characterizes the asymptotic behavior of the remainder term.

Theorem C (Young's form of Taylor's Theorem). *Let* g *have a finite* nth *derivative at the point* x. *Then*

$$g(y) - g(x) - \sum_{i=1}^{n} \frac{g^{(k)}(x)}{k!}(y - x)^k = o(|y - x|^n), \qquad \textit{as} \quad y \to x.$$

PROOF. Follows readily by induction. Or see Hardy (1952), p. 278. ■

1.12.2 Differentials

The appropriate multi-dimensional generalization of derivative of a function of one argument is given in terms of the *differential*. A function g defined on R^m is said to have a *differential*, or to be *totally differentiable*, at the point $\mathbf{x}_0$ if the partial derivatives

$$\frac{\partial g}{\partial x_1}, \ldots, \frac{\partial g}{\partial x_m}$$

all exist at $\mathbf{x} = \mathbf{x}_0$ and the function

$$g(\mathbf{x}_0; \mathbf{t}) = \sum_{i=1}^{m} \frac{\partial g}{\partial x_i}\bigg|_{\mathbf{x} = \mathbf{x}_0} \cdot t_i, \qquad \mathbf{t} \in R^m,$$

(called the "differential") satisfies the property that, for every $\varepsilon > 0$, there exists a neighborhood $N_\varepsilon(\mathbf{x}_0)$ such that

$$|g(\mathbf{x}) - g(\mathbf{x}_0) - g(\mathbf{x}_0; \mathbf{x} - \mathbf{x}_0)| \leq \varepsilon \|\mathbf{x} - \mathbf{x}_0\|, \qquad \text{all} \quad \mathbf{x} \in N_\varepsilon(\mathbf{x}_0).$$

Some interrelationships among differentials, partial derivatives, and continuity are expressed in the following result.

Lemma (Apostol (1957), pp. 110 and 118). (i) *If* g *has a differential at* $\mathbf{x}_0$, *then* g *is continuous at* $\mathbf{x}_0$.

(ii) *If the partial derivatives* $\partial g/\partial x_i$, $1 \leq i \leq m$, *exist in a neighborhood of* $\mathbf{x}_0$ *and are continuous at* $\mathbf{x}_0$, *then* g *has a differential at* $\mathbf{x}_0$.

1.13 CONDITIONS FOR DETERMINATION OF A DISTRIBUTION BY ITS MOMENTS

Let F be a distribution on the real line with moment sequence

$$\alpha_k = \int_{-\infty}^{\infty} x^k \, dF(x), \qquad k = 1, 2, \ldots.$$

The question of when an F having a given moment sequence $\{\alpha_k\}$ is the unique such distribution arises, for example, in connection with the Fréchet and Shohat Theorem (1.5.1B). Some sufficient conditions are as follows.

Theorem. *The moment sequence* $\{\alpha_k\}$ *determines the distribution* F *uniquely if the Carleman condition*

$$(*)\qquad \sum_{n=1}^{\infty} \alpha_{2n}^{-1/2n} = \infty$$

holds. Each of the following conditions is sufficient for (*):

(i) $\overline{\lim_{k\to\infty}} \frac{1}{k}\left(\int_{-\infty}^{\infty} |x|^k \, dF(x)\right)^{1/k} = \lambda < \infty;$

(ii) $\sum_{k=1}^{\infty} \frac{\alpha_k}{k!} \lambda^k$ *converges absolutely in an interval* $|\lambda| < \lambda_0$.

For proofs, discussion and references to further literature, see Feller (1966), pp. 224, 230 and 487.

An example of *nonuniqueness* consists of the class of density functions

$$\phi_\alpha(t) = \tfrac{1}{24} e^{-t^{1/4}}(1 - \alpha \sin t^{1/4}), \qquad 0 < t < \infty,$$

for $0 < \alpha < 1$, all of which possess the same moment sequence. For discussion of this and other oddities, see Feller (1966), p. 224.

1.14 CONDITIONS FOR EXISTENCE OF MOMENTS OF A DISTRIBUTION

Lemma. *For any random variable* X,

(i) $E|X| = \int_0^\infty P(|X| \geq t)dt, \qquad (\leq \infty)$

and

(ii) *if* $E|X| < \infty$, *then* $P(|X| \geq t) = o(t^{-1}),\ t \to \infty$.

PROOF. Denote by G the distribution function of $|X|$ and let c denote a (finite) continuity point of G. By integration by parts, we have

$$\text{(A)}\qquad \int_0^c x\,dG(x) = \int_0^c [1 - G(x)]dx - c[1 - G(c)],$$

and hence also

$$\text{(B)}\qquad \int_0^c x\,dG(x) \leq \int_0^c [1 - G(x)]dx.$$

Further, it is easily seen that

$$\text{(C)} \qquad c[1 - G(c)] \le \int_c^{\infty} x\, dG(x).$$

Now suppose that $E|X| = \infty$. Then (B) yields (i) for this case. On the other hand, suppose that $E|X| < \infty$. Then (C) yields (ii). Also, making use of (ii) in conjunction with (A), we obtain (i) for this case. ■

The lemma immediately yields (Problem 1.P.29) its own generalization:

Corollary. *For any random variable* X *and real number* $r > 0$,

(i) $E|X|^r = r\int_0^\infty t^{r-1}P(|X| \ge t)dt$

and

(ii) *if* $E|X|^r < \infty$, *then* $P(|X| \ge t) = t) = o(t^{-r}), t \to \infty$.

Remark. It follows that a necessary and sufficient condition for $E|X|^r < \infty$ is that $t^{r-1}P(|X| \ge t)$ be integrable. Also, if $P(|X| \ge t) = O(t^{-s})$, then $E|X|^r < \infty$ for all $r < s$. ■

1.15 ASYMPTOTIC ASPECTS OF STATISTICAL INFERENCE PROCEDURES

By "inference procedure" is usually meant a statistical procedure for estimating a parameter or testing a hypothesis about a parameter. More generally, it may be cast in decision-theoretic terms as a procedure for selecting an action in the face of risks that depend upon an unknown parameter. In the present discussion, the more general context will not be stressed but should be kept in mind nevertheless.

Let the family of possible models for the data be represented as a collection of probability spaces $\{(\Omega, \mathscr{A}, P_\theta), \theta \in \Theta\}$, indexed by the "parameter" θ. In discussing "estimation," we shall consider estimation of some parametric function $g(\theta)$. In discussing "hypothesis testing," we have in mind some "null hypothesis": $\theta \in \Theta_0(\subset \Theta)$. In either case, the relevant statistic ("estimator" or "test statistic") will be represented as a *sequence* of statistics $T_1, T_2, \ldots$. Typically, by "statistic" we mean a specified function of the sample, and T_n denotes the evaluation of the function at the first n sample observations $X_1, \ldots, X_n$. This book deals with the *asymptotic* properties of a great variety of sequences $\{T_n\}$ of proven or potential interest in statistical inference.

For such sequences $\{T_n\}$, we treat several important asymptotic features: "asymptotic unbiasedness" (in the context of estimation only); "consistency" (in estimation) and "almost sure behavior"; "asymptotic distribution theory"; "asymptotic relative efficiency." These notions are discussed in **1.15.1–1.15.4**, respectively. The concept of "asymptotic efficiency," which is related to "asymptotic relative efficiency," will be introduced in Chapter 4, in connection with the theory of maximum likelihood estimation. Some further important concepts—"deficiency," "asymptotic sufficiency," "local asymptotic normality," "local asymptotic admissibility," and "local asymptotic minimaxity"—are not treated in this book.

1.15.1 Asymptotic Unbiasedness (in Estimation)

Recall that in estimation we say that an estimator T of a parametric function $g(\theta)$ is *unbiased* if $E_\theta\{T\} = g(\theta)$, all $\theta \in \Theta$. Accordingly, we say that a sequence of estimators $\{T_n\}$ is *asymptotically unbiased* for estimation of $g(\theta)$ if

$$\lim_{n\to\infty} E_\theta\{T_n\} = g(\theta), \qquad \text{each} \quad \theta \in \Theta.$$

(In hypothesis testing, a test is unbiased if at each $\theta \notin \Theta_0$, the "power" of the test is at least as high as the "size" of the test. An asymptotic version of this concept may be defined also, but we shall not pursue it.)

1.15.2 Consistency (in Estimation) and Almost Sure Behavior

A sequence of estimators $\{T_n\}$ for a parametric function $g(\theta)$ is "consistent" if T_n converges to $g(\theta)$ in some appropriate sense. We speak of *weak consistency*,

$$T_n \xrightarrow{p} g(\theta),$$

strong consistency,

$$T_n \xrightarrow{wp1} g(\theta),$$

and *consistency in* rth *mean*,

$$T_n \xrightarrow{r\text{th}} g(\theta).$$

When the term "consistent" is used without qualification, usually the *weak* mode is meant.

(In hypothesis testing, consistency means that at each $\theta \notin \Theta_0$, the power of the test tends to 1 as $n \to \infty$. We shall not pursue this notion.)

Consistency is usually considered a minimal requirement for an inference procedure. Those procedures not having such a property are usually dropped from consideration.

A useful technique for establishing mean square consistency of an estimator T_n is to show that it is asymptotically unbiased and has variance tending to 0.

Recalling the relationships considered in Section 1.3, we see that strong consistency may be established by proving weak or rth mean consistency with a sufficiently fast rate of convergence.

There arises the question of which of these forms of consistency is of the greatest *practical* interest. To a large extent, this is a philosophical issue, the answer depending upon one's point of view. Concerning rth mean consistency versus the weak or strong versions, the issue is between "moments" and "probability concentrations" (see **1.15.4** for some further discussion). Regarding weak versus strong consistency, some remarks in support of insisting on the *strong* version follow.

(i) Many statisticians would find it distasteful to use an estimator which, if sampling were to continue indefinitely, could possibly fail to converge to the correct value. After all, there should be some pay-off for increased sampling, which advantage should be exploited by any "good" estimator.

(ii) An example presented by Stout (1974), Chapter 1, concerns a physician treating patients with a drug having unknown cure probability θ (the same for each patient). The physician intends to continue use of the drug until a superior alternative is known. Occasionally he assesses his experience by estimating θ by the proportion $\hat{\theta}_n$ of cures for the n patients treated up to that point in time. He wants to be able to estimate θ within a prescribed tolerance $\varepsilon > 0$. Moreover, he desires the reassuring feature that, with a specified high probability, he can reach a point in time such that his current estimate has become within ε of θ and no subsequent value of the estimator would misleadingly wander more than ε from θ. That is, the physician desires, for prescribed $\delta > 0$, that there exist an integer N such that

$$P\left(\max_{n \geq N} |\hat{\theta}_n - \theta| \leq \varepsilon\right) \geq 1 - \delta.$$

Weak consistency (which follows in this case by the WLLN) asserts only that

$$P(|\hat{\theta}_n - \theta| \leq \varepsilon) \to 1, \qquad n \to \infty,$$

and hence fails to supply the reassurance desired. Only by *strong* consistency (which follows in this case by the SLLN) is the existence of such an N gaurranteed.

(iii) When confronted with two competing sequences $\{T_n\}$ and $\{T_n^*\}$ of estimators or test statistics, one wishes to select the *best*. This decision calls upon knowledge of the optimum properties, whatever they may be, possessed by the two sequences. In particular, strong consistency thus becomes a useful distinguishing property.

So far we have discussed "consistency" and have focused upon the strong version. More broadly, we can retain the focus on strong consistency but widen the scope to include the precise asymptotic order of mangitude of the fluctuations $T_n - g(\theta)$, just as in **1.10** we considered the LIL as a refinement of the SLLN. In this sense, as a refinement of strong convergence, we will seek to characterize the "almost sure behavior" of sequences $\{T_n\}$. Such characterizations are of interest not only for sequences of estimators but also for sequences of test statistics. (In the latter case $g(\theta)$ represents a parameter to which the test statistic T_n converges under the model indexed by θ.)

1.15.3 The Role of Asymptotic Distribution Theory for Estimators and Test Statistics

We note that consistency of a sequence T_n for $g(\theta)$ implies convergence in distribution:

$$(*) \qquad T_n \xrightarrow{d} g(\theta).$$

However, for purposes of practical application to approximate the probability distribution of T_n, we need a result of the type which asserts that a suitably normalized version,

$$\tilde{T}_n = \frac{T_n - a_n}{b_n},$$

converges in distribution to a *nondegenerate* random variable $\tilde{T}$, that is,

$$(**) \qquad F_{\tilde{T}_n} \Rightarrow F_{\tilde{T}},$$

where $F_{\tilde{T}}$ is a nondegenerate distribution. Note that (*) is of no use in attempting to approximate the probability $P(T_n \leq t_n)$, unless one is satisfied with an approximation constrained to take only the values 0 or 1. On the other hand, writing (assuming $b_n > 0$)

$$P(T_n \leq t_n) = P\left(\tilde{T}_n \leq \frac{t_n - a_n}{b_n}\right),$$

we obtain from (**) the more realistic approximation $F_{\tilde{T}}((t_n - a_n)/b_n)$ for the probability $P(T_n \leq t_n)$.

Such considerations are relevant in calculating the approximate confidence coefficients of confidence intervals $T_n \pm d_n$ in connection with estimators T_n, and in finding critical points c_n for forming critical regions $\{T_n > c_n\}$ of approximate specified size in connection with test statistics T_n.

Thus, in developing the minimal amount of asymptotic theory regarding a sequence of statistics $\{T_n\}$, it does not suffice merely to establish a consistency property. In addition to such a property, one must also seek normalizing constants a_n and b_n such that $(T_n - a_n)/b_n$ converges in distribution to a random variable having a nondegenerate distribution (which then must be determined).

1.15.4 Asymptotic Relative Efficiency

For two competing statistical procedures A and B, suppose that a desired performance criterion is specified and let n_1 and n_2 be the respective sample sizes at which the two procedures "perform equivalently" with respect to the adopted criterion. Then the ratio

$$\frac{n_1}{n_2}$$

is usually regarded as the *relative efficiency* (in the given sense) of procedure B relative to procedure A. Suppose that the specified performance criterion is tightened in a way that causes the required sample sizes n_1 and n_2 to tend to ∞. If in this case the ratio n_1/n_2 approaches to limit L, then the value L represents the *asymptotic relative efficiency* of procedure B relative to procedure A. It is stressed that the value L obtained depends upon the particular performance criterion adopted.

As an example, consider estimation. Let $\{T_{An}\}$ and $\{T_{Bn}\}$ denote competing estimation sequences for a parametric function $g(\theta)$. Suppose that

$$T_{An} \text{ is } AN\left(g(\theta), \frac{\sigma_A^2(\theta)}{n}\right), \qquad T_{Bn} \text{ is } AN\left(g(\theta), \frac{\sigma_B^2(\theta)}{n}\right).$$

If our criterion is based upon the *variance* parameters $\sigma_A^2(\theta)$ and $\sigma_B^2(\theta)$ of the asymptotic distributions, then the two procedures "perform equivalently" at respective sample sizes n_1 and n_2 satisfying

$$\frac{\sigma_A^2(\theta)}{n_1} \sim \frac{\sigma_B^2(\theta)}{n_2},$$

in which case

$$\frac{n_1}{n_2} \to \frac{\sigma_A^2(\theta)}{\sigma_B^2(\theta)}.$$

Thus $\sigma_A^2(\theta)/\sigma_B^2(\theta)$ emerges as a measure of asymptotic relative efficiency of procedure B relative to procedure A. If, however, we adopt as performance criterion the *probability concentration* of the estimate in an ε-neighborhood of $g(\theta)$, for ε specified and fixed, then a different quantity emerges as the measure of asymptotic relative efficiency. For a comparison of $\{T_{An}\}$ and $\{T_{Bn}\}$ by this criterion, we may consider the quantities

$$P_{An}(\varepsilon, \theta) = P_\theta(|T_{An} - g(\theta)| > \varepsilon), \qquad P_{Bn}(\varepsilon, \theta) = P_\theta(|T_{Bn} - g(\theta)| > \varepsilon),$$

and compare the rates at which these quantities tend to 0 as $n \to \infty$. In typical cases, the convergence is "exponentially fast";

$$\frac{\log P_{An}(\varepsilon, \theta)}{n} \to -\gamma_A(\varepsilon, \theta), \qquad \frac{\log P_{Bn}(\varepsilon, \theta)}{n} \to \gamma_B(\varepsilon, \theta).$$

In such a case, the two procedures may be said to "perform equivalently" at respective sample sizes n_1 and n_2 satisfying

$$\frac{\log P_{Bn_2}(\varepsilon, \theta)}{\log P_{An_1}(\varepsilon, \theta)} \to 1.$$

In this case

$$\frac{n_1}{n_2} \sim \frac{n_2^{-1} \log P_{Bn_2}(\varepsilon, \theta)}{n_1^{-1} \log P_{An_1}(\varepsilon, \theta)} \to \frac{\gamma_B(\varepsilon, \theta)}{\gamma_A(\varepsilon, \theta)},$$

yielding $\gamma_B(\varepsilon, \theta)/\gamma_A(\varepsilon, \theta)$ as a measure of asymptotic relative efficiency of procedure B relative to procedure A, in the sense of the probability concentration criterion.

It is thus seen that the "asymptotic variance" and "probability concentration" criteria yield differing measures of asymptotic relative efficiency. It can happen in a given problem that these two approaches lead to *discordant* measures (one having value >1, the other <1). For an example, see Basu (1956).

The preceding discussion has been confined to asymptotic relative efficiency in *estimation*. Various examples will appear in Chapters 2–9. For the asymptotic variance criterion, the multidimensional version and the related concept of *asymptotic efficiency* (in an "absolute" sense) will be treated in Chapter 4. The notion of asymptotic relative efficiency in *testing* is deferred to Chapter 10, which is devoted wholly to the topic. (The apparent dichotomy between estimation and testing should not, however, be taken too seriously, for "testing" problems can usually be recast in the context of estimation, and vice versa.)

Further introductory discussion of asymptotic relative efficiency is found in Cramér (1946), Sections 37.3–37.5, Fraser (1957), Section 7.3, Rao (1973), Sections 5c.2 and 7a.7, and Bahadur (1967).

1.P PROBLEMS

Section 1.1

1. Prove Lemma 1.1.4.

Section 1.2

2. (a) Show that $(X_{n1}, \ldots, X_{nk}) \xrightarrow{p} (X_1, \ldots, X_k)$ if and only if $X_{nj} \xrightarrow{p} X_j$ for each $j = 1, \ldots, k$.

(b) Same problem for $\xrightarrow{wp1}$.

(c) Show that $\mathbf{X}_n = (X_{n1}, \ldots, X_{nk}) \xrightarrow{wp1} \mathbf{X}_\infty = (X_{\infty 1}, \ldots, X_{\infty k})$ if and only if, for every $\varepsilon > 0$.

$$\lim_{n\to\infty} P\{\|\mathbf{X}_m - \mathbf{X}_\infty\| < \varepsilon, \text{ all } m \geq n\} = 1.$$

3. Show that $X_n \xrightarrow{d} X$ implies $X_n = O_p(1)$.
4. Show that $U_n = o_p(V_n)$ implies $U_n = O_p(V_p)$.
5. Resolve the question posed in **1.2.6**.

Section 1.3

6. Construct a sequence $\{X_n\}$ convergent *wp*1 but *not* in *r*th mean, for any $r > 0$, by taking an appropriate subsequence of the sequence in Example 1.3.8D.

Section 1.4

7. Prove Lemma 1.4B.

8. Verify that Lemma 1.4C contains two generalizations of Theorem 1.3.6.

9. Prove Theorem 1.4D.

Section 1.5

10. Do the task assigned in the proof of Theorem 1.5.1A.

11. (a) Show that Scheffé's Theorem (1.5.1C) is indeed a criterion for convergence in distribution.

(b) Exemplify a sequence of densities f_n pointwise convergent to a function f not a density.

12. (a) Prove Pólya's Theorem (1.5.3).

(b) Give a counterexample for the case of F having discontinuities.

13. Prove part (ii) of Slutsky's Theorem (1.5.4).

14. (a) Prove Corollary 1.5.4A by direct application of Theorem 1.5.4(i).

(b) Prove Corollary 1.5.4B.

15. Prove Lemma 1.5.5A. (Hint: apply Pólya's Theorem.)

16. Prove Lemma 1.5.5B.

17. Show that $\mathbf{X}_n$ is $AN(\boldsymbol{\mu}_n, c_n^2\boldsymbol{\Sigma})$ if and only if

$$\frac{\mathbf{X}_n - \boldsymbol{\mu}_n}{c_n} \xrightarrow{d} N(\mathbf{0}, \boldsymbol{\Sigma}).$$

Here $\{c_n\}$ is a sequence of real constants and $\boldsymbol{\Sigma}$ a covariance matrix.

18. Prove or give counter-example: If $X_n \xrightarrow{d} X$ and $Y_n \xrightarrow{p} Y$, then $X_n + Y_n \xrightarrow{d} X + Y$.

19. Let X_n be $AN(\mu, \sigma^2/n)$, let Y_n be $AN(c, v/n)$, $c \neq 0$, and put $Z_n = \sqrt{n}(X_n - \mu)/Y_n$. Show that Z_n is $AN(0, \sigma^2/c^2)$. (Hint: apply Problem 1.P.20.)

20. Let X_n be $AN(\mu, \sigma_n^2)$. Show that $X_n \xrightarrow{p} \mu$ if and only if $\sigma_n \to 0$, $n \to \infty$. (See Problem 1.P.23 for a multivariate extension.)

21. Let X_n be $AN(\mu, \sigma_n^2)$ and let $Y_n = 0$ with probability $1 - n^{-1}$ and $= n$ with probability n^{-1}. Show that $X_n + Y_n$ is $AN(\mu, \sigma_n^2)$.

Section 1.7

22. Verify Application B of Corollary 1.7. (Hint: Apply Problem 1.P.17, Corollary 1.7 and then Theorem 1.7 with $g(x) = \sqrt{x}$.)

23. Verify Application C of Corollary 1.7. (Hint: Apply the Cramér–Wold device, Problem 1.P.20, and the argument used in the previous problem. Alternatively, instead of the latter, Problems 1.P.2 and 1.P.14(b) may be used.)

24. (a) Verify Application D of Corollary 1.7.
(b) Do analogues hold for convergence in distribution?

Section 1.9

25. Derive Theorem 1.9.1B from Theorem 1.9.1A.

26. Obtain Corollary 1.9.3.

27. Let X_n be a χ_n^2 random variable.
(a) Show that X_n is $AN(n, 2n)$.
(b) Evaluate the bound on the error of approximation provided by the Berry–Esseen Theorem (with van Beeck's improved constant).

Section 1.13

28. Justify that the distribution $N(\mu, \sigma^2)$ is uniquely determined by its moments.

Section 1.14

29. Obtain Corollary 1.14 from Lemma 1.14.

Section 1.15

30. Let X_n have finite mean μ_n, $n = 1, 2, \ldots$. Consider estimation of a parameter θ by X_n. Answer (with justifications):
(a) If X_n is consistent for θ, must X_n be asymptotically unbiased?
(b) If X_n is asymptotically unbiased, must X_n be consistent?
(c) If X_n is asymptotically unbiased and $\text{Var}\{X_n\} \to 0$, must X_n be consistent?
(Hint: See Problem 1.P.21.)

CHAPTER 2

The Basic Sample Statistics

This chapter considers a sample $X_1, \ldots, X_n$ of independent observations on a distribution function F and examines the most basic types of statistic usually of interest. The *sample distribution function* and the closely related *Kolmogorov–Smirnov* and *Cramér–von Mises* statistics, along with *sample density functions*, are treated in Section 2.1. The *sample moments*, the *sample quantiles*, and the *order statistics* are treated in Sections 2.2, 2.3 and 2.4, respectively.

There exist useful *asymptotic representations*, first introduced by R. R. Bahadur, by which the sample quantiles and the order statistics may be expressed in terms of the sample distribution function as simple *sums* of random variables. These relationships and their applications are examined in Section 2.5.

By way of illustration of some of the results on sample moments, sample quantiles, and order statistics, a study of confidence intervals for (population) quantiles is provided in Section 2.6.

A common form of statistical reduction of a sample consists of grouping the observations into calls. The asymptotic multivariate normality of the corresponding *cell frequency vectors* is derived in Section 2.7.

Deeper investigation of the basic sample statistics may be carried out within the framework of stochastic process theory. Some relevant *stochastic processes* associated with a sample are pointed out in Section 2.8.

Many statistics of interest may be represented as *transformations* of one or more of the "basic" sample statistics. The case of functions of several sample moments or sample quantiles, or of cell frequency vectors, and the like, is treated in Chapter 3. The case of statistics defined as functionals of the sample distribution function is dealt with in Chapter 6.

Further, many statistics of interest may be conceptualized as some sorts of *generalization* of a "basic" type. A generalization of the idea of forming a sample average consists of the *U-statistics*, introduced by W. Hoeffding. These are studied in Chapter 5. As a generalization of single order statistics, the so-called *linear functions of order statistics* are investigated in Chapter 8.

2.1 THE SAMPLE DISTRIBUTION FUNCTION

Consider an I.I.D. sequence $\{X_i\}$ with distribution function F. For each sample of size n, $\{X_1, \ldots, X_n\}$, a corresponding *sample distribution function* F_n is constructed by placing at each observation X_i a mass $1/n$. Thus F_n may be represented as

$$F_n(x) = \frac{1}{n} \sum_{i=1}^{n} I(X_i \leq x), \qquad -\infty < x < \infty.$$

(The definition for F defined on R^k is completely analogous.)

For each fixed sample $\{X_1, \ldots, X_n\}$, $F_n(\cdot)$ is a *distribution function*, considered as a function of x. On the other hand, for each fixed value of x, $F_n(x)$ is a *random variable*, considered as a function of the sample. In a view encompassing both features, $F_n(\cdot)$ is a *random distribution function* and thus may be treated as a particular *stochastic process* (a random element of a suitable function space).

The simplest aspect of F_n is that, for each fixed x, $F_n(x)$ serves as an estimator of $F(x)$. For example, note that $F_n(x)$ is *unbiased*: $E\{F_n(x)\} = F(x)$. Other properties, such as consistency and asymptotic normality, are treated in **2.1.1.**

Considered as a whole, however, the function F_n is a very *basic* sample statistic, for from it the entire set of sample values can be recovered (although their order of occurrence is lost). Therefore, it can and does play a fundamental role in statistical inference. Various aspects are discussed in **2.1.2**, and some important random variables closely related to F_n are introduced. One of these, the *Kolmogorov–Smirnov* statistic, may be formulated in two ways: as a measure of distance between F_n and F, and as a test statistic for a hypothesis $H: F = F_0$. For the Kolmogorov–Smirnov *distance*, some probability inequalities are presented in **2.1.3**, the almost sure behavior is characterized in **2.1.4**, and the asymptotic distribution theory is given in **2.1.5**. Asymptotic distribution theory for the Kolmogorov–Smirnov *test statistic* is discussed in **2.1.6**. For another such random variable, the *Cramér-von Mises* statistic, almost sure behavior and asymptotic distribution theory is discussed in **2.1.7**.

For the case of a distribution function F having a *density* f, "sample density function" estimators (of f) are of interest and play similar roles to F_n. However, their theoretical treatment is more difficult. A brief introduction is given in **2.1.8**.

Finally, in **2.1.9**, some complementary remarks are made. For a stochastic process formulation of the sample distribution function, and for related considerations, see Section 2.8.

2.1.1 $F_n(x)$ as Pointwise Estimator of $F(x)$

We have noted above that $F_n(x)$ is unbiased for estimation of $F(x)$. Moreover,

$$\text{Var}\{F_n(x)\} = \frac{F(x)[1 - F(x)]}{n} \to 0, \qquad n \to \infty,$$

so that $F_n(x) \xrightarrow{2\text{nd}} F(x)$. That is, $F_n(x)$ is *consistent in mean square* (and hence weakly consistent) for estimation of $F(x)$. Furthermore, by a direct application of the SLLN (Theorem 1.8B), $F_n(x)$ is *strongly consistent*: $F_n(x) \xrightarrow{wp1} F(x)$. Indeed, the latter convergence holds *uniformly in x* (*see* **2.1.4**).

Regarding the distribution theory of $F_n(x)$, note that the *exact* distribution of $nF_n(x)$ is simply binomial $(n, F(x))$. And, immediately from the Lindeberg–Lévy CLT (1.9.1A), the *asymptotic* distribution is given by

Theorem. *For each fixed* x, $-\infty < x < \infty$,

$$F_n(x) \quad \textit{is} \quad AN\left(F(x), \frac{F(x)[1 - F(x)]}{n}\right).$$

2.1.2 The Role of the Sample Distribution Function in Statistical Inference

We shall consider several ways in which the sample distribution function is utilized in statistical inference. Firstly, its most direct application is for estimation of the population distribution function F. Besides *pointwise* estimation of $F(x)$, each x, as considered in **2.1.1**, it is also of interest to characterize *globally* the estimation of F by F_n. To this effect, a very useful measure of closeness of F_n to F is the *Kolmogorov–Smirnov* distance

$$D_n = \sup_{-\infty < x < \infty} |F_n(x) - F(x)|.$$

A related problem is to express *confidence bands* for $F(x)$, $-\infty < x < \infty$. Thus, for selected functions $a(x)$ and $b(x)$, it is of interest to compute probabilities of the form

$$P(F_n(x) - a(x) \le F(x) \le F_n(x) + b(x), -\infty < x < \infty).$$

The general problem is quite difficult; for discussion and references, see Durbin (1973a), Section 2.5. However, in the simplest case, namely $a(x) \equiv b(x) \equiv d$, the problem reduces to computation of

$$P(D_n < d).$$

In this form, and for F continuous, the problem of confidence bands is treated in Wilks (1962), Section 11.7, as well as in Durbin (1973a).

Secondly, we consider "goodness of fit" test statistics based on the sample distribution function. The null hypothesis in the simple case is $H: F = F_0$, where F_0 is specified. A useful procedure is the *Kolmogorov–Smirnov test statistic*

$$\Delta_n = \sup_{-\infty < x < \infty} |F_n(x) - F_0(x)|,$$

which reduces to D_n under the null hypothesis. More broadly, a class of such statistics is obtained by introducing weight functions:

$$\sup_{-\infty < x < \infty} |w(x)[F_n(x) - F_0(x)]|.$$

(Similarly, more general versions of D_n may be formulated.) There are also one-sided versions of Δ_n:

$$\Delta_n^+ = \sup_{-\infty < x < \infty} [F_n(x) - F_0(x)],$$

$$\Delta_n^- = \sup_{-\infty < x < \infty} [F_0(x) - F_n(x)].$$

Another important class of statistics is based on the *Cramér–von Mises test statistic*

$$C_n = n \int_{-\infty}^{\infty} [F_n(x) - F_0(x)]^2 \, dF_0(x)$$

and takes the general form $n \int w(F_0(x))[F_n(x) - F_0(x)]^2 \, dF_0(x)$. For example, for $w(t) = [t(1 - t)]^{-1/2}$, each discrepancy $F_n(x) - F_0(x)$ becomes weighted by the reciprocal of its standard deviation (under H_0), yielding the Anderson–Darling statistic.

Thirdly, some so-called "tests on the circle" are based on F_n. The context concerns data in the form of directions, and the null hypothesis of randomness of directions is formulated as randomness of n points distributed on the circumference of the unit circle. With appropriately defined X_i's, a suitable test statistic is the Kuiper statistic

$$V_n = \Delta_n^+ - \Delta_n^-.$$

This statistic also happens to have useful properties when used as an alternative to Δ_n in the goodness-of-fit problem.

Finally, we mention that the theoretical investigation of many statistics of interest can advantageously be carried out by representing the statistics, either exactly or approximately, as functionals of the sample distribution function, or as functionals of a stochastic process based on the sample distribution function. (See Section 2.8 and Chapter 6). In this respect, metrics such as D_n play a useful role.

In light of the foregoing remarks, it is seen that the random variable D_n and related random variables merit extensive investigation. Thus we devote **2.1.3–2.1.6** to this purpose.

An excellent introduction to the theory underlying statistical tests based on F_n is the monograph by Durbin (1973a). An excellent overview of the probabilistic theory for F_n considered as a stochastic process, and with attention to multidimensional F, is the survey paper of Gaenssler and Stute (1979). Useful further reading is provided by the references in these manuscripts. Also, further elementary reading of general scope consists of Bickel and Doksum (1977), Section 9.6, Cramér (1946), Section 25.3, Lindgren (1968), Section 6.4, Noether (1967), Chapter 4, Rao (1973), Section 6f.1, and Wilks (1962), Chapters 11 and 14.

2.1.3 Probability Inequalities for the Kolmogorov–Smirnov Distance

Consider an I.I.D. sequence $\{\mathbf{X}_i\}$ of elements of R^k, let F and F_n denote the corresponding population and sample distribution functions, and put

$$D_n = \sup_{\mathbf{x} \in R_k} |F_n(\mathbf{x}) - F(\mathbf{x})|.$$

For the case $k = 1$, an exponential-type probability inequality for D_n was established by Dvoretzky, Kiefer, and Wolfowitz (1956).

Theorem A (Dvoretzky, Kiefer, and Wolfowitz). *Let* F *be defined on* R. *There exists a finite positive constant* C (*not depending on* F) *such that*

$$P(D_n > d) \le Ce^{-2nd^2}, \qquad d > 0,$$

for all $n = 1, 2, \ldots$.

Remarks. (i) DKW actually prove this result only for F uniform on $[0, 1]$, extension to the general case being left implicit. The extension may be seen as follows. Given independent observations X_i having distribution F and defined on a common probability space, one can construct independent uniform $[0, 1]$ variates Y_i such that $P[X_i = F^{-1}(Y_i)] = 1$, $1 \le i \le n$. Let G denote the uniform $[0, 1]$ distribution and G_n the sample distribution function of the Y_i's. Then $F(x) = G(F(x))$ and, by Lemma 1.1.4(iii), ($wp1$) $F_n(x) = G_n(F(x))$. Thus

$$\begin{aligned} D_n^F = \sup_{-\infty < x < \infty} |F_n(x) - F(x)| &= \sup_{-\infty < x < \infty} |G_n(F(x)) - G(F(x))| \\ &\le \sup_{0 < t < 1} |G_n(t) - G(t)| = D_n^G, \end{aligned}$$

so that $P(D_n^F > d) \le P(D_n^G > d)$.

Alternatively, reduction to the uniform case may be carried out as follows. Let $Y_1, \ldots, Y_n$ be independent uniform [0, 1] random variables. Then $\mathscr{L}\{(X_1, \ldots, X_n)\} = \mathscr{L}\{(F^{-1}(Y_1), \ldots, F^{-1}(Y_n))\}$. Thus

$$\mathscr{L}\{D_n(X_1, \ldots, X_n)\} = \mathscr{L}\{D_n(F^{-1}(Y_1), \ldots, F^{-1}(Y_n))\}.$$

But

$$\begin{aligned} D_n(F^{-1}(Y_1), \ldots, F^{-1}(Y_n)) &= \sup_{-\infty<x<\infty} \left| n^{-1}\sum_{i=1}^{n} I(F^{-1}(Y_i) \le x) - F(x) \right| \\ &= \sup_{-\infty<x<\infty} \left| n^{-1}\sum_{i=1}^{n} I(Y_i \le F(x)) - F(x) \right| \\ &\le \sup_{0<t<1} \left| n^{-1}\sum_{i=1}^{n} I(Y_i \le t) - t \right| \\ &= D_n(Y_1, \ldots, Y_n). \end{aligned}$$

(ii) The foregoing construction does not retain the distribution-free property in generalizing to multidimensional F. For F in R^k, let F_j denote the jth marginal of F, $1 \le j \le k$, and put $\vec{F}(\mathbf{x}) = (F_1(x_1), \ldots, F_k(x_k))$ for $\mathbf{x} = x_1, \ldots, x_k$). Putting $\mathbf{Y}_i = \vec{F}(\mathbf{X}_i)$, $1 \le i \le n$, and letting G^F denote the distribution of each $\mathbf{Y}_i$, and G_n the sample distribution function of the $\mathbf{Y}_i$'s, we have $F(\mathbf{x}) = G_F(\vec{F}(\mathbf{x}))$ and $F_n(\mathbf{x}) = G_n(\vec{F}(\mathbf{x}))$, so that

$$F_n(\mathbf{x}) - F(\mathbf{x}) = G_n(\vec{F}(\mathbf{x})) - G^F(\vec{F}(\mathbf{x})).$$

Again, we have achieved a reduction to the case of distributions on the k-dimensional unit cube, but in some cases the distribution G^F depends on F. (Also, see Kiefer and Wolfowitz (1958).)

(iii) The inequality in Theorem A may be expressed in the form:

$$P(n^{1/2}D_n > d) \le C\exp(-2d^2).$$

In **2.1.5** a limit distribution for $n^{1/2}D_n$ will be given. Thus the present result augments the limit distribution result by providing a useful bound for probabilities of large deviations.

(iv) Theorem A also yields important results on the almost sure behavior of D_n. See **2.1.4**. ■

The exponential bound of Theorem A is quite powerful, as the following corollary shows.

Corollary. *Let* F *and* C *be as in Theorem A. Then, for every* $\varepsilon > 0$,

$$P\left(\sup_{m\ge n} D_n > \varepsilon\right) \le \frac{C}{1-\rho_\varepsilon}\rho_\varepsilon^n.$$

where $\rho_\varepsilon = \exp(-2\varepsilon^2)$.

PROOF. Let $\varepsilon > 0$.

$$P\left(\sup_{m \geq n} D_n > \varepsilon\right) \leq \sum_{m=n}^{\infty} P(D_m > \varepsilon) \leq C \sum_{m=n}^{\infty} \rho_\varepsilon^m = \frac{C}{1 - \rho_\varepsilon} \rho_\varepsilon^n. \quad \blacksquare$$

The extension of Theorem A to multidimensional F was established by Kiefer (1961):

Theorem B (Kiefer). *Let* F *be defined on* R^k, $k \geq 2$. *For each* $\varepsilon > 0$, *there exists a finite positive constant* $C = C(\varepsilon, k)$ (*not depending on* F) *such that*

$$P(D_n > d) \leq Ce^{-(2-\varepsilon)nd^2}, \qquad d > 0,$$

for all $n = 1, 2, \ldots$.

As a counter-example to the possibility of extending the result to the case $\varepsilon = 0$, as was possible for the 1-dimensional case, Kiefer cites a 2-dimensional F satisfying $P(n^{1/2}D_n \geq d) \to 8d^2 \exp(-2d^2)$, $n \to \infty$.

(An analogue of the corollary to Theorem A follows from Theorem B.)

2.1.4 Almost Sure Behavior of the Kolmogorov–Smirnov Distance

(We continue the notation of **2.1.3**.) The simplest almost sure property of D_n is that it converges to 0 with probability 1:

Theorem A (Glivenko–Cantelli). $D_n \xrightarrow{wp1} 0$.

PROOF. For the 1-dimensional case, this result was proved by Glivenko (1933) for continuous F and by Cantelli (1933) for general F. See Loève (1977), p. 21, or Gnedenko (1962), Section 67, for a proof based on application of the SLLN. Alternatively, simply apply the Dvoretzky–Kiefer–Wolfowitz probability inequality (Theorem 2.1.3A) in conjunction with Theorem 1.3.4 to obtain

$$\sum_{n=1}^{\infty} P(D_n > \varepsilon) < \infty \qquad \text{for every} \quad \varepsilon > 0,$$

showing thus that D_n *converges completely* to 0. Even more strongly, we can utilize Corollary 2.1.3 in similar fashion to establish that $\sup_{m \geq n} D_n$ converges completely to 0.

Likewise, the multidimensional case of the above theorem may be deduced from Theorem 2.1.3B. $\blacksquare$

The extreme fluctuations in the convergence of D_n to 0 are characterized by the following LIL.

Theorem B. *With probability* 1,

$$\overline{\lim_{n\to\infty}} \frac{n^{1/2}D_n}{(2 \log \log n)^{1/2}} = c(F).$$

where

$$c(F) = \sup_{\mathbf{x} \in R^k} \{F(\mathbf{x})[1 - F(\mathbf{x})]\}^{1/2}.$$

(Note that $c(F) = \frac{1}{2}$ if F is continuous.)

For F 1-dimensional and continuous, proofs are contained in the papers of Smirnov (1944), Chung (1949), and Csáki (1968). Kiefer (1961) extended to multidimensional continuous F and Richter (1974) to general multidimensional F.

2.1.5 Asymptotic Distribution Theory for the Kolmogorov–Smirnov Distance

We confine attention to the case of F *1-dimensional.*

The exact distribution of D_n is complicated to express. See Durbin (1973a), Section 2.4, for discussion of various computational approaches. On the other hand, the asymptotic distribution theory, for continuous F, is easy to state:

Theorem A (Kolmogorov). *Let* F *be* 1-*dimensional and continuous. Then*

$$\lim_{n\to\infty} P(n^{1/2}D_n \le d) = 1 - 2\sum_{j=1}^{\infty}(-1)^{j+1}e^{-2j^2d^2}, \qquad d > 0.$$

The proposition was originally established by Kolmogorov (1933), using a representation of F_n as a conditioned Poisson process (see **2.1.9**). Later writers have found other approaches. For proof via convergence in distribution in $C[0, 1]$, see Hájek and Šidák (1967), Section V.3, or Billingsley (1968), Section 13. Alternatively, see Breiman (1968) or Brillinger (1969) for proof via Skorokhod constructions.

A convenient feature of the preceding approximation is that it does not depend upon F. In fact, this is true also of the exact distribution of D_n for the class of continuous F's (see, e.g., Lindgren (1968), Section 8.1.)

In the case of F having discontinuities, $n^{1/2}D_n$ still has a limit distribution, but it depends on F (through the values of F at the points of discontinuity). Extension to the case of F having finitely many discontinuities and not being purely atomic was obtained by Schmid (1958), who gives the limit distribution explicitly. The general case is treated in Billingsley (1968), Section 16. Here there is only implicit characterization of the limit distribution, namely, as that of a specified functional of a specified Gaussian stochastic process (see Section 2.8 for details).

For multidimensional F, also, $n^{1/2}D_n$ has a limit distribution. This has been established by Kiefer and Wolfowitz (1958) primarily as an *existence* result, the limit distribution not being characterized in general. For dimension ≥ 2, the limit distribution depends on F even in the continuous case.

Let us also consider *one-sided* Kolmogorov–Smirnov distances, typified by

$$D_n^+ = \sup_{-\infty < x < \infty} [F_n(x) - F(x)].$$

For continuous F, the distribution of D_n^+ does not depend on F. The exact distribution is somewhat more tractable than that of D_n (see Durbin (1973a) for details). The asymptotic distribution, due to Smirnov (1944) (or see Billingsley (1968), p. 85), is quite simple:

Theorem B (Smirnov). *Let* F *be* 1-*dimensional and continuous. Then*

$$\lim_{n\to\infty} P(n^{1/2}D_n^+ \leq d) = \lim_{n\to\infty} P(n^{1/2}D_n^- \leq -d) = 1 - e^{-2d^2}, \qquad d > 0.$$

An associated Berry–Esséen bound of order $O(n^{-1/2} \log n)$ has been established by Komlós, Major and Tusnády (1975). Asymptotic expansions in powers of $n^{-1/2}$ are discussed in Durbin (1973a) and Gaenssler and Stute (1979).

2.1.6 Asymptotic Distribution Theory for the Kolmogorov–Smirnov Test Statistic

Let $X_1, X_2, \ldots$ be I.I.D. with (1-dimensional) *continuous* distribution function F, and let F_0 be a specified *hypothetical* continuous distribution. For the null hypothesis H: $F = F_0$, the Kolmogorov–Smirnov test statistic was introduced in **2.1.2**:

$$\Delta_n = \sup_{-\infty < x < \infty} |F_n(x) - F_0(x)|.$$

The asymptotic distribution of Δ_n under the *null* hypothesis is given by Theorem 2.1.5A, for in this case $\Delta_n = D_n$. Under the *alternative* hypothesis H^*: $F \neq F_0$, the parameter

$$\Delta = \sup_{-\infty < x < \infty} |F(x) - F_0(x)|$$

is relevant. Raghavachari (1973) obtains the limit distribution of $n^{1/2}(\Delta_n - \Delta)$, expressed as the distribution of a specified functional of a specified Gaussian stochastic process, both specifications depending on F and F_0 (see Section 2.8 for details). He also obtains analogous results for other Kolmogorov–Smirnov type statistics considered in **2.1.2**.

2.1.7 Almost Sure Behavior and Asymptotic Distribution Theory of the Cramér–von Mises Test Statistic

Let $\{X_i\}$, F and F_0 be as in **2.1.6**. We confine attention to the *null hypothesis* situation, in which case $F = F_0$ and the test statistic introduced in **2.1.2** may be viewed and written as a measure of disparity between F_n and F:

$$C_n = n \int_{-\infty}^{\infty} [F_n(x) - F(x)]^2 \, dF(x).$$

In this respect, we present analogues of results for D_n established in **2.1.4** and **2.1.5**. We also remark that in the present context C_n, like D_n, has a distribution not depending on F.

Theorem A (Finkelstein). *With probability* 1,

$$\overline{\lim_{n\to\infty}} \frac{C_n}{2 \log \log n} = \frac{1}{\pi^2}.$$

Finkelstein (1971) obtains this as a corollary of her general theorem on the LIL for the sample distribution function.

Theorem B. *Let ξ be a random variable representable as*

$$\xi = \sum_{j=1}^{\infty} \frac{\chi_{1j}^2}{j^2\pi^2},$$

where $\chi_{1j}^2, \chi_{2j}^2, \ldots$ are independent χ_1^2 variates. Then

$$\lim_{n\to\infty} P(C_n \le c) = P(\xi \le c), \qquad c > 0.$$

For details of proof, and of computation of $P(\xi < c)$, see Durbin (1973a), Section 4.4.

2.1.8 Sample Density Functions

Let $X_1, X_2, \ldots$ be I.I.D. with (1-dimensional) absolutely continuous F having density $f = F'$. A natural way to estimate $f(x)$ is by a difference quotient

$$f_n(x) = \frac{F_n(x + b_n) - F_n(x - b_n)}{2b_n}.$$

Here $\{b_n\}$ is a sequence of constants selected to $\to 0$ at a suitable rate. Noting that $2nb_n f_n(x)$ has the binomial $(n, F(x + b_n) - F(x - b_n))$ distribution, one finds (Problem 2.P.3) that

$$E\{f_n(x)\} \to f(x) \quad \text{if} \quad b_n \to 0, \qquad n \to \infty,$$

and

$$\mathrm{Var}\{f_n(x)\} \to 0 \quad \text{if} \quad b_n \to 0 \quad \text{and} \quad nb_n \to \infty, \qquad n \to \infty,$$

Thus one wants b_n to converge to 0 slower than n^{-1}. (Further options on the choice of $\{b_n\}$ are based on *a priori* knowledge of f and on the actual sample size n.) In this case $f_n(x)$ is *consistent in mean square* for estimation of f. Further, under suitable smoothness restrictions on f at x and additional convergence restrictions on $\{b_n\}$, it is found (Problems 2.P.4–5) that $f_n(x)$ is $AN(f(x), f(x)/nb_n)$. See Bickel and Doksum (1977) for practical discussion regarding the estimator $f_n(x)$.

A popular alternative estimator of similar type is the *histogram*

$$f_n^*(x) = \frac{F_n(a + (j + 1)b_n) - F_n(a + jb_n)}{2b_n}, \qquad x \in [a + jb_n, a + (j + 1)b_n).$$

Its asymptotic properties are similar to those of $f_n(x)$.

A useful class of estimators generalizing $f_n(x)$ is defined by the form

$$f_n(x) = \int_{-\infty}^{\infty} b_n^{-1} W\left(\frac{x - y}{b_n}\right) dF_n(y) = \frac{1}{nb_n} \sum_{i=1}^{n} W\left(\frac{x - X_i}{b_n}\right),$$

where $W(\cdot)$ is an integrable nonnegative weight function. (The case $W(z) = \frac{1}{2}$, $|z| \le 1$, and $= 0$ otherwise, gives essentially the simple estimator considered above.) Under restrictions on $W(\cdot)$, $f(\cdot)$ and $\{b_n\}$, the almost sure behavior of the distance

$$\sup_{-\infty < x < \infty} |f_n(x) - f(x)|$$

is characterized by Silverman (1978). For two other such global measures,

$$\sup_x \left| \frac{f_n(x) - f(x)}{f^{1/2}(x)} \right| \quad \text{and} \quad \int \frac{[f_n(x) - f(x)]^2}{f(x)} dx,$$

asymptotic distributions are determined by Bickel and Rosenblatt (1973). Regarding pointwise estimation of $f(x)$ by $f_n(x)$, asymptotic normality results are given by Rosenblatt (1971).

2.1.9 Complements

(i) The problem of estimation of F is treated from a *decision-theoretic* standpoint by Ferguson (1967), Section 4.8. For best "invariant" estimation, and in connection with various loss functions, some forms of sample distribution function other than F_n arise for consideration. They weight the X_i's differently than simply n^{-1} uniformly.

(ii) The speed of the Glivenko–Cantelli convergence is characterized stochastically by an LIL, as seen in **2.1.4**. In terms of *nonstochastic* quantities, such as $E\{D_n\}$ and $E\{\int |F_n(x) - F(x)|\,dx\}$, Dudley (1969) establishes rates $O(n^{-1/2})$.

(iii) The Kolmogorov–Smirnov test statistic Δ_n considered in **2.1.2** and **2.1.6** is also of interest when the hypothesized F_0 involves unknown *nuisance parameters* which have to be estimated from the data in order to formulate Δ_n. See Durbin (1973a, b) for development of the relevant theory. For further development, see Neuhaus (1976).

(iv) For theory of Kolmogorov–Smirnov type test statistics generalized to include *regression constants*, for power against regression alternatives, see Hájek and Šidák (1967).

(v) Consider *continuous* F and thus reduce without loss of generality to F uniform on [0, 1]. Then (see Durbin (1973a))

(a) $\{F_n(t)\}$ is a *Markov process*: for any $0 < t_1 < \cdots < t_k < t_k < 1$, the conditional distribution of $F_n(t_k)$ given $F_n(t_1), \ldots, F_n(t_k)$ depends only on $F_n(t_{k-1})$.

(b) $\{F_n(t)\}$ is a *conditioned Poisson process*: it has the same distribution as the stochastic process $\{P_n(t)\}$ given $P_n(1) = 1$, where $\{P_n(t)\}$ is the Poisson process with occurrence rate n and jumps of n^{-1}.

(vi) In **2.1.3** we stated large deviation probability inequalities for D_n. "*Large deviation*" *probabilities* for F_n may also be characterized. For suitable types of set $\mathscr{F}_0$ of distribution functions, and for F not in $\mathscr{F}_0$, there exist numbers $c(F, \mathscr{F}_0)$ such that

$$\frac{\log P(F_n \in \mathscr{F}_0)}{n} \to c(F, \mathscr{F}_0), \qquad n \to \infty.$$

See Hoadley (1967), Bahadur (1971), Bahadur and Zabell (1979), and Chapter 10.

2.2 THE SAMPLE MOMENTS

Let $X_1, X_2, \ldots$ be I.I.D. with distribution function F. For a positive integer k, the kth *moment* of F is defined as

$$\alpha_k = \int_{-\infty}^{\infty} x^k \, dF(x) = E\{X_1^k\}.$$

The first moment α_1 is also called the *mean* and denoted by μ when convenient. likewise, the kth *central moment* of F is defined as

$$\mu_k = \int_{-\infty}^{\infty} (x - \mu)^k \, dF(x) = E\{(X_1 - \mu)^k\}.$$

Note that $\mu_1 = 0$. The $\{\alpha_i\}$ and $\{\mu_i\}$ represent important parameters in terms of which the description of F, or manipulations with F, can sometimes be greatly simplified. Natural estimators of these parameters are given by the corresponding moments of the sample distribution function F_n. Thus α_k may be estimated by

$$a_k = \int_{-\infty}^{\infty} x^k \, dF_n(x) = \frac{1}{n} \sum_{i=1}^{n} X_i^k \qquad (k = 1, 2, \ldots)$$

(and let a_1 also be denoted by $\bar{X}$), and μ_k may be estimated by

$$m_k = \int_{-\infty}^{\infty} (x - \bar{X})^k \, dF_n(x) = \frac{1}{n} \sum_{i=1}^{n} (X_i - \bar{X})^k \qquad (k = 2, 3, \ldots).$$

Since F_n possesses desirable properties as an estimator of F, as seen in Section 2.1, it might be expected that the sample moments a_k and the sample central moments m_k possess desirable features as estimators of α_k and μ_k. Indeed, we shall establish that these estimators are *consistent* in the usual senses and jointly are *asymptotically multivariate normal* in distribution. Further, we shall examine *bias* and *variance* quantities. The estimates a_k are treated in **2.2.1**. Following some preliminaries in **2.2.2**, the estimates m_k are treated in **2.2.3**. The results include treatment of the joint asymptotic distribution of the a_k's and m_k's taken together. In **2.2.4** some complements are presented.

2.2.1 The Estimates a_k

Note that a_k is a mean of I.I.D. random variables having mean α_k and variance $\alpha_{2k} - \alpha_k^2$. Thus by trivial computations and the SLLN (Theorem 1.8B), we have

Theorem A.

(i) $a_k \xrightarrow{wp1} \alpha_k$;

(ii) $E\{a_k\} = \alpha_k$:

(iii) $\mathrm{Var}\{a_k\} = \dfrac{\alpha_{2k} - \alpha_k^2}{n}$.

(It is implicitly assumed that all stated moments are finite.) Note that (i) implies *strong consistency* and (ii) and (iii) together yield *mean square consistency*.

More comprehensively, the vector $(a_1, a_2, \ldots, a_k)$ is the mean of the I.I.D. vectors $(X_i, X_i^2, \ldots, X_i^k)$, $1 \le i \le n$. Thus a direct application of the multivariate Lindeberg–Lévy CLT (Theorem 1.9.1B) yields that $(a_1, \ldots, a_k)$ is *asymptotically normal* with mean vector $(\alpha_1, \ldots, \alpha_k)$ and covariances $(\alpha_{i+j} - \alpha_i\alpha_j)/n$. Formally:

Theorem B. *If* $\alpha_{2k} < \infty$, *the random vector* $n^{1/2}(a_1 - \alpha_1, \ldots, a_k - \alpha_k)$ *converges in distribution to* k*-variate normal with mean vector* $(0, \ldots, 0)$ *and covariance matrix* $[\sigma_{ij}]_{k \times k}$, *where* $\sigma_{ij} = \alpha_{i+j} - \alpha_i \alpha_j$.

2.2.2 Some Preliminary and Auxiliary Results

Preliminary to deriving properties of the estimates m_k, it is advantageous to consider the closely related random variables

$$b_k = \frac{1}{n} \sum_{i=1}^{n} (X_i - \mu)^k \qquad (k = 1, 2, \ldots).$$

Properties of the m_k's will be deduced from those of the b_k's.

The same arguments employed in dealing with the a_k's immediately yield

Lemma A.

(i) $b_k \xrightarrow{wp1} \mu_k$;

(ii) $E\{b_k\} = \mu_k$;

(iii) *For* $\mu_{2k} < \infty$, $\mathrm{Var}\{b_k\} = \dfrac{\mu_{2k} - \mu_k^2}{n}$;

(iv) *For* $\mu_{2k} < \infty$, *the random vector* $(b_1, \ldots, b_k)$ *is asymptotically normal with mean vector* $(\mu_1, \ldots, \mu_k)$ *and covariances* $(\mu_{i+j} - \mu_i \mu_j)/n$.

Note that b_k and m_k represent alternate ways of estimating μ_k by moment statistics. The use of b_k presupposes knowledge of μ, whereas m_k employs the sample mean $\bar{X}$ in place of μ. This makes m_k of greater practical utility than b_k, but more cumbersome to analyze theoretically.

As another preliminary, we state

Lemma B. *Let* $\{Z_i\}$ *be I.I.D. with* $E\{Z_1\} = 0$ *and with* $E|Z_1|^\nu < \infty$, *where* $\nu \geq 2$. *Then*

$$E\left|\sum_{i=1}^{n} Z_i\right|^\nu = O(n^{1/2\nu}), \qquad n \to \infty.$$

For proof and more general results, see Loève (1977), p. 276, or Marcinkiewicz and Zygmund (1937). See also Lemma 9.2.6A.

We shall utilize Lemma B through the implication

$$E\{b_1^j\} = E\{(\bar{X} - \mu)^j\} = O(n^{-(1/2)j}), \qquad n \to \infty,$$

for $j \geq 2$.

2.2.3 The Estimates m_k

Although analogous in form to b_k, the random variable m_k differs crucially in *not* being expressible as an average of I.I.D. random variables. Therefore, instead of dealing with m_k directly, we exploit the connection between m_k and the b_j's. Writing

$$m_k = \frac{1}{n}\sum_{i=1}^{n}(X_i - \bar{X})^k = \frac{1}{n}\sum_{i=1}^{n}\sum_{j=0}^{k}\binom{k}{j}(X_i - \mu)^j(\mu - \bar{X})^{k-j},$$

we obtain

$$(*) \qquad m_k = \sum_{j=0}^{k}\binom{k}{j}(-1)^{k-j}b_j b_1^{k-j},$$

where we define $b_0 = 1$.

The following result treats the *bias*, *mean square consistency*, *mean square error*, and *strong consistency* of m_k.

Theorem A.

(i) $m_k \xrightarrow{wp1} \mu_k$;

(ii) *The bias of* m_k *satisfies*

$$E\{m_k\} - \mu_k = \frac{\frac{1}{2}k(k-1)\mu_{k-1}\mu_2 - k\mu_k}{n} + O(n^{-2}), \qquad n \to \infty;$$

(iii) *The variance of* m_k *satisfies*

$$\mathrm{Var}\{m_k\} = \frac{\mu_{2k} - \mu_k^2 - 2k\mu_{k-1}\mu_{k+1} + k^2\mu_2\mu_{k-1}^2}{n} + O(n^{-2}), \qquad n \to \infty;$$

(iv) *Hence* $E(m_k - \mu_k)^2 \sim \mathrm{Var}\{m_k\} = O(n^{-1}), \qquad n \to \infty.$

PROOF. (i) In relation (*), apply Lemma 2.2.2A(i) in conjunction with Application D of Corollary 1.7, and note that $\mu_1 = 0$.

(ii) Again utilize (*) to write

$$E\{m_k\} - \mu_k = \sum_{j=1}^{k}\binom{k}{j}(-1)^j E\{b_{k-j}b_1^j\}.$$

Now, making use of the independence of the X_i's,

$$E\{b_{k-1}b_1\} = \frac{1}{n^2}E\left\{\sum_{i=1}^{n}(X_i - \mu)^{k-1}\sum_{j=1}^{n}(X_j - \mu)\right\}$$

$$= \frac{1}{n^2}\sum_{i=1}^{n}E\{(X_i - \mu)^k\} = \frac{\mu_k}{n}.$$

Similarly,

$$E\{b_{k-2}b_1^2\} = \frac{1}{n^3} E\left\{\sum_{i_1=1}^n \sum_{i_2=1}^n \sum_{i_3=1}^n (X_{i_1} - \mu)^{k-2}(X_{i_2} - \mu)(X_{i_3} - \mu)\right\}$$
$$= \frac{1}{n^3} \sum_{i_1=1}^n \sum_{i_2=1}^n E\{(X_{i_1} - \mu)^{k-2}(X_{i_2} - \mu)^2\},$$

since the expectation of a term in the triple summation is 0 if $i_3 \neq i_2$. Hence

$$E\{b_{k-2}b_1^2\} = \frac{n\mu_k + n(n-1)\mu_{k-2}\mu_2}{n^3} = \frac{\mu_{k-2}\mu_2}{n} + O(n^{-2}), \qquad n \to \infty.$$

Similarly (exercise),

$$E\{b_{k-3}b_1^3\} = O(n^{-2}), \qquad n \to \infty.$$

For $j > 3$, use Hölder's inequality (Appendix)

$$|E\{b_{k-j}b_1^j\}| \leq [E|b_{k-j}|^{k/(k-j)}]^{(k-j)/k}[E|b_1|^k]^{j/k}.$$

By application of Minkowski's inequality (Appendix) in connection with the first factor on the right, and Lemma 2.2.2B in connection with the second factor, we obtain

$$E\{b_{k-j}b_1^j\} = O(1)[O(n^{-(1/2)k})]^{j/k} = O(n^{-(1/2)j})$$
$$= O(n^{-2}), \qquad n \to \infty \ (j > 3).$$

Collecting these results, we have

$$E\{m_k\} - \mu_k = \binom{k}{1}(-1)\frac{\mu_k}{n} + \binom{k}{2}(-1)^2\frac{\mu_{k-2}\mu_2}{n} + O(n^{-2}), \qquad n \to \infty.$$

(iii) Writing $\mathrm{Var}\{m_k\} = E\{m_k^2\} - [E\{m_k\}]^2$, we seek to compute $E\{m_k^2\}$ and combine with the result in (ii). For $E\{m_k^2\}$, we need to compute quantities of the form

$$E\{b_{k-j_1}b_1^{j_1}b_{k-j_2}b_1^{j_2}\} = E\{b_{k-j_1}b_{k-j_2}n_1^{j_1+j_2}\},$$

for $0 \leq j_1, j_2 \leq k$. For $j_1 = j_2 = 0$, we have

$$E\{b_k^2\} = \frac{1}{n^2} E\left\{\sum_{i=1}^n (X_i - \mu)^k \sum_{j=1}^n (X_j - \mu)^k\right\}$$
$$= \frac{n\mu_{2k} + n(n-1)\mu_k^2}{n^2} = \mu_k^2 + \frac{\mu_{2k} - \mu_k^2}{n}.$$

For $(j_1, j_2) = (0, 1)$ or $(1, 0)$, we have

$$E\{b_{k-1}b_k b_1\} = \frac{1}{n^3} E\left\{\sum_{i_1=1}^{n} (X_{i_1} - \mu)^{k-1} \sum_{i_2=1}^{n} (X_{i_2} - \mu)^k \sum_{i_3=1}^{n} (X_{i_3} - \mu)\right\}$$

$$= \frac{1}{n^3}\left[\sum_{i_1=1}^{n} E(X_{i_1} - \mu)^{2k} + \sum_{i_1 \neq i_2}\sum E\{(X_{i_1} - \mu)^k (X_{i_2} - \mu)^k\}\right.$$

$$\left. + \sum_{i_1 \neq i_2} E\{(X_{i_1} - \mu)^{k-1}(X_{i_2} - \mu)^{k+1}\}\right]$$

$$= \frac{1}{n^3}[n\mu_{2k} + n(n-1)\mu_k^2 + n(n-1)\mu_{k-1}\mu_{k+1}]$$

$$= \frac{\mu_k^2 + \mu_{k-1}\mu_{k+1}}{n} + O(n^{-2}), \qquad n \to \infty.$$

For $j_1 = j_2 = 1$, we have (exercise)

$$E\{b_{k-1}^2 b_1^2\} = \frac{\mu_{k-1}^2 \mu_2}{n} + O(n^{-2}), \qquad n \to \infty,$$

and

$$E\{b_k b_{k-2} b_1^2\} = \frac{\mu_k \mu_{k-2} \mu_2}{n} + O(n^{-2}), \qquad n \to \infty.$$

Finally, for $j_1 + j_2 > 2$, we have (exercise)

$$E\{b_{k-j_1} b_{k-j_2} b_1^{j_1+j_2}\} = O(n^{-2}), \qquad n \to \infty.$$

Consequently, by (*),

$$E\{m_k^2\} = E\{b_k^2\} - 2kE\{b_k b_{k-1} b_1\} + k^2 E\{b_{k-1}^2 b_1^2\}$$
$$+ k(k-1)E\{b_k b_{k-2} b_1^2\} + O(n^{-2}), \qquad n \to \infty,$$

$$= \mu_k^2 + \frac{\mu_{2k} - \mu_k^2 - 2k(\mu_k^2 + \mu_{k-1}\mu_{k-1}) + k^2\mu_{k-1}^2\mu_2 + k(k-1)\mu_k\mu_{k-2}\mu_2}{n}$$
$$+ O(n^{-2}), \qquad n \to \infty.$$

(iv) trivial. ■

Next we establish *asymptotic normality* of the vector $(m_2, \ldots, m_k)$. The following lemma is useful.

Lemma. *For each* k,

$$n^{1/2}(m_k - \mu_k) = n^{1/2}(b_k - \mu_k - k\mu_{k-1}b_1) + o_p(1), \qquad n \to \infty.$$

PROOF. By (*), write

$$n^{1/2}(m_k - \mu_k) = n^{1/2}(b_k - \mu_k - k\mu_{k-1}b_1)$$
$$+ n^{1/2}b_1\left[k(\mu_{k-1} - b_{k-1}) + \sum_{j=0}^{k-2}\binom{k}{j}(-1)^{k-j}b_j b_1^{k-j-1}\right].$$

The second term on the right is a product of two factors, the first converging in distribution and the second converging to 0 *wp*1, these properties following from Lemma 2.2.2A. Therefore, by Slutsky's Theorem (1.5.4), the product converges to 0 in probability. ■

Theorem B. *If* $\mu_{2k} < \infty$, *the random vector* $n^{1/2}(m_2 - \mu_2, \ldots, m_k - \mu_k)$ *converges in distribution to* (k − 1)*-variate normal with mean vector* (0, . . . , 0) *and covariance matrix* $[\sigma^*_{ij}]_{(k-1)\times(k-1)}$, *where*

$$\sigma^*_{ij} = \mu_{i+j+2} - \mu_{i+1}\mu_{j+1} - (i+1)\mu_i\mu_{j+2} - (j+1)\mu_{i+2}\mu_j$$
$$+ (i+1)(j+1)\mu_i\mu_j\mu_2.$$

PROOF. By the preceding lemma, in conjunction with the Cramér–Wold device (Theorem 1.5.2) and Slutsky's Theorem, the random vector

$$n^{1/2}(m_2 - \mu_2, \ldots, m_k - \mu_k)$$

has the same limit distribution (if any) as the vector $n^{1/2}(b_2 - \mu_2 - 2\mu_1 b_1, \ldots, b_k - \mu_k - k\mu_{k-1}b_1)$. But the latter is simply $n^{1/2}$ times the average of the I.I.D. vectors

$$[(X_i - \mu)^2 - \mu_2 - 2\mu_1(X_i - \mu), \ldots, (X_i - \mu)^k - \mu_k - k\mu_{k-1}(X_i - \mu)],$$
$$1 \le i \le n.$$

Application of the CLT (Theorem 1.9.1B) gives the desired result. ■

By similar techniques, we can obtain asymptotic normality of any vector $(a_1, \ldots, a_{k_1}, m_2, \ldots, m_{k_2})$. In particular, let us consider (a_1, m_2) = (sample mean, sample variance) = $(\bar{X}, s^2)$. It is readily seen (Problem 2.P.8) that

$$n^{1/2}(\bar{X} - \mu, s^2 - \sigma^2) \xrightarrow{d} N\left((0, 0), \begin{bmatrix} \sigma^2 & \mu_3 \\ \mu_3 & \mu_4 - \sigma^4 \end{bmatrix}\right).$$

Here we have denoted μ_2 by σ^2, as usual.

2.2.4 Complements

(i) *Examples: the sample mean and the sample variance.* The joint asymptotic distribution of $\bar{X}$ and s^2 was expressed at the conclusion of **2.2.3.** From this, or directly from Theorems 2.2.1B and 2.2.3B, it is seen that each of these statistics is asymptotically normal:

$$\bar{X} \quad \text{is} \quad AN\left(\mu, \frac{\sigma^2}{n}\right)$$

and

$$s^2 \quad \text{is} \quad AN\left(\sigma^2, \frac{\mu_4 - \sigma^4}{n}\right).$$

(ii) *Rates of convergence in connection with the asymptotic normality of the sample mean and sample variance.* Regarding $\bar{X}$, the rate of convergence to 0 of the normal approximation error follows from the Berry–Esséen Theorem (1.9.5). For s^2, the rate of this convergence is found via consideration of s^2 as a U-statistic (Section 5.5).

(iii) *Efficiency of "moment" estimators.* Despite the good properties of the moment estimators, there typically are more efficient estimators available when the distribution F is known to belong to a parametric family. Further, the "method of moments" is inapplicable if F fails to possess the relevant moments, as in the case of the Cauchy distribution. (See additional discussion in **2.3.5** and **4.3.**)

(iv) *The case* $\mu = 0$. In this case the relations $\alpha_k = \mu_k$ hold ($k = 2, 3, \ldots$) and so the two sets of estimates $\{a_2, a_3, \ldots\}$ and $\{m_2, m_3, \ldots\}$ offer alternative ways to estimate the parameters $\{\alpha_2 = \mu_2, \alpha_3 = \mu_3, \ldots\}$. In this situation, Lemma 2.2.3 shows that

$$m_k - \mu_k = a_k - \mu_k - k\mu_{k-1}\bar{X} + o_p(n^{-1/2}).$$

That is, the errors of estimation using a_k and m_k differ by a nonnegligible component, except in the case $k = 2$.

(v) *Correction factors to achieve unbiased estimators.* If desired, correction factors may be introduced to convert the m_k's into *unbiased* consistent estimators

$$M_2 = \frac{n}{n-1} m_2,$$

$$M_3 = \frac{n}{(n-1)(n-2)} m_3,$$

$$M_4 = \frac{n(n^2 - 2n + 3)}{(n-1)(n-2)(n-3)} m_4 - \frac{3n(2n-3)}{(n-1)(n-2)(n-3)} m_2^2,$$

etc., for $\mu_2, \mu_3, \mu_4, \ldots$. However, as seen from Theorem 2.2.3A, the bias of the unadjusted m_k's is asymptotically negligible. Its contribution to the mean square error is $O(n^{-2})$, while that of the variance is of order n^{-1}.

(vi) *Rates of convergence in connection with the strong convergence of* a_k *and* m_k. This topic is treated in **5.1.5**.

(vii) *Further reading.* Cramér (1946), Sections 27.1–6 and 28.1–3, and Rao (1973), Section 6*h*. ∎

2.3 THE SAMPLE QUANTILES

Let F be a distribution function (continuous from the right, as usual). For $0 < p < 1$, the pth *quantile* or *fractile* of F is defined (recall **1.1.4**) as

$$\xi_p = \inf\{x : F(x) \geq p\}$$

and is alternately denoted by $F^{-1}(p)$. Note that ξ_p satisfies

$$F(\xi_p-) \leq p \leq F(\xi_p).$$

Other useful properties have been presented in Lemmas 1.1.4 and 1.5.6.

Corresponding to a sample $\{X_1, \ldots, X_n\}$ of observations on F, the *sample* pth *quantile* is defined as the pth quantile of the sample distribution function F_n, that is, as $F_n^{-1}(p)$. Regarding the sample pth quantile as an estimator of ξ_p, we denote it by $\hat{\xi}_{pn}$, or simply by $\hat{\xi}_p$ when convenient.

It will be seen (**2.3.1**) that $\hat{\xi}_p$ is *strongly consistent* for estimation of ξ_p, under mild restrictions on F in the neighborhood of ξ_p. We exhibit (**2.3.2**) bounds on the related probability

$$P\left(\sup_{m \geq n} |\hat{\xi}_{pm} - \xi_p| > \varepsilon\right),$$

showing that it converges to 0 at an *exponential* rate.

The asymptotic distribution theory of $\hat{\xi}_p$ is treated in **2.3.3.** In particular, under mild smoothness requirements on F in the neighborhoods of the points $\xi_{p_1}, \ldots, \xi_{p_k}$, the vector of sample quantiles $(\hat{\xi}_{p_1}, \ldots, \hat{\xi}_{p_k})$ is *asymptotically normal.* Also, several complementary results will be given, including a rate of convergence for the asymptotic normality.

If F has a density, then so does the distribution of $\hat{\xi}_p$. This result and its application are discussed in **2.3.4**.

Comparison of *quantiles versus moments* as estimators is made in **2.3.5**, and the mean and median are compared for illustration. In **2.3.6** a *measure of dispersion* based on quantiles is examined. Finally, in **2.7.7**, brief discussion of nonparametric tests based on quantiles is provided.

Further background reading may be found in Cramér (1946), Section 28.5, and Rao (1973), Section 6f.2.

2.3.1 Strong Consistency of $\hat{\xi}_p$

The following result asserts that $\hat{\xi}_p$ is *strongly consistent* for estimation of ξ_p, unless *both* $F(\xi_p) = p$ and F is flat in a right-neighborhood of ξ_p.

Theorem. *Let* $0 < p < 1$. *If* ξ_p *is the unique solution* x *of* $F(x-) \le p \le F(x)$, *then* $\hat{\xi}_{pn} \xrightarrow{wp1} \xi_p$.

PROOF. Let $\varepsilon > 0$. By the uniqueness condition and the definition of ξ_p, we have

$$F(\xi_p - \varepsilon) < p < F(\xi_p + \varepsilon).$$

It was seen in **2.1.1** that $F_n(\xi_p - \varepsilon) \xrightarrow{wp1} F(\xi_p - \varepsilon)$ and $F_n(\xi_p + \varepsilon) \xrightarrow{wp1} F(\xi_p + \varepsilon)$. Hence (review **1.2.2**)

$$P(F_m(\xi_p - \varepsilon) < p < F_m(\xi_p + \varepsilon), \text{ all } m \ge n) \to 1, \qquad n \to \infty.$$

Thus, by Lemma 1.1.4(iii),

$$P(\xi_p - \varepsilon < \hat{\xi}_{pm} \le \xi_p + \varepsilon, \text{ all } m \ge n) \to 1, \qquad n \to \infty.$$

That is,

$$P\left(\sup_{m \ge n} |\hat{\xi}_{pm} - \xi_p| > \varepsilon\right) \to 0, \qquad n \to \infty.$$

As an exercise, show that the uniqueness requirement on ξ_p cannot be dropped (Problem 2.P.11).

In the following subsection, we obtain results which contain the preceding theorem, but which require more powerful techniques of proof.

2.3.2 A Probability Inequality for $|\hat{\xi}_{pn} - \xi_p|$

We shall use the following result of Hoeffding (1963).

Lemma (Hoeffding). *Let* $Y_1, \ldots, Y_n$ *be independent random variables satisfying* $P(a \le Y_i \le b) = 1$, *each* i, *where* $a < b$. *Then, for* $t > 0$,

$$P\left(\sum_{i=1}^{n} Y_i - \sum_{i=1}^{n} E\{Y_i\} \ge nt\right) \le e^{-2nt^2/(b-a)^2}.$$

Theorem. *Let* $0 < p < 1$. *Suppose that* ξ_p *is the unique solution* x *of* $F(x-) \le p \le F(x)$. *Then, for every* $\varepsilon > 0$,

$$P(|\hat{\xi}_{pn} - \xi_p| > \varepsilon) \le 2e^{-2n\delta_\varepsilon^2}, \qquad \textit{all } n,$$

where $\delta_\varepsilon = \min\{F(\xi_p + \varepsilon) - p, p - F(\xi_p - \varepsilon)\}$.

PROOF. (We apply Hoeffding's lemma in conjunction with a technique of proof of Smirnov (1952).) Let $\varepsilon > 0$. Write

$$P(|\hat{\xi}_{pn} - \xi_p| > \varepsilon) = P(\hat{\xi}_{pn} > \xi_p + \varepsilon) + P(\hat{\xi}_{pn} < \xi_p + \varepsilon).$$

By Lemma 1.1.4,

$$P(\hat{\xi}_{pn} > \xi_p + \varepsilon) = P(p > F_n(\xi_p + \varepsilon))$$
$$= P\left(\sum_{i=1}^{n} I(X_i > \xi_p + \varepsilon) > n(1 - p)\right)$$
$$= P\left(\sum_{i=1}^{n} V_i - \sum_{i=1}^{n} E\{V_i\} > n\delta_1\right),$$

where $V_i = I(X_i > \xi_p + \varepsilon)$ and $\delta_1 = F(\xi_p + \varepsilon) - p$. Likewise,

$$P(\hat{\xi}_{pn} < \xi_p + \varepsilon) \le P(p \le F_n(\xi_p - \varepsilon))$$
$$= P\left(\sum_{i=1}^{n} W_i - \sum_{i=1}^{n} E\{W_i\} \ge n\delta_2\right),$$

where $W_i = I(X_i \le \xi_p - \varepsilon)$ and $\delta_2 = p - F(\xi_p - \varepsilon)$. Therefore, utilizing Hoeffding's lemma, we have

$$P(\hat{\xi}_{pn} > \xi_p + \varepsilon) \le e^{-2n\delta_1^2}$$

and

$$P(\hat{\xi}_{pn} < \xi_p - \varepsilon) \le e^{-2n\delta_2^2}.$$

Putting $\delta_\varepsilon = \min\{\delta_1, \delta_2\}$, the proof is complete. ∎

Thus $P(|\hat{\xi}_{pn} - \xi_p| > \varepsilon) \to 0$ *exponentially fast*, which implies (via Theorem 1.3.4) that $\hat{\xi}_{pn}$ *converges completely* to ξ_p. Even more strongly, we have

Corollary. *Under the assumptions of the theorem, for every* $\varepsilon > 0$,

$$P\left(\sup_{m \ge n} |\hat{\xi}_{pm} - \xi_p| > \varepsilon\right) \le \frac{2}{1 - \rho_\varepsilon} \rho_\varepsilon^n, \qquad \textit{all } n,$$

where $\rho_\varepsilon = \exp(-2\delta_\varepsilon^2)$ *and* $\delta_\varepsilon = \min\{F(\xi_p + \varepsilon) - p, p - F(\xi_p - \varepsilon)\}$.

(derived the same way as the corollary to Theorem 2.1.3A)

Remarks. (i) The value of ε (>0) in the preceding results *may depend upon n* if desired.

(ii) The bounds established in these results are *exact*. They hold for each $n = 1, 2, \ldots$ and so may be applied for any fixed n as well as for asymptotic analyses.

(iii) A slightly modified version of the preceding theorem, asserting the same exponential rate, may be obtained by using the Dvoretzky–Kiefer–Wolfowitz probability inequality for D_n, instead of the Hoeffding lemma, in the proof. (Problem 2.P.12). ∎

2.3.3 Asymptotic Normality of $\hat{\xi}_p$

The *exact* distribution of $\hat{\xi}_p$ will be examined in **2.3.4**. Here we prove *asymptotic normality* of $\hat{\xi}_p$ in the case that F possesses left- or right-hand derivatives at the point ξ_p. If F lacks this degree of smoothness at ξ_p, the limit distribution of $\hat{\xi}_p$ (suitably normalized) need not be normal (no pun intended). The various possibilities are all covered in Theorem 4 of Smirnov (1952). In the present treatment we confine attention to the case of chief importance, that in which a normal law arises as limit.

The following theorem slightly extends Smirnov's result for the case of a normal law as limit. However, the corollaries we state are included in Smirnov's result also.

When assumed to exist, the left- and right-hand derivatives of F at δ_p will be denoted by $F'(\xi_p-)$ and $F'(\xi_p+)$, respectively.

Theorem A. *Let* $0 < p < 1$. *Suppose that* F *is continuous at* ξ_p.

(i) *If there exists* $F'(\xi_p-) > 0$, *then for* $t < 0$,

$$\lim_{n\to\infty} P\left(\frac{n^{1/2}(\hat{\xi}_{pn} - \xi_p)}{[p(1-p)]^{1/2}/F'(\xi_p-)} \le t\right) = \Phi(t).$$

(ii) *If there exists* $F'(\xi_p+) > 0$, *then for* $t > 0$,

$$\lim_{n\to\infty} P\left(\frac{n^{1/2}(\hat{\xi}_{pn} - \xi_p)}{[p(1-p)]^{1/2}/F'(\xi_p+)} \le t\right) = \Phi(t).$$

(iii) *In any case,*

$$\lim_{n\to\infty} P(n^{1/2}(\hat{\xi}_{pn} - \xi_p) \le 0) = \Phi(0) = \tfrac{1}{2}.$$

Corollary A. *Let* $0 < p < 1$. *If* F *is differentiable at* ξ_p *and* $F'(\xi_p) > 0$, *then*

$$\hat{\xi}_{pn} \quad \textit{is} \quad AN\left(\xi_p, \frac{p(1-p)}{[F'(\xi_p)]^2 n}\right).$$

Corollary B. *Let* $0 < p < 1$. *If* F *possesses a density* f *in a neighborhood of* ξ_p *and* f *is positive and continuous at* ξ_p, *then*

$$\hat{\xi}_{pn} \quad \textit{is} \quad AN\left(\xi_p, \frac{p(1-p)}{f^2(\xi_p)n}\right).$$

These corollaries follow immediately from Theorem A. Firstly, if F is differentiable at ξ_p, then $F'(\xi_p-) = F'(\xi_p+) = F'(\xi_p)$. Thus Corollary A follows. Secondly, if f is a density of F, it is not necessary that $f \equiv F'$. However, if f is *continuous* at x_0, then $f(x_0) = F'(x_0)$. (See **1.1.8**.) Thus Corollary B follows from Corollary A. Among these three results, it is Corollary B that is typically used in practice.

PROOF OF THEOREM A. Fix t. Let $A > 0$ be a normalizing constant to be specified later, and put

$$G_n(t) = P\left(\frac{n^{1/2}(\hat{\xi}_{pn} - \xi_p)}{A} \le t\right).$$

Applying Lemma 1.1.4 (iii), we have

$$\begin{aligned} G_n(t) &= P(\hat{\xi}_{pn} \le \xi_p + tAn^{-1/2}) = P(p \le F_n(\xi_p + tAn^{-1/2})) \\ &= P[np \le Z_n(F(\xi_p + tAn^{-1/2}))], \end{aligned}$$

where $Z_n(\Delta)$ denotes a binomial (n, Δ) random variable. In terms of the standardized form of $Z_n(\Delta)$,

$$Z_n^*(\Delta) = \frac{Z_n(\Delta) - n\Delta}{[n\Delta(1 - \Delta)]^{1/2}},$$

we have

$$(*) \qquad G_n(t) = P(Z_n^*(\Delta_{nt}) \ge -c_{nt}),$$

where

$$\Delta_{nt} = F(\xi_p + tAn^{-1/2})$$

and

$$c_{nt} = \frac{n^{1/2}(\Delta_{nt} - p)}{[\Delta_{nt}(1 - \Delta_{nt})]^{1/2}}.$$

At this point we may easily obtain (iii). Putting $t = 0$ in (*), we have $G_n(0) = P(Z_n^*(p) \ge 0) \to \Phi(0) = \frac{1}{2}$, $n \to \infty$, by the Lindeberg–Lévy CLT.

Now utilize the Berry–Esséen Theorem (1.9.5) to write

$$(**) \qquad \sup_{-\infty < x < \infty} |P(Z_n^*(\Delta) < x) - \Phi(x)| \le C \frac{\rho_\Delta}{\sigma_\Delta^3 n^{1/2}} = C \frac{\gamma(\Delta)}{n^{1/2}},$$

where C is a universal constant, $\sigma_\Delta^2 = \text{Var}\{Z_1(\Delta)\} = \Delta(1 - \Delta)$, $\rho_\Delta = E|Z_1(\Delta) - \Delta|^3 = \Delta(1 - \Delta)[(1 - \Delta)^2 + \Delta^2]$, and thus

$$\gamma(\Delta) = \frac{\rho_\Delta}{\sigma_\Delta^3} = \frac{(1 - \Delta)^2 + \Delta^2}{[\Delta(1 - \Delta)]^{1/2}}.$$

Using (*) to write

$$\begin{aligned} \Phi(t) - G_n(t) &= P(Z_n^*(\Delta_{nt}) < -c_{nt}) - [1 - \Phi(t)] \\ &= P(Z_n^*(\Delta_{nt}) < -c_{nt}) - \Phi(-c_{nt}) + \Phi(t) - \Phi(c_{nt}), \end{aligned}$$

we have by (**) that

$$|G_n(t) - \Phi(t)| \le C\frac{\gamma(\Delta_{nt})}{n^{1/2}} + |\Phi(t) - \Phi(c_{nt})|.$$

Since F is continuous at ξ_p, we have $\Delta_{nt}(1 - \Delta_{nt}) \to p(1 - p) > 0$, and thus $\gamma(\Delta_{nt})n^{-1/2} \to 0$, $n \to \infty$. It remains to investigate whether $c_{nt} \to t$. Writing

$$c_{nt} = t \cdot \frac{A}{[\Delta_{nt}(1 - \Delta_{nt})]^{1/2}} \cdot \frac{F(\xi_p + tAn^{-1/2}) - F(\xi_p)}{tAn^{-1/2}},$$

we see that, if $t > 0$, then

$$c_{nt} \to \frac{tA}{[p(1 - p)]^{1/2}} \cdot F'(\xi_p +),$$

and, if $t < 0$, then

$$c_{nt} \to \frac{tA}{[p(1 - p)]^{1/2}} \cdot F'(\xi_p -).$$

Thus $c_{nt} \to t$ if either

$$t > 0 \quad \text{and} \quad A = [p(1 - p)]^{1/2}/F'(\xi_p +)$$

or

$$t < 0 \quad \text{and} \quad A = [p(1 - p)]^{1/2}/F'(\xi_p -).$$

This establishes (i) and (ii). ■

Remark. The specific rate $O(n^{-1/2})$ provided by the Berry–Esséen Theorem was not actually utilized in the preceding proof. However, in proving Theorem C we do make application of this specific order of magnitude. ■

Corollaries A and B cover typical cases in which $\hat{\xi}_p$ is asymptotically normal, that is, a suitably normalized version of $\hat{\xi}_p$ converges in distribution to $N(0, 1)$. However, more generally, Theorem A may provide a normal approximation even when no limit distribution exists. That is, $n^{1/2}(\hat{\xi}_{pn} - \xi_p)$ may fail to have a limit distribution, but its distribution may nevertheless be approximated, as a function of t, by normal distribution probabilities: for $t < 0$, based on the distribution $N(0, p(1 - p)/[F'(\xi_p -)]^2)$; for $t > 0$, based on the distribution $N(0, p(1 - p)/[F'(\xi_p +)]^2)$. The various possibilities are illustrated in the following example.

Example. *Estimation of the median.* Consider estimation of the median $\xi_{1/2}$ of F by the sample median $\hat{\xi}_{1/2}$.

(i) If F has a positive derivative $F'(\xi_{1/2})$ at $x = \xi_{1/2}$, then

$$\hat{\xi}_{1/2} \quad \text{is} \quad AN\left(\xi_{1/2}, \frac{1}{4[F'(\xi_{1/2})]^2 n}\right).$$

(ii) If, further, F has a density f *continuous* at $\xi_{1/2}$, then equivalently we may write

$$\hat{\xi}_{1/2} \quad \text{is} \quad AN\left(\xi_{1/2}, \frac{1}{4f^2(\xi_{1/2})n}\right).$$

(iii) However, suppose that F has a density f which is *discontinuous* at $\xi_{1/2}$. For example, consider the distribution

$$F(x) = \begin{cases} x, & 0 \le x \le \frac{1}{2}, \\ 2x - \frac{1}{2}, & \frac{1}{2} \le x \le \frac{3}{4}. \end{cases}$$

A density for F is

$$f(x) = \begin{cases} 1, & 0 \le x \le \frac{1}{2} \\ 2, & \frac{1}{2} \le x \le \frac{3}{4}, \end{cases}$$

which is discontinuous at $\xi_{1/2} = \frac{1}{2}$. Thus the sample median $\hat{\xi}_{1/2}$ is *not* asymptotically normal in the strict sense, but nevertheless we can approximate the probability

$$P(n^{1/2}(\hat{\xi}_{1/2} - \xi_{1/2}) \le t).$$

We use the distribution $N(0, \frac{1}{4})$ if $t < 0$ and the distribution $N(0, \frac{1}{16})$ if $t > 0$. For $t = 0$, we use the value $\frac{1}{2}$ as an approximation. ■

The multivariate generalization of Corollary B is

Theorem B. *Let* $0 < p_1 < \cdots < p_k < 1$. *Suppose that* F *has a density* f *in neighborhoods of* $\xi_{p_1}, \ldots, \xi_{p_k}$ *and that* f *is positive and continuous at* $\xi_{p_1}, \ldots, \xi_{p_k}$. *Then* $(\hat{\xi}_{p_1}, \ldots, \hat{\xi}_{p_k})$ *is asymptotically normal with mean vector* $(\xi_{p_1}, \ldots, \xi_{p_k})$ *and covariances* σ_{ij}/n, *where*

$$\sigma_{ij} = \frac{p_i(1 - p_j)}{f(\xi_{p_i})f(\xi_{p_j})} \quad \textit{for} \quad i \le j$$

and $\sigma_{ij} = \sigma_{ji}$ *for* $i > j$.

One method of proof will be seen in **2.3.4**, another in **2.5.1**. Or see Cramér (1946), p. 369.

We now consider the *rate of convergence* in connection with the asymptotic normality of $\hat{\xi}_p$. Theorem C below provides the rate $O(n^{-1/2})$. Although left implicit here, an explicit constant of proportionality could be determined by careful scrutiny of the details of proof (Problem 2.P.13). In proving the theorem, we shall utilize the probability inequality for $|\hat{\xi}_{pn} - \xi_p|$ given by Theorem 2.3.2, as well as the following lemma.

Lemma. *For* $|ax| \le \frac{1}{2}$.

$$|\Phi(x + ax^2) - \Phi(x)| \le 5|a|\sup_x[x^2\phi(x)].$$

PROOF. By the mean value theorem,

$$\Phi(x + ax^2) - \Phi(x) = ax^2\phi(x^*),$$

where x^* lies between x and $x + ax^2$, both of which have the same sign. Since $\phi(x)$ is increasing on $(-\infty, 0)$ and decreasing on $(0, \infty)$, we have

$$\phi(x^*) \le \phi(x + \tfrac{1}{2}x) + \phi(x - \tfrac{1}{2}x)$$

and hence

$$\begin{aligned} x^2\phi(x^*) &\le x^2\phi(x + \tfrac{1}{2}x) + x^2\phi(x - \tfrac{1}{2}x) \\ &= \tfrac{4}{9}(\tfrac{3}{2}x)^2\phi(\tfrac{3}{2}x) + 4(\tfrac{1}{2}x)^2\phi(\tfrac{1}{2}x) \\ &\le 5\sup_x[x^2\phi(x)]. \quad \blacksquare \end{aligned}$$

Theorem C. *Let* $0 < p < 1$. *Suppose that in a neighborhood of* ξ_p, F *possesses a positive continuous density* f *and a bounded second derivative* F″. *Then*

$$\sup_{-\infty<t<\infty}\left|P\left(\frac{n^{1/2}(\hat{\xi}_{pn} - \xi_p)}{[p(1 - p)]^{1/2}/f(\xi_p)} \le t\right) - \Phi(t)\right| = O(n^{-1/2}), \qquad n \to \infty.$$

PROOF. Put $A = [p(1 - p)]^{1/2}/f(\xi_p)$ and

$$G_n(t) = P(n^{1/2}(\hat{\xi}_{pn} - \xi_p)/A \le t).$$

Let $L_n = B(\log n)^{1/2}$. We shall introduce restrictions on the constant B as needed in the course of the proof. Now note that

(1)

$$\begin{aligned} \sup_{|t|>L_n}|G_n(t) - \Phi(t)| &= \max\left\{\sup_{t<-L_n}|G_n(t) - \Phi(t)|, \sup_{t>L_n}|G_n(t) - \Phi(t)|\right\} \\ &\le \max\{G_n(-L_n) + \Phi(-L_n), 1 - G_n(L_n) + 1 - \Phi(L_n)\} \\ &\le G_n(-L_n) + 1 - G_n(L_n) + 1 - \Phi(L_n) \\ &\le P(|\hat{\xi}_{pn} - \xi_p| \ge AL_n n^{-1/2}) + 1 - \Phi(L_n). \end{aligned}$$

As is well-known and easily checked (or see Gnedenko (1962), p. 134),

$$1 - \Phi(x) \le \frac{(2\pi)^{-1/2}}{x}e^{-(1/2)x^2}, \qquad x > 0,$$

so that

(2) $$1 - \Phi(L_n) \le \frac{(2\pi)^{-1/2}}{L_n}n^{-(1/2)B^2} = O(n^{-1/2}),$$

provided that

(3) $$B^2 \geq 1.$$

To obtain a similar result for the other term in (1), we use the probability inequality given by Theorem 2.3.2, with ε given by

$$\varepsilon_n = (A - \varepsilon_0)L_n n^{-1/2},$$

where ε_0 is arbitrarily chosen subject to $0 < \varepsilon_0 < A$. In order to deal with $\delta_{\varepsilon_n} = \min\{F(\delta_p + \varepsilon_n) - p,\ p - F(\xi_p - \varepsilon_n)\}$, we utilize Taylor's Theorem (1.12.1A) to write

$$F(\xi_p + \varepsilon_n) - p = f(\xi_p)\varepsilon_n + \tfrac{1}{2}F''(z^*)\varepsilon_n^2,$$

where z^* lies between ξ_p and $\xi_p + \varepsilon_n$, and

$$p - F(\xi_p - \varepsilon_n) = f(\xi_p)\varepsilon_n - \tfrac{1}{2}F''(z^{**})\varepsilon_n^2,$$

where z^{**} lies between ξ_p and $\xi_p - \varepsilon_n$. Then

$$\begin{aligned}\delta_{\varepsilon_n}^2 = \min\{&f^2(\xi_p)\varepsilon_n^2 + f(\xi_p)F''(z^*)\varepsilon_n^3 + \tfrac{1}{4}[F''(z^*)]^2\varepsilon_n^4, \\ &f^2(\xi_p)\varepsilon_n^2 - f(\xi_p)F''(z^{**})\varepsilon_n^3 + \tfrac{1}{4}[F''(z^{**})]^2\varepsilon_n^4 \\ \geq \varepsilon_n^2 f(\xi_p)&[f(\xi_p) - M\varepsilon_n],\end{aligned}$$

where M satisfies

(4) $$\sup_{|z| \leq \varepsilon_n} |F''(\xi_p + z)| \leq M < \infty$$

for all n under consideration. Hence

$$\begin{aligned}-2n\delta_{\varepsilon_n}^2 &\leq -2n\varepsilon_n^2 f(\xi_p)[f(\xi_p) - M\varepsilon_n] \\ &= -2L_n^2(A - \varepsilon_0)^2 f(\xi_p)[f(\xi_p) - M\varepsilon_n].\end{aligned}$$

For convenience let us now put

$$\varepsilon_0 = \tfrac{1}{2}A.$$

Recalling the definition of A, we thus have

$$-2n\delta_{\varepsilon_n}^2 \leq -\tfrac{1}{2}p(1 - p)B^2\left(1 - \frac{[p(1 - p)]^{1/2}MB(\log n)^{1/2}}{2f^2(\xi_p)n^{1/2}}\right)\log n.$$

Hence

(5)

$$\begin{aligned}P(|\hat{\xi}_{pn} - \xi_p| \geq AL_n n^{-1/2}) &\leq P(|\hat{\xi}_{pn} - \xi_p| > \varepsilon_n) \\ &\leq 2n^{-(1/2)p(1-p)B^2\left(1 - \frac{[p(1 - p)]^{1/2}MB(\log n)^{1/2}}{2f^2(\xi_p)n^{1/2}}\right)} \\ &= O(n^{-1/2}),\end{aligned}$$

provided that

$$B^2 > \frac{1}{p(1-p)}. \tag{6}$$

We now treat

$$\sup_{|t| \le L_n} |G_n(t) - \Phi(t)|.$$

From the proof of Theorem A, it is seen that

$$\sup_{|t| \le L_n} |G_n(t) - \Phi(t)| \le n^{-1/2} C \sup_{|t| \le L_n} \gamma(\Delta_{nt}) + \sup_{|t| \le L_n} |\Phi(t) - \Phi(c_{nt})|, \tag{7}$$

where C is a universal constant, $\Delta_{nt} = F(\xi_p + Atn^{-1/2})$,

$$\gamma(\Delta_{nt}) = \frac{(1 - \Delta_{nt})^2 + \Delta_{nt}^2}{[\Delta_{nt}(1 - \Delta_{nt})]^{1/2}},$$

and

$$c_{nt} = \frac{n^{1/2}(\Delta_{nt} - p)}{[\Delta_{nt}(1 - \Delta_{nt})]^{1/2}}.$$

Defining $g(z) = [F(\xi_p + z) - F^2(\xi_p + z)]^{-1/2}$, we have $g(0) = [p(1-p)]^{-1/2}$ and

$$g'(z) = -\tfrac{1}{2} f(\xi_p + z)[1 - 2F(\xi_p + z)][F(\xi_p + z) - F^2(\xi_p + z)]^{-3/2},$$

and thus

$$[\Delta_{nt}(1 - \Delta_{nt})]^{-1/2} = [p(1-p)]^{-1/2} + g'(z_{nt})Atn^{-1/2},$$

where z_{nt} lies between ξ_p and $\xi_p + Atn^{-1/2}$. Inspection of $g'(z)$ shows that the quantity

$$w_n = \sup_{|z - \xi_p| \le AL_n n^{-1/2}} |g'(z)|$$

is finite; in fact

$$w_n \to w_\infty = f(\xi_p)|\tfrac{1}{2} - p|[p(1-p)]^{-3/2}$$

as $n \to \infty$. Hence also the quantity

$$\gamma_n = \sup_{|t| \le L_n} \gamma(\Delta_{nt})$$

is finite; in fact

$$\gamma_n \to \gamma_\infty = [(1-p)^2 + p^2][p(1-p)]^{-3/2} \tag{8}$$

as $n \to \infty$.

Finally, by Taylor's Theorem again,

$$n^{1/2}(\Delta_{nt} - p) = At\,\frac{F(\xi_p + Atn^{-1/2}) - F(\xi_p)}{Atn^{-1/2}}$$

$$= At[f(\xi_p) + F''(\xi_{pt})Atn^{-1/2}],$$

where ξ_{pt} lies between ξ_p and $\xi_p + Atn^{-1/2}$. Thus

$$c_{nt} = t\left(1 + \frac{F''(\xi_{pt})}{f(\xi_p)}\,Atn^{-1/2}\right)(1 + [p(1-p)]^{1/2}g'(z_{nt})Atn^{-1/2})$$

$$= t(1 + h_{nt}tn^{-1/2}), \text{ say,}$$

where we have

(9) $$\sup_{|t| \le L_n} |h_{nt}| = H_n = O(1),$$

since F'' is bounded in a neighborhood of ξ_p. Thus, for n large enough that

(10) $$H_n L_n n^{-1/2} \le \tfrac{1}{2},$$

application of the lemma preceding the theorem, with $a = h_{nt}n^{-1/2}$, yields

(11) $$|\Phi(t) - \Phi(c_{nt})| \le 5H_n n^{-1/2} \sup_x [x^2\phi(x)], \qquad |t| \le L_n.$$

Since $\sup_x[x^2\phi(x)] < \infty$, it follows by (7), (8) and (11) that

(12) $$\sup_{|t| \le L_n} |G_n(t) - \Phi(t)| = O(n^{-1/2}).$$

Combining (1), (2), (5) and (12), the proof is complete. ■

A theorem similar to the preceding result has also been established, independently, by Reiss (1974).

2.3.4 The Density of $\hat{\xi}_{pn}$

If F has a density f, then the distribution G_n of $\hat{\xi}_{pn}$ also has a density, $g_n(t) = G_n'(t)$, for which we now derive an expression.

By Lemma 1.1.4 and the fact that $nF_n(\Delta)$ is binomial $(n, F(\Delta))$, we have

$$G_n(t) = P(\hat{\xi}_{pn} \le t) = P(F_n(t) \ge p) = P(nF_n(t) \ge np)$$

$$= \sum_{i=m}^{n} \binom{n}{i} [F(t)]^i [1 - F(t)]^{n-i},$$

where

$$m = \begin{cases} np & \text{if } np \text{ is an integer} \\ [np] + 1 & \text{if } np \text{ is not an integer.} \end{cases}$$

Taking derivatives, it is found (Problem 2.P.14) that

$$g_n(t) = n\binom{n-1}{m-1}[F(t)]^{m-1}[1 - F(t)]^{n-m}f(t).$$

Incidentally, this result provides another way to prove the asymptotic normality of $\hat{\xi}_{pn}$, as stated in Corollary 2.3.3B. The density of the random variable $n^{1/2}(\hat{\xi}_{pn} - \xi_p)$ is

$$h_n(t) = n^{-1/2}g_n(\xi_p + tn^{-1/2}).$$

Using the expression just derived, it may be shown that

$$(*) \qquad \lim_{n\to\infty} h_n(t) = \phi(tf(\xi_p)[p(1-p)]^{-1/2}), \qquad \text{each } t,$$

that is, $h_n(t)$ converges pointwise to the density of $N(0, p(1-p)/f^2(\xi_p))$. Then Scheffé's Theorem (1.5.1C) yields the desired conclusion. For details of proof of (*), see Cramér (1946) or Rao (1973). Moreover, this technique of proof generalizes easily for the multivariate extension, Theorem 2.3.3B.

Finally, we comment that from the above expression for $G_n(t)$ one can establish that if F has a finite mean, then for each k, $\hat{\xi}_{pn}$ has finite kth moment for all sufficiently large n (Problem 2.P.15).

2.3.5 Quantiles Versus Moments

In some instances the quantile approach is feasible and useful when other approaches are out of the question. For example, to estimate the parameter of a Cauchy distribution, with density $f(x) = 1/\pi[1 + (x - \mu)^2]$, $-\infty < x < \infty$, the sample mean $\bar{X}$ is *not* a consistent estimate of the location parameter μ. However, the sample median $\hat{\xi}_{1/2}$ is $AN(\mu, \pi^2/4n)$ and thus quite well-behaved.

When *both* the quantile and moment approaches are feasible, it is of interest to examine their relative efficiency. For example, consider a symmetric distribution F having finite variance σ^2 and mean (= median) μ $(= \xi_{1/2})$. In this case both $\bar{X}$ and $\hat{\xi}_{1/2}$ are competitors for estimation of μ. Assume that F has a density f positive and continuous at μ. Then, according to the theorems we have established,

$$\bar{X} \quad \text{is} \quad AN\left(\mu, \frac{\sigma^2}{n}\right)$$

and

$$\hat{\xi}_{1/2} \quad \text{is} \quad AN\left(\mu, \frac{1}{4f^2(\mu)n}\right).$$

If we consider asymptotic relative efficiency in the sense of the criterion of small asymptotic variance in the normal approximation, then the asymptotic relative efficiency of $\hat{\xi}_{1/2}$ relative to $\overline{X}$ is (recall **1.15.4**)

$$e(\hat{\xi}_{1/2}, \overline{X}) = 4\sigma^2 f^2(\mu),$$

that is, the limiting ratio of sample sizes (of $\overline{X}$ and $\hat{\xi}_{1/2}$, respectively) at which performance is "equivalent." For a *normal* distribution F, this relative efficiency is $2/\pi$, indicating the degree of superiority of $\overline{X}$ over $\hat{\xi}_{1/2}$. As an exercise (Problem 1.P.16), evaluate $e(\hat{\xi}_{1/2}, \overline{X})$ for some other distributions F. Discover some cases when $\hat{\xi}_{1/2}$ is superior to $\overline{X}$.

2.3.6 A Measure of Dispersion Based on Quantiles

An alternative to the standard deviation σ of F, as a measure of dispersion, is the *semi-interquartile range*

$$R = \tfrac{1}{2}(\xi_{3/4} - \xi_{1/4}).$$

A natural estimator of R is the sample analogue

$$\hat{R} = \tfrac{1}{2}(\hat{\xi}_{3/4} - \hat{\xi}_{1/4}).$$

By Theorem 2.3.3B and the Cramér–Wold device (Theorem 1.5.2), it follows that (Problem 2.P.17)

$$\hat{R} \text{ is } AN\left(R, \frac{1}{64n}\left(\frac{3}{f^2(\xi_{3/4})} - \frac{2}{f(\xi_{1/4})f(\xi_{3/4})} + \frac{3}{f^2(\xi_{1/4})}\right)\right).$$

For $F = N(\mu, \sigma^2)$, we have

$$\hat{R} \text{ is } AN\left(0.6745\sigma, \frac{(0.7867)^2\sigma^2}{n}\right).$$

See Cramér (1946), pp. 181 and 370.)

2.3.7 Nonparametric Tests Based on Quantiles

A number of hypothesis-testing problems in nonparametric inference may be formulated suitably in terms of quantiles (see Fraser (1957), Chapter 3). Among these are:

(i) *single sample location problem*
  Hypothesis: $\xi_p = v_0$
  Alternative: $\xi_0 > v_0$
  (Here p and v_0 are to be specified. Of course, other types of alternative may be considered.

(ii) *single sample location and symmetry problem*
  Hypothesis: $\xi_{1/2} = v_0$ and F symmetric
  Alternative: $\xi_{1/2} \neq v_0$ or F not symmetric

(iii) *two-sample scale problem*
(Given $X_1, \ldots, X_{n_1}$ I.I.D. F and $Y_1, \ldots, Y_{n_2}$ I.I.D. G)
Hypothesis: $F(x) = G(x + c)$, all x
Alternative: $\xi_{p_2}(F) - \xi_{p_1}(F) < \xi_{p_2}(G) - \xi_{p_1}(G)$, all $p_1 < p_2$.

The most widely known "quantile" test arising for these problems is the *sign* test for problem (i). Most texts provide some discussion of it, often in the context of *order* statistics, which we shall examine in the forthcoming section.

2.4 THE ORDER STATISTICS

For a sample of independent observations $X_1, \ldots, X_n$ on a distribution F, the ordered sample values

$$X_{(1)} \le X_{(2)} \le \cdots \le X_{(n)},$$

or, in more explicit notation,

$$X_{n1} \le X_{n2} \le \cdots \le X_{nn},$$

are called the *order statistics* and the vector

$$\mathbf{X}_{(n)} = (X_{n1}, \ldots, X_{nn})$$

is called the *order statistic* of the sample. If F is continuous, then with probability 1 the order statistics of the sample take distinct values (and conversely).

The *exact* distribution of the kth order statistic X_{nk} is easily found, but cumbersome to use:

$$P(X_{nk} \le x) = \sum_{i=k}^{n} \binom{n}{i} [F(x)]^i [1 - F(x)]^{n-i}, \qquad -\infty < x < \infty.$$

The *asymptotic* theory of sequences $\{X_{nk_n}\}$ of order statistics is discussed in **2.4.3**, with some particular results exhibited in **2.4.4.** We further discuss asymptotic theory of order statistics in **2.5**, **3.6** and Chapter 8.

Comments on the fundamental role of order statistics and their connection with sample quantiles are provided in **2.4.1**, and on their scope of application in **2.4.2**.

Useful general reading on order statistics consists of David (1970), Galambos (1978), Renyi (1953), Sarhan and Greenberg (1962), and Wilks (1948, 1962).

2.4.1 Fundamental Role of the Order Statistics; Connection with the Sample Quantiles

Since the order statistic $\mathbf{X}_{(n)}$ is equivalent to the sample distribution function F_n, its role is fundamental even if not always explicit. Thus, for example, the sample mean $\bar{X}$ may be regarded as the mean of the order statistics, and the sample pth quantile may be expressed as

$$(*) \qquad \hat{\xi}_{pn} = \begin{cases} X_{n,np} & \text{if } np \text{ is an integer,} \\ X_{n,[np]+1} & \text{if } np \text{ is not an integer.} \end{cases}$$

The representations of $\bar{X}$ and $\hat{\xi}_{pn}$ in terms of order statistics are a bit artificial. On the other hand, for many useful statistics, the most natural and effective representations are in terms of order statistics. Examples are the *extreme values* X_{n1} and X_{nn}, and the *sample range* $X_{nn} - X_{n1}$. (In **2.4.4** it is seen that these latter examples have asymptotic behavior quite different from asymptotic normality.)

The relation (*) may be inverted:

$$(**) \qquad X_{nk} = \hat{\xi}_{k/n,n}, \qquad 1 \le k \le n.$$

In view of (*) and (**), the entire discussion of order statistics could be carried out formally in terms of sample quantiles, and vice versa. The choice of formulation depends upon the point of view which is most relevant and convenient to the particular purpose or application at hand. Together, therefore, the previous section (**2.3**) and the present section (**2.4**) comprise the two basic elements of a single general theory. The cohesion of these basic elements will be viewed more fully in a complementary analysis developed in **2.5**.

2.4.2 Remarks on Applications of Order Statistics

The *extreme values*, X_{n1} and X_{nn}, arise quite naturally in the study of floods or droughts, and in problems of breaking strength or fatigue failure.

A quick measure of dispersion is provided by the *sample range*, suitably normalized. More generally, a variety of *short-cut* procedures for quick estimates of location or dispersion, or for quick tests of hypotheses about location or dispersion, are provided in the form of *linear functions of order statistics*, that is, statistics of the form $\sum_{i=1}^{n} c_{ni} X_{ni}$. The class of such statistics is important also in the context of *robust inference*. We shall study these statistics technically in Chapter 8.

Order statistics are clearly relevant in problems with *censored* data. A typical situation arises in connection with *life-testing* experiments, in which a fixed number n of items are placed on test and the experiment is terminated as soon as a prescribed number r have failed. The observed lifetimes are thus $X_{n1} \le \cdots \le X_{nr}$, whereas the lifetimes $X_{n,r+1} \le \cdots \le X_{nn}$ remain unobserved. For a survey of some important results on order statistics and their role in estimation and hypothesis testing in life testing and *reliability* problems, see Gupta and Panchapakesan (1974). A useful methodological text consists of Mann, Schafer and Singpurwalla (1974).

Pairs of order statistics, such as (X_{nr}, X_{ns}), serve to provide *distribution-free tolerance limits* (see Wilks (1962), p. 334) and *distribution-free confidence intervals for quantiles* (see **2.6**).

Some further discussion of applications of order statistics is provided in **3.6**.

2.4.3 Asymptotic Behavior of Sequences $\{X_{nk_n}\}$.

The discussion here is general. Particular results are given in **2.2.4** and **2.5** (and in principle in **2.3**).

For an order statistic X_{nk}, the ratio k/n is called its *rank*. Consider a sequence of order statistics, $\{X_{nk_n}\}_{n=1}^{\infty}$, for which k_n/n has a limit L (called the *limiting rank*). Three cases are distinguished: sequences of *central terms* $(0 < L < 1)$, sequences of *intermediate terms* ($L = 0$ and $k_n \to \infty$, or $L = 1$ $n - k_n \to \infty$), and sequences of *extreme terms* ($L = 0$ and k_n bounded, or $L = 1$ and $n - k_n$ bounded).

A typical example of a sequence of central terms having limiting rank p, where $0 < p < 1$, is the sequence of sample pth quantiles $\{\hat{\xi}_{pn}\}_{n=1}^{\infty}$. On the basic elements of a single general theory. The cohesion of these basic elements quences of central terms in general have asymptotically normal behavior and converge strongly to appropriate limits. This will be corroborated in **2.5**.

An example of a sequence of extreme terms having limiting rank 1 is $\{X_{nn}\}_{n=1}^{\infty}$.

Generalizing work of Gnedenko (1943), Smirnov (1952) provided the asymptotic distribution theory for both central and extreme sequences. For each case, he established the class of possible limit distributions and for each limit distribution the corresponding domain of attraction. For extension to the case of independent but nonidentically distributed random variables, see Mejzler and Weissman (1969). For investigation of intermediate sequences, see Kawata (1951), Cheng (1965) and Watts (1977).

2.4.4 Asymptotic Behavior of X_{nn}

If the random variable $(X_{nn} - a_n)/b_n$ has a limit distribution for some choice of constants $\{a_n\}$, $\{b_n\}$, then the limit distribution must be of the form G_1, G_2, or G_3, where

$$G_1(t) = \begin{cases} 0, & t \le 0, \\ e^{-t^{-\alpha}}, & t \ge 0, \end{cases}$$

$$G_2(t) = \begin{cases} e^{-(-t)^{\alpha}} & t \le 0, \\ 1, & t \ge 0, \end{cases}$$

and

$$G_3(t) = e^{-e^{-t}}, \qquad -\infty < t < \infty.$$

(In G_1 and G_2, α is a positive constant.) This result was established by Gnedenko (1943), following less rigorous treatments by earlier authors. Each of the three types G_1, G_2, and G_3 arises in practice, but G_3 occupies the pre-eminent position. Typical cases are illustrated by the following examples.

Example A. F is *exponential* : $F(x) = 1 - \exp(-x)$, $x > 0$. Putting $a_n = \log n$ and $b_n = 1$, we have

$$\begin{aligned} P\left(\frac{X_{nn} - a_n}{b_n} \le t\right) &= P(X_{nn} - \log n \le t) \\ &= (1 - e^{-\log n - t})^n \\ &\to e^{-e^{-t}}, \qquad n \to \infty. \quad \blacksquare \end{aligned}$$

Example B. F is *logistic*: $F(x) = [1 + \exp(-x)]^{-1}$, $-\infty < x < \infty$. Again taking $a_n = \log n$ and $b_n = 1$, we may obtain (Problem 2.P.18)

$$P(X_{nn} - \log n \le t) \to e^{-e^{-t}}, \qquad n \to \infty. \quad \blacksquare$$

Example C. F is *normal*: $F = \Phi$. With

$$a_n = (2 \log n)^{1/2} - \tfrac{1}{2}(\log \log n + \log 4\pi)(2 \log n)^{-1/2}$$

and

$$b_n = (2 \log n)^{-1/2}$$

it is found (Cramér (1946), p. 374) that

$$P\left(\frac{X_{nn} - a_n}{b_n} \le t\right) \to e^{-e^{-t}}, \qquad n \to \infty. \quad \blacksquare$$

In Examples A and B, the *rate* of the convergence in distribution is quite fast. In Example C, however, the error of approximation tends to 0 at the rate $O((\log n)^{-\beta})$, for some $\beta > 0$, but not faster. For discussion and pictorial illustration, see Cramér (1946), Section 28.6. The lack of agreement between the exact and limit distributions is seen to be in the tails of the distributions. Further literature on the issue is cited in David (1970), p. 209. See also Galambos (1978), Section 2.10.

Statistics closely related to X_{nn} include the *range* $X_{nn} - X_{n1}$ and the *studentized extreme deviate*, whose asymptotic distributions are discussed in Section 3.6.

The *almost sure* asymptotic properties of X_{nn} can also be characterized. For example, in connection with F *normal*, we anticipate by Example C above that

X_{nn} is close to $(2 \log n)^{1/2}$ in appropriate stochastic senses. This is in fact true: both

$$P\left(\lim_{n\to\infty} [X_{nn} - (2 \log n)^{1/2}] = 0\right) = 1$$

and

$$P\left(\lim_{n\to\infty} \frac{X_{nn}}{(2 \log n)^{1/2}} = 1\right) = 1.$$

Thus X_{nn} satisfies both additive and multiplicative forms of strong convergence. For a treatment of the almost sure behavior of X_{nn} for arbitrary F, see Galambos (1978), Chapter 4.

2.5 ASYMPTOTIC REPRESENTATION THEORY FOR SAMPLE QUANTILES, ORDER STATISTICS, AND SAMPLE DISTRIBUTION FUNCTIONS

Throughout we deal as usual with a sequence of I.I.D. observations $X_1, X_2, \ldots$ having distribution function F.

We shall see that it is possible to express sample quantiles and "central" order statistics asymptotically as *sums*, via representation as a linear transform of the sample distribution function evaluated at the relevant quantile. From these representations, a number of important insights and properties follow.

The representations were pre-figured in Wilks (1962), Section 9.6. However, they were first presented in their own right, and with a full view of their significance, by Bahadur (1966). His work gave impetus to a number of important additional studies, as will be noted.

Bahadur's representations for sample quantiles and order statistics are presented in **2.5.1** and **2.5.2**, respectively, with discussion of their implications. A sketch of the proof is presented in general terms in **2.5.3**, and the full details of proof are given in **2.5.4**. Further properties of the errors of approximation in the representations are examined in **2.5.5**. An application of the representation theory will be made in Section 2.6, in connection with the problem of confidence intervals for quantiles.

Besides references cited herein, further discussion and bibliography may be found in Kiefer (1970b).

2.5.1 Sample Quantiles as Sums Via the Sample Distribution Function

Theorem (Bahadur (1966)). *Let* $0 < p < 1$. *Suppose that* F *is twice differentiable at* ξ_p, *with* $F'(\xi_p) = f(\xi_p) > 0$. *Then*

$$\hat{\xi}_{pn} = \xi_p + \frac{p - F_n(\xi_p)}{f(\xi_p)} + R_n,$$

where with probability 1

$$R_n = O(n^{-3/4}(\log n)^{3/4}), \qquad n \to \infty.$$

Details of proof are given in **2.5.3** and **2.5.4**, and the random variable R_n is examined somewhat further in **2.5.5**.

Remarks. (i) By the statement "with probability 1, $Y_n = O(g(n))$ as $n \to \infty$" is meant that there exists a set Ω_0 such that $P(\Omega_0) = 1$ and for each $\omega \in \Omega_0$ there exists a constant $B(\omega)$ such that

$$|Y_n(\omega)| \geq B(\omega)g(n), \qquad \text{all } n \text{ sufficiently large.}$$

(For Y_n given by the R_n of the theorem, it can be seen from the proof that the constants $B(\omega)$ may be chosen not to depend upon ω.)

(ii) Bahadur (1966) actually assumes in addition that F'' exists and is bounded in a neighborhood of ξ_p. However, by substituting in his argument the use of Young's form of Taylor's Theorem instead of the standard version, the extra requirements on F'' may be dropped.

(iii) Actually, Bahadur established that

$$R_n = O(n^{-3/4}(\log n)^{1/2}(\log \log n)^{1/4}, \qquad n \to \infty,$$

with probability 1. (See Remark 2.5.4D.) Further, Kiefer (1967) obtained the *exact* order for R_n, namely $O(n^{-3/4}(\log \log n)^{3/4})$. See **2.5.5** for precise details.

(iv) (continuation) However, for many statistical applications, it suffices merely to have $R_n = o_p(n^{-1/2})$. Ghosh (1971) has obtained this weaker conclusion by a simpler proof requiring only that F be *once* differentiable at ξ_p with $F'(\xi_p) > 0$.

(v) The conclusion stated in the theorem may alternatively be expressed as follows: with probability 1

$$n^{1/2}(\hat{\xi}_{pn} - \xi_p) = \frac{n^{1/2}(p - F_n(\xi_p))}{f(\xi_p)} + O(n^{-1/4}(\log n)^{3/4}), \qquad n \to \infty.$$

(vi) (continuation) The theorem thus provides a link between two asymptotic normality results, that of $\hat{\xi}_{pn}$ and that of $F_n(\xi_p)$. We have seen previously as separate results (Corollary 2.3.3B and Theorem 2.1.1, respectively) that the random variables

$$n^{1/2}(\hat{\xi}_{pn} - \xi_p), \frac{n^{1/2}(p - F_n(\xi_p))}{f(\xi_p)}$$

each converge in distribution to $N(0, p(1 - p)/f^2(\xi_p))$. The theorem of Bahadur goes much further, by revealing that the actual difference between these random variables tends to 0 *wp*1, and this at a rate $O(n^{-3/4}(\log n)^{1/2})$.

(vii) *Representation of a sample quantile as a sample mean.* Let

$$Y_i = \xi_p + \frac{p - I(X_i \le \xi_p)}{f(\xi_p)}, \qquad i = 1, 2, \ldots.$$

Then the conclusion of the theorem may be expressed as follows: *wp*1

$$\hat{\xi}_{pn} = \frac{1}{n}\sum_{i=1}^{n} Y_i + O(n^{-3/4}(\log n)^{3/4}), \qquad n \to \infty.$$

That is, *wp*1 $\hat{\xi}_{pn}$ is asymptotically (*but not exactly*) the mean of the first n members of the I.I.D. sequence $\{Y_i\}$.

(viii) *Law of the iterated logarithm for sample quantiles* (under the conditions of the theorem). As a consequence of the preceding remark, in conjunction with the classical LIL (Theorem 1.10A), we have: *wp*1

$$\overline{\lim_{n\to\infty}} \pm \frac{n^{1/2}(\hat{\xi}_{pn} - \xi_p)}{(2 \log \log n)^{1/2}} = \frac{[p(1-p)]^{1/2}}{f(\xi_p)},$$

for either choice of sign (Problem 2.P.20). This result has been extended to a larger class of distributions F by de Haan (1974).

(ix) *Asymptotic multivariate normality of sample quantiles* (under the conditions of the theorem). As another consequence of remark (vii), the conclusion of Theorem 2.3.3B is obtained (Problem 2.P.21). ■

2.5.2 Central Order Statistics as Sums Via the Sample Distribution Function

The following theorem applies to a sequence of "central" order statistics $\{X_{nk_n}\}$ as considered in **2.4.3**. It is required, in addition, that the convergence of k_n/n to p be at a sufficiently fast rate.

Theorem (Bahadur (1966)). *Let* $0 < p < 1$. *Suppose that* F *is twice differentiable at* ξ_p, *with* $F'(\xi_p) = f(\xi_p) > 0$. *Let* $\{k_n\}$ *be a sequence of positive integers* $(1 \le k_n \le n)$ *such that*

$$\frac{k_n}{n} = p + o\left(\frac{(\log n)^{\Delta}}{n^{1/2}}\right), \qquad n \to \infty,$$

for some $\Delta \ge \frac{1}{2}$. *Then*

$$X_{nk_n} = \xi_p + \frac{(k_n/n) - F_n(\xi_p)}{f(\xi_p)} + \tilde{R}_n,$$

where with probability 1

$$\tilde{R}_n = O(n^{-3/4}(\log n)^{(1/2)(\Delta+1)}), \qquad n \to \infty.$$

Remarks. (i) Bahadur (1966) actually assumes in addition that F'' exists and is bounded in a neighborhood of ξ_p. Refer to discussion in Remark 2.5.1(ii).

(ii) Extension to certain cases of "intermediate" order statistics has been carried out by Watts (1977). ■

This theorem, taken in conjunction with Theorem 2.5.1, shows that the order statistic X_{nk_n} and the sample pth quantile $\hat{\xi}_{pn}$ are roughly equivalent as estimates of ξ_p, provided that the rank k_n/n tends to p sufficiently fast. More precisely, we have (Problem 2.P.22) the following useful and interesting result.

Corollary. *Assume the conditions of the preceding theorem and suppose that*

$$\frac{k_n}{n} = p + \frac{k}{n^{1/2}} + o\left(\frac{1}{n^{1/2}}\right), \qquad n \to \infty.$$

Then

$$(*) \qquad n^{1/2}(X_{nk_n} - \hat{\xi}_{pn}) \xrightarrow{wp1} \frac{k}{f(\xi_p)}$$

and

$$(**) \qquad n^{1/2}(X_{nk_n} - \xi_p) \xrightarrow{d} N\left(\frac{k}{f(\xi_p)}, \frac{p(1-p)}{f^2(\xi_p)}\right).$$

By (*) it is seen that X_{nk_n} trails along with $\hat{\xi}_{pn}$ as a *strongly consistent* estimator of ξ_p. We also see from (*) that the closeness of X_{nk_n} to $\hat{\xi}_{pn}$ is regulated rigidly by the exact rate of the convergence of k_n/n to p. Further, despite the consistency of X_{nk_n} for estimation of ξ_p, it is seen by (**) that, on the other hand, the normalized estimator has a limit normal distribution *not* centered at 0 (unless $k = 0$), but rather at a constant determined by the exact rate of convergence of k_n/n to p. These aspects will be of particular interest in our treatment of confidence intervals for quantiles (Section 2.6).

2.5.3 Sketch of Bahadur's Method of Proof

Here we sketch, in general form, the line of argument used to establish Theorems 2.5.1 and 2.5.2. Complete details of proof are provided in **2.5.4**.

Objective. Suppose that we have an estimator T_n satisfying $T_n \xrightarrow{wp1} \theta$ and that we seek to represent T_n asymptotically as simply a linear transformation of $G_n(\theta)$, where $G_n(\cdot)$ is a random function which pointwise has the structure of a sample mean. (For example, G_n might be the sample distribution function.)

Approach. (i) Let $G(\cdot)$ be the function that G_n estimates. Assume that G is sufficiently regular at ξ_p to apply Taylor's Theorem (in Young's form) and obtain a linearization of $G(T_n)$:

$$(1) \qquad G(T_n) - G(\theta) = G'(\theta)(T_n - \theta) + \Delta_n,$$

where *wp*1 $\Delta_n = O((T_n - \theta)^2)$, $n \to \infty$.

(ii) In the left-hand side of (1), switch from G to G_n, subject to adding another component Δ_n' to the remainder. This yields

$$(2) \qquad G_n(T_n) - G_n(\theta) = G'(\theta)(T_n - \theta) + \Delta_n + \Delta_n'.$$

(iii) Express $G_n(T_n)$ in the form

$$G_n(T_n) = c_n + \Delta_n'',$$

where c_n is a constant and Δ_n'' is suitably negligible. Introduce into (2) and solve for T_n, obtaining:

$$(3) \qquad T_n = \theta + \frac{c_n - G_n(\theta)}{G'(\theta)} + O(\Delta_n) + O(\Delta_n') + O(\Delta_n'').$$

Clearly, the usefulness of (3) depends upon the $O(\cdot)$ terms. This requires judicious choices of T_n and G_n. In **2.5.4** we take G_n to be F_n and T_n to be either $\hat{\xi}_{pn}$ or X_{nk_n}. In this case $\Delta_n'' = O(n^{-1})$. Regarding Δ_n, it will be shown that for these T_n we have that *wp*1 $|T_n - \theta| = O(n^{-1/2}(\log n)^{1/2})$, yielding $\Delta_n = O(n^{-1}\log n)$. Finally, regarding Δ_n', Bahadur proves a unique and interesting lemma showing that *wp*1 $\Delta_n' = O(n^{-3/4}(\log n)^{(1/2)(q+1)})$, under the condition that $|T_n - \theta| = O(n^{-1/2}(\log n)^q)$, where $q \geq \frac{1}{2}$.

2.5.4 Basic Lemmas and Proofs for Theorems 2.5.1 and 2.5.2

As a preliminary, we consider the following probability inequality, one of many attributed to S. N. Bernstein. For proof, see Uspensky (1937).

Lemma A (Bernstein). *Let* $Y_1, \ldots, Y_n$ *be independent random variables satisfying* $P(|Y_i - E\{Y_i\}| \leq m) = 1$, *each* i, *where* $m < \infty$. *Then, for* $t > 0$,

$$P\left(\left|\sum_{i=1}^{n} Y_i - \sum_{i=1}^{n} E\{Y_i\}\right| \geq nt\right) \leq 2\exp\left(-\frac{n^2t^2}{2\sum_{i=1}^{n} \mathrm{Var}\{Y_i\} + \frac{2}{3}mnt}\right),$$

for all $n = 1, 2, \ldots$.

Remarks A. (i) For the case $\mathrm{Var}\{Y_i\} \equiv \sigma^2$, the bound reduces to

$$2\exp\left(-\frac{nt^2}{2\sigma^2 + \frac{2}{3}mt}\right).$$

(ii) For Y_i binomial $(1, p)$, the bound may be replaced by

$$2\exp\left(-\frac{nt^2}{2(p+t)}\right).$$

This version will serve our purposes in the proof of Lemma E below. ■

The next two lemmas give conditions under which $\hat{\xi}_{pn}$ and X_{nk_n} are contained in a suitably small neighborhood of ξ_p for all sufficiently large n, $wp1$.

Lemma B. *Let* $0 < p < 1$. *Suppose that* F *is differentiable at* ξ_p, *with* $F'(\xi_p) = f(\xi_p) > 0$. *Then with probability* 1

$$|\hat{\xi}_{pn} - \xi_p| \le \frac{2(\log n)^{1/2}}{f(\xi_p)n^{1/2}}, \qquad \textit{for all n sufficiently large.}$$

PROOF. Since F is continuous at ξ_p with $F'(\xi_p) > 0$, ξ_p is the unique solution of $F(x-) \le p \le F(x)$ and $F(\xi_p) = p$. Thus we may apply Theorem 2.3.2. Put

$$\varepsilon_n = \frac{2(\log n)^{1/2}}{f(\xi_p)n^{1/2}}.$$

We then have

$$\begin{aligned} F(\xi_p + \varepsilon_n) - p &= F(\xi_p + \varepsilon_n) - F(\xi_p) \\ &= f(\xi_p)\varepsilon_n + o(\varepsilon_n) \\ &\ge \frac{(\log n)^{1/2}}{n^{1/2}}, \qquad \text{for all } n \text{ sufficiently large.} \end{aligned}$$

Likewise, $p - F(\xi_p - \varepsilon_n)$ satisfies a similar relation. Thus, for $\delta_{\varepsilon_n} = \min\{F(\xi_p + \varepsilon_n) - p, p - F(\xi_p - \varepsilon_n)\}$, we have

$$2n\delta_{\varepsilon_n}^2 \ge 2\log n, \qquad \text{for all } n \text{ sufficiently large.}$$

Hence, by Theorem 2.3.2,

$$P(|\hat{\xi}_{pn} - \xi_p| > \varepsilon_n) \le \frac{2}{n^2}, \qquad \text{for all } n \text{ sufficiently large.}$$

By the Borel–Cantelli Lemma (Appendix), it follows that $wp1$ the relations $|\hat{\xi}_{pn} - \xi_p| > \varepsilon_n$ hold for only finitely many n. ■

Remark B. Note from Remark 2.5.1 (viii) that if in addition $F''(\xi_p)$ exists, then we may assert: with probability 1,

$$|\hat{\xi}_{pn} - \xi_p| \le \frac{(\log\log n)^{1/2}}{f(\xi_p)n^{1/2}}, \qquad \text{for all sufficiently large } n.$$

Lemma C. *Let* $0 < \mathrm{p} < 1$. *Suppose that in a neighborhood of* ξ_p, $F'(x) = f(x)$ *exists, is positive, and is continuous at* ξ_p. *Let* $\{k_n\}$ *be a sequence of positive integers* $(1 \le k_n \le n)$ *such that*

$$\frac{k_n}{n} = p + o\left(\frac{(\log n)^\Delta}{n^{1/2}}\right), \qquad n \to \infty,$$

for some $\Delta \ge \frac{1}{2}$. *Then with probability* 1

$$|X_{nk_n} - \xi_p| \le \frac{2(\log n)^\Delta}{f(\xi_p)n^{1/2}}, \qquad \textit{for all n sufficiently large.}$$

PROOF. Define

$$\varepsilon_n = \frac{2(\log n)^\Delta}{f(\xi_p)n^{1/2}}.$$

Following the proof of Theorem 2.3.2, we can establish

$$P(|X_{nk_n} - \xi_p| > \varepsilon_n) \le 2e^{-2n\delta_{\varepsilon_n}^2}, \qquad \text{all } n,$$

where $\delta_{\varepsilon_n} = \min\{F(\xi_p + \varepsilon_n) - k_n/n, k_n/n - F(\xi_p - \varepsilon_n)\}$. Then, following the proof of Lemma B above, we can establish

$$2n\delta_{\varepsilon_n}^2 \ge 2 \log n, \qquad \text{for all } n \text{ sufficiently large,}$$

and obtain the desired conclusion. (Complete all details as an exercise, Problem 2.P.24.) ■

As an exercise (Problem 2.P.25), verify

Lemma D. *Let* $0 < p < 1$. *Suppose that* F *is twice differentiable at* ξ_p. *Let* $T_n \xrightarrow{wp1} \xi_p$. *Then with probability* 1

$$F(T_n) - F(\xi_p) = F'(\xi_p)(T_n - \xi_p) + O((T_n - \xi_p)^2), \qquad n \to \infty.$$

As our final preparation, we establish the following ingenious result of Bahadur (1966).

Lemma E (Bahadur). *Let* $0 < p < 1$. *Suppose that* F *is twice differentiable at* ξ_p, *with* $F'(\xi_p) = f(\xi_p) > 0$. *Let* $\{a_n\}$ *be a sequence of positive constants such that*

$$a_n \sim c_0 n^{-1/2}(\log n)^q, \qquad n \to \infty,$$

for some constants $c_0 > 0$ *and* $q \ge \frac{1}{2}$. *Put*

$$H_{pn} = \sup_{|x| \le a_n} |[F_n(\xi_p + x) - F_n(\xi_p)] - [F(\xi_p + x) - F(\xi_p)]|.$$

Then with probability 1

$$H_{pn} = O(n^{-3/4}(\log n)^{(1/2)(q+1)}), \qquad n \to \infty.$$

PROOF. Let $\{b_n\}$ be a sequence of positive integers such that $b_n \sim c_0 n^{1/4}(\log n)^q$, $n \to \infty$. For integers $r = -b_n, \ldots, b_n$, put

$$\eta_{r,n} = \xi_p + a_n b_n^{-1} r, \; \alpha_{r,n} = F(\eta_{r+1,n}) - F(\eta_{r,n}),$$

and

$$G_{r,n} = |[F_n(\eta_{r,n}) - F_n(\xi_p)] - [F(\eta_{r,n}) - F(\xi_p)]|.$$

Using the monotonicity of F_n and F, it is easily seen (*exercise*) that

$$H_{pn} \le K_n + \beta_n$$

where

$$K_n = \max\{G_{r,n}: -b_n \le r \le b_n\}$$

and

$$\beta_n = \max\{\alpha_{r,n}: -b_n \le r \le b_n - 1\}.$$

Since $\eta_{r+1,n} - \eta_{r,n} = a_n b_n^{-1} = n^{-3/4}$, $-b_n \le r \le b_n - 1$, we have by the Mean Value Theorem that

$$\alpha_{r,n} \le \left[\sup_{|x| \le a_n} F'(\xi_p + x)\right](\eta_{r+1,n} - \eta_{r,n}) = \left[\sup_{|x| \le a_n} F'(\xi_p + x)\right] n^{-3/4},$$

$-b_n \le r \le b_n - 1$, and thus

(1) $$\beta_n = O(n^{-3/4}), \qquad n \to \infty.$$

We now establish that with probability 1

(2) $$K_n = O(n^{-3/4}(\log n)^{(1/2)(q+1)}), \qquad n \to \infty.$$

For this it suffices by the Borel–Cantelli Lemma to show that

(3) $$\sum_{n=1}^{\infty} P(K_n \ge \gamma_n) < \infty,$$

where $\gamma_n = c_1 n^{-3/4}(\log n)^{(1/2)(q+1)}$, for a constant $c_1 > 0$ to be specified later. Now, crudely but nevertheless effectively, we use

(4) $$P(K_n \ge \gamma_n) \le \sum_{r=-b_n}^{b_n} P(G_{r,n} \ge \gamma_n).$$

Note that $nG_{r,n}$ is distributed as $|\sum_1^n Y_i - \sum_1^n E\{Y_i\}|$, where the Y_i's are independent binomial $(1, z_{r,n})$, with $z_{r,n} = |F(\eta_{r,n}) - F(\xi_p)|$. Therefore, by Bernstein's Inequality (Lemma A, Remark A(ii)), we have

$$P(G_{r,n} \ge \gamma_n) \le 2e^{-\theta_{r,n}},$$

where

$$\theta_{r,n} = \frac{n\gamma_n^2}{2(z_{r,n} + \gamma_n)}.$$

Let c_2 be a constant $> f(\xi_p)$. Then (*justify*) there exists N such that

$$F(\xi_p + a_n) - F(\xi_p) < c_2 a_n$$

and

$$F(\xi_p) - F(\xi_p - a_n) < c_2 a_n$$

for all $n > N$. Then $z_{r,n} \le c_2 a_n$ for $|r| \le b_n$ and $n > N$. Hence $\theta_{r,n} \ge \delta_n$ for $|r| \le b_n$ and $n > N$, where

$$\delta_n = \frac{n\gamma_n^2}{2(c_2 a_n + \gamma_n)}.$$

Note that

$$\delta_n \ge \frac{c_1^2}{4c_0 c_2}(\log n)$$

for all n sufficiently large. Given c_0 and c_2, we may choose c_1 large enough that $c_1^2/4c_0c_2 > 2$. It then follows that there exists N^* such that

$$P(G_{r,n} \ge \gamma_n) \le 2n^{-2}$$

for $|r| \le b_n$ and $n > N^*$. Consequently, for $n \ge N^*$,

$$P(K_n \ge \gamma_n) \le 8b_n n^{-2}.$$

That is,

$$P(K_n \ge \gamma_n) = O(n^{-3/2}).$$

Hence (3) holds and (2) is valid. Combining with (1), the proof is complete. ∎

Remark C. For an extension of the preceding result to the random variable $H_n = \sup_{0<p<1} H_{pn}$, see Sen and Ghosh (1971), pp. 192–194. ∎

PROOF OF THEOREM 2.5.1. Under the conditions of the theorem, we may apply Lemma B. Therefore, Lemma D is applicable with $T_n = \hat{\xi}_{pn}$, and we have: *wp*1

$$(*) \qquad F(\hat{\xi}_{pn}) - F(\xi_p) = f(\xi_p)(\hat{\xi}_{pn} - \xi_p) + O(n^{-1}\log n), \qquad n \to \infty.$$

Utilizing Lemma E with $q = \frac{1}{2}$, and again appealing to Lemma B, we may pass from (*) to: *wp*1

$$(**) \quad F_n(\hat{\xi}_{pn}) - F_n(\xi_p) = f(\xi_p)(\hat{\xi}_{pn} - \xi_p) + O(n^{-3/4}(\log n)^{3/4}), \qquad n \to \infty.$$

Finally, since $wp1$ $F_n(\hat{\xi}_{pn}) = p + O(n^{-1})$, $n \to \infty$, we have: $wp1$

$$p - F_n(\xi_p) = f(\xi_p)(\hat{\xi}_{pn} - \xi_p) + O(n^{-3/4}(\log n)^{3/4}), \qquad n \to \infty.$$

This completes the proof. ■

A similar argument (Problem 2.P.27) yields Theorem 2.5.2.

Remark D. As a corollary of Theorem 2.5.1, we have the option of replacing Lemma B by Remark B, in the proof of Theorem 2.5.1. Therefore, instead of requiring

$$a_n = O(n^{-1/2}(\log n)^q)$$

in Lemma E, we could for this purpose assume merely

$$a_n = O(n^{-1/2}(\log \log n)^{1/2}).$$

In this case a revised form of Lemma E would assert the rate

$$O(n^{-3/4}(\log n)^{1/2}(\log \log n)^{1/4}).$$

Consequently, this same rate could be asserted in Theorem 2.5.1. ■

2.5.5 The Precise Behavior of the Remainder Term R_n

Bahadur (1966) showed (see Theorem 2.5.1 and Remark 2.5.4D) that $wp1$ $R_n = O(n^{-3/4}(\log n)^{1/2}(\log \log n)^{1/4})$, $n \to \infty$. Further analysis by Eicker (1966) revealed that $R_n = o_p(n^{-3/4}g(n))$ if and only if $g(n) \to \infty$. Kiefer (1967) obtained very precise details, given by the following two theorems.

Concerning the precise order of magnitude of the deviations R_n, we have

Theorem A (Kiefer). *With probability* 1

$$\overline{\lim_{n\to\infty}} \pm \frac{n^{3/4}R_n}{(\log \log n)^{3/4}} = \frac{2^{5/4}[p(1-p)]^{1/4}}{3^{3/4}},$$

for either choice of sign.

Concerning the asymptotic distribution theory of R_n, we have that $n^{3/4}R_n$ has a nondegenerate limit distribution:

Theorem B (Kiefer).

$$\lim_{n\to\infty} P(n^{3/4}f(\xi_p)R_n \le z) = \frac{2}{[p(1-p)]^{1/2}} \int_0^\infty \Phi\left(\frac{z}{u^{1/2}}\right)\phi\left(\frac{u}{[p(1-p)]^{1/2}}\right)du.$$

(Here Φ and ϕ denote, as usual, the $N(0, 1)$ distribution function and density.)

The limit distribution in the preceding theorem has mean 0 and variance $[2p(1-p)/\pi]^{1/2}$. A complementary result has been given by Duttweiler (1973), as follows.

Theorem C (Duttweiler). *For any* $\varepsilon > 0$,

$$E\{(n^{3/4}f(\xi_p)R_n)^2\} = [2p(1-p)/\pi]^{1/2} + o(n^{-1/4+\varepsilon}), \qquad n \to \infty.$$

It is also of interest and of value to describe the behavior of the *worst deviation* of the form R_n, for p taking values $0 < p < 1$. For such a discussion, the quantity R_n defined in Theorem 2.5.1 is denoted more explicitly as a function of p, by $R_n(p)$. We thus are concerned now with

$$R_n^* = \sup_{0<p<1} f(\xi_p)|R_n(p)|.$$

This and some related random variables are investigated very thoroughly by Kiefer (1970a).

Concerning the precise order of magnitude of R_n^*, we have

Theorem D (Kiefer). *With probability* 1

$$\overline{\lim_{n\to\infty}} \frac{n^{3/4}R_n^*}{(\log n)^{1/2}(\log\log n)^{1/4}} = 2^{-1/4}.$$

Concerning the asymptotic distribution theory of R_n^*, we have that $n^{3/4}(\log n)^{-1/2}R_n^*$ has a nondegenerate limit distribution:

Theorem E (Kiefer).

$$\lim_{n\to\infty} P\left(\frac{n^{3/4}R_n^*}{(\log n)^{1/2}} \le z\right) = 1 - 2\sum_{j=1}^{\infty}(-1)^{j+1}e^{-2j^2z^4}, \qquad z > 0.$$

It is interesting that the limit distribution appearing in the preceding result happens to be the same as that of the random variable $n^{1/4}D_n^{1/2}$ considered in Section 2.1 (see Theorem 2.1.5A). That is, the random variables

$$n^{1/4}D_n^{1/2}, \frac{n^{3/4}R_n^*}{(\log n)^{1/2}}$$

have the same limit distribution. This is, in fact, more than a mere coincidence. For the following result shows that these random variables are closely related to each other, in the sense of a multiplicative form of the WLLN.

Theorem F (Kiefer).

$$\frac{n^{1/2}}{(\log n)^{1/2}} \frac{R_n^*}{D_n^{1/2}} \xrightarrow{p} 1.$$

Note that Theorem E then follows from Theorem F in conjunction with Theorem 2.1.5A (Kolmogorov) and Theorem 1.5.4 (Slutsky).

2.6 CONFIDENCE INTERVALS FOR QUANTILES

Here we consider various methods of determining a confidence interval for a given quantile ξ_p of a distribution function F. It is assumed that $0 < p < 1$ and that F is *continuous* and *strictly increasing* in a neighborhood of ξ_p. Additional regularity properties for F, such as introduced in Sections 2.3–2.5, will be postulated as needed, either explicitly or implicitly. Throughout, we deal with I.I.D. observations $X_1, X_2, \ldots$ on F. As usual, Φ denotes $N(0, 1)$. Also, K_α will denote $\Phi^{-1}(1 - \alpha)$, the $(1 - \alpha)$-quantile of Φ.

An *exact* (that is, fixed sample size) distribution-free confidence interval approach is described in **2.6.1**. Then we examine four *asymptotic* approaches: one based on *sample quantiles* in **2.6.2**, one based on *order statistics* in **2.6.3** (an equivalence between these two procedures is shown in **2.6.4**), one based on order statistics in terms of the *Wilcoxon one-sample statistic* in **2.6.5**, and one based on the *sample mean* in **2.6.6** (in each of the latter two approaches, attention is confined to the case of the *median*, i.e., the case $p = \frac{1}{2}$). Finally, in **2.6.7** the asymptotic *relative efficiencies* of the four asymptotic procedures are derived according to one criterion of comparison, and also an alternate criterion is discussed.

2.6.1 An Exact Distribution-Free Approach Based on Order Statistics

Form a confidence interval for ξ_p by using as endpoints two order statistics, X_{nk_1} and X_{nk_2}, where k_1 and k_2 are integers, $1 \le k_1 < k_2 \le n$. The interval thus defined,

$$(X_{nk_1}, X_{nk_2}),$$

has confidence coefficient *not* depending on F. For it is easily justified (exercise) that

$$\begin{aligned} P(X_{nk_1} < \xi_p < X_{nk_2}) &= P(F(X_{nk_1}) < p < F(X_{nk_2})) \\ &= P(U_{nk_1} < p < U_{nk_2}), \end{aligned}$$

where $U_{n1} \le \cdots \le U_{nn}$ denote the order statistics for a sample of size n from the uniform (0, 1) distribution. The computation of the confidence coefficient may thus be carried out via

$$P(U_{nk_1} < p < U_{nk_2}) = I_p(k, n - k_1 + 1) - I_p(k_2, n - k_2 + 1),$$

where $I_p(v_1, v_2)$ is the incomplete beta function,

$$I_p(v_1, v_2) = \frac{\Gamma(v_1 + v_2)}{\Gamma(v_1)\Gamma(v_2)} \int_0^p t^{v_1 - 1}(1 - t)^{v_2 - 1}\, dt.$$

Tables of $I_p(v_1, v_2)$ may be used to select values of k_1 and k_2 to achieve a specified confidence coefficient. Ordinarily, one chooses k_1 and k_2 as close together as possible. See Wilks (1962), Section 11.2, for further details.

Alternatively, the computations can be carried out via tables of binomial probabilities, since $P(U_{nk_1} < p < U_{nk_2})$ may be represented as the probability that a binomial (n, p) variable takes a value at least k_1 but less than k_2.

The *asymptotic* approaches in the following subsections provide for avoiding these cumbersome computations.

2.6.2 An Asymptotic Approach Based on the Sample *p*th Quantile

We utilize the asymptotic distribution theory for $\hat{\xi}_{pn}$, which was seen in **2.3.3** to be $AN(\xi_p, p(1 - p)/f^2(\xi_p)n)$. Therefore, the confidence interval

$$I_{Qn} = \left(\hat{\xi}_{pn} - \frac{K_\alpha[p(1 - p)]^{1/2}}{f(\xi_p)n^{1/2}}, \hat{\xi}_{pn} + \frac{K_\alpha[p(1 - p)]^{1/2}}{f(\xi_p)n^{1/2}}\right)$$

satisfies

(*) confidence coefficient of $I_{Qn} \to 1 - 2\alpha,\ n \to \infty$,

and

(**) length of interval $I_{Qn} = \dfrac{2K_\alpha[p(1 - p]^{1/2}}{f(\xi_p)n^{1/2}}$, all n.

A drawback of this procedure is that $f(\xi_p)$ must be *known* in order to express the interval I_{Qn}. Of course, a modified procedure replacing $f(\xi_p)$ by a consistent estimator would eliminate this difficulty. In effect, this is accomplished by the procedure we consider next.

2.6.3 An Asymptotic Approach Based on Order Statistics

An asymptotic version of the distribution-free approach of **2.6.1** is obtained by choosing k_1 and k_2 to be appropriate functions of n. Let $\{k_{1n}\}$ and $\{k_{2n}\}$ be sequences of integers satisfying $1 \le k_{1n} < k_{2n} \le n$ and

$$\frac{k_{1n}}{n} - p \sim -\frac{K_\alpha[p(1 - p)]^{1/2}}{n^{1/2}},$$

$$\frac{k_{2n}}{n} - p \sim \frac{K_\alpha[p(1 - p)]^{1/2}}{n^{1/2}},$$

$n \to \infty$. Then the intervals

$$I_{S_n} = (X_{nk_{1n}}, X_{nk_{2n}}), \qquad n = 1, 2, \ldots,$$

are distribution-free and, we shall show, satisfy

(*) confidence coefficient of $I_{S_n} \to 1 - 2\alpha, \qquad n \to \infty,$

and

(**) with probability 1,

$$\text{length of interval } I_{S_n} \sim \frac{2K_\alpha[p(1-p)]^{1/2}}{f(\xi_p)n^{1/2}}, \qquad n \to \infty.$$

It follows from (*) and (**) that the interval I_{S_n} is asymptotically equivalent to the interval I_{Q_n}, in a sense discussed precisely in **2.6.4**. Yet $f(\xi_p)$ need *not* be known for use of the interval I_{S_n}.

To establish (*) and (**), we first show that I_{Qn} and I_{Sn} in fact *coincide asymptotically*, in the sense that *wp*1 the nonoverlapping portions have length negligible relative to that of the overlap, as $n \to \infty$. Write

$$X_{nk_{1n}} - \left(\hat{\xi}_{pn} - \frac{K_\alpha[p(1-p)]^{1/2}}{f(\xi_p)n^{1/2}}\right)$$
$$= n^{-1/2}\left[n^{1/2}(X_{nk_{1n}} - \hat{\xi}_{pn}) + \frac{K_\alpha[p(1-p)]^{1/2}}{f(\xi_p)}\right]$$

Applying Corollary 2.5.2 to the right-hand side, we obtain that *wp*1

$$X_{nk_{1n}} - \left(\hat{\xi}_{pn} - \frac{K_\alpha[p(1-p)]^{1/2}}{f(\xi_p)n^{1/2}}\right) = o(n^{-1/2}), \qquad n \to \infty. \tag{1}$$

That is, the lower endpoints of I_{Q_n} and I_{S_n} are separated by an amount which *wp*1 is $o(n^{-1/2})$, $n \to \infty$. The same is true for the upper endpoints. Since the length of I_{Q_n} is of exact order $n^{-1/2}$, (**) follows. Further, utilizing (1) to write

$$P(X_{nk_{1n}} > \xi_p) = P\left(\hat{\xi}_{pn} - \frac{K_\alpha[p(1-p)]^{1/2}}{f(\xi_p)n^{1/2}} + o_p(n^{-1/2}) > \xi_p\right)$$
$$= P\left(n^{1/2}(\hat{\xi}_{pn} - \xi_p) + o_p(1) > \frac{K_\alpha[p(1-p)]^{1/2}}{f(\xi_p)}\right),$$

we have from Theorem 1.5.4 (Slutsky) that $P(X_{nk_{1n}} > \xi_p) \to \alpha$. Similarly, $P(X_{nk_{2n}} < \xi_p) \to \alpha$. Hence (*) is valid.

2.6.4 Asymptotic Equivalence of I_{Qn} and I_{Sn}.

Let us formalize the notion of relative efficiency suggested by the preceding discussion. Take as an efficiency criterion the (almost sure) rate at which the *length* of the confidence interval tends to 0 while the confidence coefficient tends to a limit γ, $0 < \gamma < 1$. In this sense, for sample sizes n_1 and n_2 respectively, procedures I_{Qn_1} and I_{Sn_2} perform "equivalently" if $n_1/n_2 \to 1$ as n_1 and $n_2 \to \infty$. Thus, in this sense, the asymptotic relative efficiency of the sequence $\{I_{Qn}\}$ to the sequence $\{I_{Sn}\}$ is 1. (A different approach toward asymptotic relative efficiency is mentioned in **2.6.7**.)

2.6.5 An Asymptotic Approach Based on the Wilcoxon One-Sample Statistic

Here we restrict to the important case $p = \frac{1}{2}$ and develop a procedure based on a sequential procedure introduced by Geertsema (1970).

Assume that F is symmetric about $\xi_{1/2}$ and has a density f satisfying

$$\int_{-\infty}^{\infty} f^2(x)dx < \infty.$$

Denote by G the distribution function of $\frac{1}{2}(X_1 + X_2)$, where X_1 and X_2 are independent observations on F. Assume that in a neighborhood of $\xi_{1/2}$, G has a positive derivative $G' = g$ and a bounded second derivative G''. (It is found that $g(\xi_{1/2}) = 2\int_{-\infty}^{\infty} f^2(x)dx$.)

The role played by the order statistics $X_{n1} \le \cdots \le X_{nn}$ in the approach given in **2.6.3** will be handed over, in the present development, to the ordered values

$$W_{n1} \le W_{n2} \le \cdots \le W_{nN_n}$$

of the $N_n = \frac{1}{2}n(n-1)$ averages

$$\tfrac{1}{2}(X_i + X_j), \qquad 1 \le i < j \le n,$$

that may be formed from $X_1, \ldots, X_n$. Geertsema proves for the W_{ni}'s an analogue of the Bahadur representation (Theorem 2.5.2) for the X_{ni}'s. The relevant theorems fall properly within the context of the theory of U-statistics and will thus be provided in Chapter 5. On the basis of these results, an interval of the form (W_{na_n}, W_{nb_n}) may be utilized as a confidence interval for $\xi_{1/2}$. In particular, if $\{a_n\}$ and $\{b_n\}$ are sequences of integers satisfying $1 \le a_n < b_n \le N_n = \frac{1}{2}n(n-1)$ and

$$\frac{a_n}{N_n} - \frac{1}{2} \sim -\frac{K_\alpha}{(3n)^{1/2}},$$

$$\frac{b_n}{N_n} - \frac{1}{2} \sim \frac{K_\alpha}{(3n)^{1/2}},$$

as $n \to \infty$, then the intervals

$$I_{Wn} = (W_{na_n}, W_{nb_n})$$

satisfy

(*) confidence coefficient of $I_{Wn} \to 1 - 2\alpha, \qquad n \to \infty,$

and

(**) with probability 1,

$$\text{length of interval } I_{Wn} \sim \frac{K_\alpha}{3^{1/2}(\int_{-\infty}^{\infty} f^2(x)dx)n^{1/2}}, \qquad n \to \infty.$$

These assertions will be justified in Chapter 5.

2.6.6 An Asymptotic Approach Based on the Sample Mean

Still another confidence interval for $\xi_{1/2}$ is given by

$$I_{Mn} = \left(\overline{X}_n - \frac{K_\alpha s_n}{n^{1/2}}, \overline{X}_n + \frac{K_\alpha s_n}{n^{1/2}}\right),$$

where $\overline{X}_n = n^{-1}\sum_1^n X_i$, $s_n^2 = n^{-1}\sum_1^n(X_i - \overline{X}_n)^2$, and it is assumed that F is symmetric about $\xi_{1/2}$ and has finite variance σ^2. Verify (Problem 2.P.28) that the intervals $\{I_{Mn}\}$ satisfy

(*) confidence coefficient of $I_{Mn} \to 1 - 2\alpha, \qquad n \to \infty,$

and

(**) with probability 1,

$$\text{length of interval } I_{Mn} \sim \frac{2K_\alpha \sigma}{n^{1/2}}, \qquad n \to \infty.$$

2.6.7 Relative Efficiency Comparisons

Let us make comparisons in the same sense as formulated in **2.6.4**. Denote by $e(A, B)$ the asymptotic relative efficiency of procedure A relative to procedure B. We have seen already that

$$e(Q, S) = 1.$$

Further, it is readily seen from **2.6.3**, **2.6.5** and **2.6.6** that, for confidence intervals for the *median* $\xi_{1/2}$,

$$e(Q, W) = e(S, W) = \frac{f^2(\xi_{1/2})}{3(\int_{-\infty}^{\infty} f^2(x)dx)^2};$$

$$e(Q, M) = e(S, M) = 4\sigma^2 f^2(\xi_{1/2});$$

$$e(W, M) = 12\sigma^2\left(\int_{-\infty}^{\infty} f^2(x)dx\right)^2.$$

As an exercise, examine these asymptotic relative efficiencies for various choices of distribution F meeting the relevant assumptions.

The asymptotic relative efficiencies just listed are identical with the *Pitman* asymptotic relative efficiencies of the corresponding test procedures, as will be seen from developments in Chapter 10. This relationship is due to a direct correspondence between consideration of a confidence interval as the length tends to 0 while the confidence coefficient tends to a constant γ, $0 < \gamma < 1$, and consideration of a test procedure as the "distance" between the alternative and the null hypothesis tends to 0 while the power tends to a limit Δ, $0 < \Delta < 1$.

Other notions of asymptotic comparison of confidence intervals are possible. For example, we may formulate the sequences of intervals in such a way that the lengths tend to a specified limit L while the confidence coefficients tend to 1. In this case, efficiency is measured by the *rate* at which the confidence coefficients tend to 1, or, more precisely, by the rate at which the noncoverage probability tends to 0. (The asymptotic relative efficiences obtained in this way correspond to the notion of *Hodges–Lehmann* asymptotic relative efficiency of test procedures, as will be seen in Chapter 10.)

The two notions of asymptotic comparison lead to differing measures of relative efficiency. In the context of *sequential* confidence interval procedures, the notion in which length $\to 0$ while confidence coefficient $\to 1 - 2\alpha\ (<1)$ has been used by Geertsema (1970) in comparing confidence interval procedures based on the *sign* test, the *Wilcoxon* test, and the *mean* test (i.e., basically the intervals $\{I_{Sn}\}$, $\{I_{Wn}\}$, and $\{I_{Mn}\}$ which we have considered). The other notion, in which length $\to L\ (>0)$ while confidence coefficient $\to 1$, has been employed by Serfling and Wackerly (1976) for an alternate comparison of sequential confidence intervals related to the *sign* test and *mean* test. (Extension to the *Wilcoxon* test remains open.)

In these two approaches toward asymptotic relative efficiency of confidence interval procedures, differing probabilistic tools are utilized. In the case of length $\to 0$ while confidence coefficient $\to 1 - 2\alpha\ (<1)$, the main tool is central limit theory. In the other case, large deviation theory is the key.

2.7 ASYMPTOTIC MULTIVARIATE NORMALITY OF CELL FREQUENCY VECTORS

Consider a sequence of n independent trials, with k possible outcomes for each trial. Let p_j denote the probability of occurrence of the jth outcome in any given trial ($\sum_1^k p_j = 1$). To avoid trivialities, we assume that $p_j > 0$, each j. Let n_j denote the number of occurrences of the jth outcome in the series of n trials ($\sum_1^k n_j = n$). We call $(n_1, \ldots, n_k)$ the "cell frequency vector" associated with the n trials.

Example. Such "cell frequency vectors" may arise in connection with general data consisting of I.I.D. random vectors $\mathbf{X}_1, \ldots, \mathbf{X}_n$ defined on $(\Omega, \mathscr{A}, P)$, as follows. Suppose that the $\mathbf{X}_i$'s take values in R^m and let $\{B_1, \ldots, B_k\}$ be a partition of R^m into "cells" of particular interest. The probability that a given observation $\mathbf{X}_i$ falls in the cell B_j is $p_j = P(\mathbf{X}_i^{-1}(B_j))$, $1 \le j \le k$. With n_j denoting the total number of observations falling in cell B_j, $1 \le j \le k$, the associated "cell frequency vector" is $(n_1, \ldots, n_k)$.

In particular, let $X_1, \ldots, X_n$ be independent $N(\theta, 1)$ random variables. For a specified constant $c > 0$, let $B_1 = (-\infty, c)$, $B_2 = [-c, c]$, and $B_3 = (c, \infty)$. Then $\{B_1, B_2, B_3\}$ partitions R into 3 cells with associated probabilities

$$p_1 = P(X_1 \le -c) = P(X_1 - \theta \le -c - \theta) = \Phi(-c - \theta),$$
$$p_3 = \Phi(-c + \theta),$$

and

$$p_2 = 1 - \Phi(-c - \theta) - \Phi(-c + \theta) = \Phi(\theta + c) - \Phi(\theta - c).$$

Note thus that the probabilities $(p_1, \ldots, p_k)$ which are associated with the vector $(n_1, \ldots, n_k)$ as parameters may arise as functions of parameters of the distribution of the $\mathbf{X}_i$'s. ■

The *exact* distribution of $(n_1, \ldots, n_k)$ is multinomial $(n; p_1, \ldots, p_k)$:

$$P((n_1, \ldots, n_k) = (r_1, \ldots, r_k)) = \binom{n}{r_1, \ldots, r_k} p_1^{r_1} \cdots p_k^{r_k},$$

for all choices of integers $r_1 \ge 0, \ldots, r_k \ge 0$, $r_1 + \cdots + r_k = n$.

We now show that $(n_1, \ldots, n_k)$ is *asymptotically k-variate normal*. Associate with the ith trial a random k-vector

$$\mathbf{Y}_i = (0, \ldots, 0, 1, 0, \ldots, 0),$$

where the single nonzero component 1 is located in the jth position if the ith trial yields the jth outcome. Then

$$(n_1, \ldots, n_k) = \sum_{i=1}^{n} \mathbf{Y}_i.$$

Further, the $\mathbf{Y}_i$'s are I.I.D. with mean vector $(p_1, \ldots, p_k)$ and (check) covariance matrix $\mathbf{\Sigma} = [\sigma_{ij}]_{k \times k}$, where

$$(*) \qquad \sigma_{ij} = \begin{cases} p_i(1 - p_i) & \text{if } i = j \\ -p_i p_j & \text{if } i \ne j. \end{cases}$$

From this formulation it follows, by the multivariate Lindeberg–Lévy CLT (Theorem 1.9.1B), that the vector of *relative frequencies* $(n_1/n, \ldots, n_k/n)$ is $AN((p_1, \ldots, p_k), n^{-1}\mathbf{\Sigma})$:

Theorem. *The random vector*

$$n^{1/2}\left(\frac{n_1}{n} - p_1, \ldots, \frac{n_k}{n} - p_k\right)$$

converges in distribution to k*-variate normal with mean* 0 *and covariance matrix* $\mathbf{\Sigma} = [\sigma_{ij}]$ *given by* (*).

2.8 STOCHASTIC PROCESSES ASSOCIATED WITH A SAMPLE

In **1.11.4** we considered a stochastic process on the unit interval [0, 1] associated in a natural way with the first n partial sums generated by a sequence of I.I.D. random variables $X_1, X_2, \ldots$. That is, for each n, a process was defined in terms of $X_1, \ldots, X_n$. For the sequence of such processes obtained as $n \to \infty$, we saw in Donsker's Theorem a useful generalization of the CLT. Thus the convergence in distribution of normalized sums to $N(0, 1)$ was seen to be a corollary of the convergence in distribution of partial sum processes to the Wiener process. Other corollaries of the generalization were indicated also.

We now consider various other stochastic processes which may be associated with a sample $X_1, \ldots, X_n$, in connection with the various types of statistic we have been considering. We introduce in **2.8.1** processes of "partial sum" type associated with the sample moments, in **2.8.2** a "sample distribution function process," or "empirical process," and in **2.8.3** a "sample quantile process." Miscellaneous other processes are mentioned in **2.8.4**. In subsequent chapters, further stochastic processes of interest will be introduced as their relevance becomes apparent.

2.8.1 Partial Sum Processes Associated with Sample Moments

In connection with the sample kth moment,

$$a_k = \frac{1}{n}\sum_{i=1}^{n} X_i^k,$$

we associate a partial sum process based on the random variables

$$\xi_i = X_i^k - \alpha_k, \qquad 1 \le i \le n.$$

The relevant theory is obtained as a special case of Donsker's Theorem.

2.8.2 The Sample Distribution Function (or "Empirical") Process

The asymptotic normality of the sample distribution function F_n is viewed more deeply by considering the stochastic process

$$n^{1/2}[F_n(x) - F(x)], \qquad -\infty < x < \infty.$$

Let us assume that F is *continuous*, so that we may equivalently consider the process

$$Y_n(t) = n^{1/2}[F_n(F^{-1}(t)) - t], \qquad 0 \le t \le 1,$$

obtained by transforming the domain from $(-\infty, \infty)$ to $[0, 1]$, by putting $x = F^{-1}(t)$, $0 < t < 1$, and defining $Y_n(0) = Y_n(1) = 0$.

The random function $\{Y_n(t), 0 \le t \le 1\}$ is not an element of the function space $C[0, 1]$ considered in Section 1.11. Rather, the natural setting here is the space $D[0, 1]$ of functions on $[0, 1]$ which are right-continuous and have left-hand limits. Suitably metrizing $D[0, 1]$ and utilizing the concept of weak convergence of probability measures on $D[0, 1]$, we have

$$Y_n \xrightarrow{d} W^0 \text{ (in } D[0, 1] \text{ suitably metrized)},$$

where W^0 denotes a random element of $D[0, 1]$ having the unique Gaussian measure determined by the mean function

$$E\{W^0(t)\} \equiv 0$$

and the covariance function

$$\text{Cov}\{W^0(s), W^0(t)\} = s(1 - t), \qquad 0 \le s \le t \le 1.$$

We shall use the notation W^0 also for the measure just defined.

The stochastic process W^0 is essentially a random element of $C[0, 1]$, in fact. That is, $W^0(C[0, 1]) = 1$. Thus, with probability 1, the sample path of the process W^0 is a continuous function on $[0, 1]$. Further, with probability 1, $W^0(0) = 0$ and $W^0(1) = 0$, that is, the random function takes the value 0 at each endpoint of the interval $[0, 1]$. Thus W^0 is picturesquely termed the "Brownian bridge," or the "tied-down Wiener" process.

The convergence $Y_n \xrightarrow{d} W^0$ is proved in Billingsley (1968). An immediate corollary is that for each fixed x, $F_n(x)$ is asymptotically normal as given by Theorem 2.1.1. Another corollary is the asymptotic distribution of the (normalized) Kolmogorov–Smirnov *distance* $n^{1/2}D_n$, which may be written in terms of the process $Y_n(\cdot)$ as

$$n^{1/2}D_n = \sup_{0<t<1} |Y_n(t)|.$$

It follows from $Y_n \xrightarrow{d} W^0$ that

(1) $$\lim_{n\to\infty} P(n^{1/2}D_n \le d) = W^0\left(\left\{x(\cdot)\colon \sup_{0\le t\le 1} |x(t)| \le d\right\}\right),$$

since $\{x(\cdot)\colon \sup_{0\le t\le 1} |x(t)| \le d\}$ can be shown to be a W^0-continuity set. Also,

$$W^0\left(\left\{x(\cdot)\colon \sup_{0\le t\le 1} |x(t)| \le d\right\}\right) = 1 - 2\sum_{j=1}^{\infty}(-1)^{j+1}e^{-2j^2d^2}, \qquad d > 0.$$

(See Billingsley (1968) for proofs of these details.) Thus follows Theorem 2.1.5A (Kolmogorov).

As discussed in **2.1.6**, the result just stated may be recast as the null-hypothesis asymptotic distribution of the Kolmogorov–Smirnov *test statistic*

$$\Delta_n = \sup_{-\infty < x < \infty} |F_n(x) - F_0(x)|.$$

That is, for the process

$$\tilde{Y}_n(t) = n^{1/2}[F_n(F_0^{-1}(t)) - t], \qquad 0 \le t \le 1,$$

we have

$$n^{1/2}\Delta_n = \sup_{0 < t < 1} |\tilde{Y}_n(t)|$$

and thus, under H_0: $F = F_0$, we have $n^{1/2}\Delta_n \xrightarrow{d} \sup_t |W^0(t)|$. Thus, under H_0, $n^{1/2}\Delta_n$ has the limit distribution (1) above.

It is also of interest to have asymptotic distribution theory for Δ_n under a fixed *alternative* hypothesis, that is, in the case $F \neq F_0$. This has been obtained by Raghavachari (1973). To state the result we introduce some further notation. Put

$$\Delta = \sup_{-\infty < x < \infty} |F(x) - F_0(x)|$$

and

$$C_1 = \{x\colon F(x) - F_0(x) = \Delta\}, C_2 = \{x\colon F(x) - F_0(x) = -\Delta\}.$$

It is convenient to switch from $-\infty < x < \infty$ to $0 < t < 1$. Noting that

$$\Delta = \sup_{0 < t < 1} |F(F_0^{-1}(t)) - t|,$$

we accordingly put

$$K_i = F_0(C_i) = \{t\colon F_0^{-1}(t) \in C_i\}, \qquad i = 1, 2.$$

Finally, on the measurable space $(C[0, 1], \mathscr{B})$ considered in **1.11**, denote by $\tilde{W}^0$ a random element having the unique Gaussian measure determined by the mean function

$$E\{\tilde{W}^0(t)\} \equiv 0$$

and the covariance function

$$\text{Cov}\{\tilde{W}^0(s), \tilde{W}^0(t)\} = F(F_0^{-1}(s))[1 - F(F_0^{-1}(t))], \qquad 0 \le s \le t \le 1.$$

We shall use the notation $\tilde{W}^0$ also for the measure just defined.

Theorem (Raghavachari). *Let* F *be continuous. Then*

$$\lim_{n\to\infty} P(n^{1/2}(\Delta_n - \Delta) \le d) = \tilde{W}^0\left(\left\{x(\cdot): \sup_{t\in K_1} x(t) \le d; \sup_{t\in K_2} x(t) \ge -d\right\}\right),$$

for $-\infty < d < \infty$.

The preceding result contains (1) above as the special case corresponding to $\Delta = 0$, $\Delta_n = D_n$ and $K_1 = K_2 = [0, 1]$, in which case the measure $\tilde{W}^0$ reduces to W^0.

It is also of interest to investigate $n^{1/2}\Delta_n$ under a sequence of *local* alternatives converging weakly to F_0 at a suitable rate. In this context, Chibisov (1965) has derived the limit behavior of the process $\tilde{Y}_n(\cdot)$.

For further discussion of empirical processes, see **2.8.3** below.

2.8.3 The Sample Quantile Process

The asymptotic normality of sample quantiles, established in **2.3** and **2.5**, may be viewed from more general perspective by considering the stochastic process

$$Z_n(p) = n^{1/2}(\hat{\xi}_{pn} - \xi_p), \qquad 0 < p < 1,$$

with $Z_n(0) = Z_n(1) = 0$. We may equivalently write

$$Z_n(p) = n^{1/2}[F_n^{-1}(p) - F^{-1}(p)], \qquad 0 < p < 1.$$

There is a close relationship between the empirical process $Y_n(\cdot)$ considered in **2.8.2** above and the quantile process $Z_n(\cdot)$. This is seen heuristically as follows (we assume that F is *absolutely continuous*):

$$\begin{aligned} Y_n(t) &= n^{1/2}[F_n(F^{-1}(t)) - t] \\ &= n^{1/2}\left[F_n\left(F_n^{-1}(t) - \frac{Z_n(t)}{n^{1/2}}\right) - t\right] \\ &\doteq n^{1/2}\left[F\left(F^{-1}(t) - \frac{Zn(t)}{n^{1/2}}\right) - t\right] \\ &\doteq -n^{1/2}F'(F^{-1}(t))\frac{Z_n(t)}{n^{1/2}} \\ &= -f(F^{-1}(t))Z_n(t). \end{aligned}$$

That is, there holds the approximate relationship

$$Y_n(p) \doteq -f(\xi_p)Z_n(p), \qquad 0 \le p \le 1.$$

For the case of F uniform [0, 1], this becomes $Y_n(p) \doteq -Z_n(p)$, $0 \le p \le 1$, which suggests $Z_n \xrightarrow{d} -W^0$, which is the same as $Z_n \xrightarrow{d} W^0$.

A precise and illuminating technical discussion of the empirical and quantile processes taken together has been given in the appendix of a paper by Shorack (1972). Another way to see the relationship between the $Y_n(\cdot)$ and

$Z_n(\cdot)$ processes is through the Bahadur representation (recall **2.5**), which gives *exactly*

$$Z_n(p) = \frac{-n^{1/2}[F_n(\xi_p) - p]}{f(\xi_p)} + n^{1/2}R_n(p)$$

$$= -\frac{Y_n(p)}{f(\xi_p)} + n^{1/2}R_n(p),$$

where for each fixed p, *wp*1 $n^{1/2}R_n(p) = O(n^{-1/4}\log n)$, $n \to \infty$.

2.8.4 Miscellaneous Other Processes

(i) *The remainder process in the Bahadur representation.* This process, $\{R_n(p), 0 \le p \le 1\}$, has just been discussed in **2.8.3** above and has also been considered in **2.5.5**. Its fundamental role is evident.

(ii) *Empirical processes with random perturbations.* A modified empirical process based on a sample distribution function subject to random perturbations and scale factors is treated by Rao and Sethuraman (1975).

(iii) *Empirical processes with estimated parameters.* It is of interest to consider modifications of the process $Y_n(\cdot)$ in connection with *composite* goodness-of-fit hypotheses, where the stated null hypothesis distributions may depend on parameters which are unknown and thus must be estimated from the data. In this regard, see Durbin (1973b), Wood (1975), and Neuhaus (1976).

(iv) "*Extremal processes.*" A stochastic process associated with the extreme order statistics $\{X_{nk}\}$, k fixed, is defined by

$$Q_{kn}(t) = \frac{X_{[nt],k} - a_n}{b_n}, \qquad \frac{k}{n} \le t \le 1,$$

where a_n and b_n are suitable normalizing constants. See Dwass (1964), Lamperti (1964), and Galambos (1978).

(v) "*Spacings*" *processes.* Another type of process based on order statistics is noted in Section 3.6.

2.P PROBLEMS

Miscellaneous

1. Let $\{a_n\}$ be a sequence of constants. Does there exist a sequence of random variables $\{X_n\}$ satisfying

(a) $X_n \xrightarrow{d} X$ for some random variable X

and

(b) $E\{X_n\} = a_n$, all n?

Justify your answer.

2. Let $\{a_n\}$ be a sequence of constants and Y a random variable. Does there exist a sequence of random variables $\{X_n\}$ satisfying

(a) $$X_n \xrightarrow{d} X \text{ for some random variable } X,$$

(b) $$X_n - a_n \xrightarrow{d} Y,$$

and

(c) $$E\{X_n\} = 0, \qquad \text{all } n?$$

If "yes," prove. Otherwise give counter-example.

Section 2.1

3. For the density estimator

$$f_n(x) = \frac{F_n(x + b_n) - F_n(x - b_n)}{2b_n},$$

(a) show that $2nb_n f_n(x)$ is distributed binomial $(n, F(x + b_n) - F(x - b_n))$,
(b) show that $E\{f_n(x)\} \to f(x)$ if $b_n \to 0$,
(c) show that $\text{Var}\{f_n(x)\} \to 0$ if $b_n \to 0$ and $nb_n \to \infty$.

4. (continuation) Apply the Berry–Esséen Theorem (1.9.5) to show that if f is continuous and positive at x, then there exists a constant K depending on $f(x)$ but not on n, such that

$$\left| P\left(\frac{f_n(x) - E\{f_n(x)\}}{[\text{Var}\{f_n(x)\}]^{1/2}} < x \right) - \Phi(x) \right| \le \frac{K}{(nb_n)^{1/2}}, \qquad \text{all } n.$$

5. (continuation) (a) Deduce from the preceding results that $(2nb_n)^{1/2}[f_n(x) - E\{f_n(x)\}]/f^{1/2}(x) \xrightarrow{d} N(0, 1)$;
(b) Apply Taylor's Theorem to obtain $(nb_n)^{1/2}[E\{f_n(x)\} - f(x)] \to 0$, $n \to \infty$, under suitable smoothness restrictions on f and rate of convergence restrictions on $\{b_n\}$;
(c) From (a) and (b), establish $(2nb_n)^{1/2}[f_n(x) - f(x)]/f^{1/2}(x) \xrightarrow{d} N(0, 1)$ under suitable (stated explicitly) conditions on f and $\{b_n\}$.

6. Justify that $n^{1/2}D_n/(\log \log n)^{1/2}$ converges to 0 in probability but not with probability 1.

Section 2.2

7. Do some of the exercises assigned in the proof of Theorem 2.2.3A.
8. Show that

$$n^{1/2}(\overline{X} - \mu, s^2 - \sigma^2) \xrightarrow{d} N\left((0, 0), \begin{bmatrix} \sigma^2 & \mu_3 \\ \mu_3 & \mu_4 - \sigma^4 \end{bmatrix}\right).$$

(Hint: Use Lemma 2.2.3 to determine that the off-diagonal element of the covariance matrix is given by $\text{Cov}\{X_1, (X_1 - \mu)^2\}$.)

9. Show that $(\bar{X}, m_2, m_3, \ldots, m_k)$ is asymptotically k-variate normal with mean $(\mu, \sigma^2, \mu_3, \ldots, \mu_k)$, and find the asymptotic covariance matrix $n^{-1}\Sigma$.

10. Let $\{X_1, \ldots, X_n\}$ be I.I.D. with mean μ and variance $\sigma^2 < \infty$. The "Student's t-statistic" for the sample is

$$T_n = \frac{n^{1/2}(\bar{X}_n - \mu)}{s_n},$$

where $\bar{X}_n = n^{-1}\sum_1^n X_i$ and $s_n^2 = (n-1)^{-1}\sum_1^n (X_i - \bar{X}_n)^2$. Derive the limit distribution of T_n.

Section 2.3

11. Show that the uniqueness assumption on ξ_p in Theorem 2.3.1 cannot be dropped.

12. Prove Theorem 2.3.2 as an application of Theorem 2.1.3A. (Hint: see Remark 2.3.2 (iii).)

13. Obtain an explicit constant of proportionality in the term $O(n^{-1/2})$ of Theorem 2.3.3C.

14. Complete the details of derivation of the density of $\hat{\xi}_{pn}$ in **2.3.4**.

15. Let F be a distribution function posessing a finite mean. Let $0 < p < 1$. Show that for any k the sample pth quantile $\hat{\xi}_{pn}$ possesses a finite kth moment for all n sufficiently large. (Hint: apply **1.14**.)

16. Evaluate the asymptotic relative efficiency of $\hat{\xi}_{1/2}$ relative to $\bar{X}$ by the criterion of asymptotic variance, for various choices of underlying distribution F. Follow the guidelines of **2.3.5**.

17. Check the asymptotic normality parameters for the sample semi-interquartile range, considered in **2.3.6**.

Section 2.4

18. Check the details of Example 2.4.4B.

Section 2.5

19. Show that $X_n \overset{wp1}{=} O(g(n))$ implies $X_n = O_p(g(n))$.

20. Verify Remark 2.5.1 (viii).

21. Verify Remark 2.5.1 (ix).

22. Prove Corollary 2.5.2.

23. Derive from Theorem 2.5.2 an LIL for sequences of central order statistics $\{X_{nk_n}\}$ for which $k_n/n \to p$ sufficiently fast.

24. Complete the details of proof of Lemma 2.5.4C.

25. Prove Lemma 2.5.4D.

26. Provide missing details for the proof of Lemma 2.5.4E.

Section 2.6

27. Verify the distribution-free property of the confidence interval procedure of **2.6.1**.

28. Verify the properties claimed for the confidence interval procedure of **2.6.6**.

29. Evaluate the asymptotic relative efficiencies of the confidence interval procedures $\{I_{S_n}\}$, $\{I_{W_n}\}$ and $\{I_{M_n}\}$, for various choices of F. Use the formulas of **2.6.7**. Be sure that your choices of F meet the relevant assumptions that underlie the derivation of these formulas.

30. Investigate the confidence interval approach

$$\left(F_n^{-1}\left(p - \frac{K_\alpha[p(1-p)]^{1/2}}{n^{1/2}}\right), F_n^{-1}\left(p + \frac{K_\alpha[p(1-p)]^{1/2}}{n^{1/2}}\right)\right)$$

for the pth quantile. Develop the asymptotic properties of this interval.

Section 2.7

31. Check the covariance matrix Σ given for the multinomial $(1; p_1, \ldots, p_k)$ random k-vector.

Section 2.8

32. Formulate explicitly the stochastic process referred to in **2.8.1** and state the relevant weak convergence result.

CHAPTER 3

Transformations of Given Statistics

In Chapter 2 we examined a variety of statistics which arise fundamentally, in connection with a sample $X_1, \ldots, X_n$. Several instances of asymptotically normal vectors of statistics were seen. A broad class of statistics of interest, such as the *sample coefficient of variation* $s/\bar{X}$, may be expressed as a smooth function of a vector of the basic sample statistics. This chapter provides methodology for deriving the asymptotic behavior of such statistics and considers various examples.

More precisely, suppose that a statistic of interest T_n is given by $g(\mathbf{X}_n)$, where $\mathbf{X}_n$ is a vector of "basic" statistics about which the asymptotic behavior is already known, and g is a function satisfying some mild regularity conditions. The aim is to deduce the asymptotic behavior of T_n.

It suffices for many applications to consider the situations

(a) $\mathbf{X}_n \xrightarrow{wp1} \mathbf{c}$, or $\mathbf{X}_n \xrightarrow{p} \mathbf{c}$;

(b) $\mathbf{X}_n \xrightarrow{d} \mathbf{X}$;

(c) $\mathbf{X}_n\, AN(\boldsymbol{\mu}, \boldsymbol{\Sigma}_n)$, where $\boldsymbol{\Sigma}_n \to \mathbf{0}$.

For situations (a) and (b), under mild continuity requirements on $g(\cdot)$, we may apply Theorem 1.7 to obtain conclusions such as $T_n \xrightarrow{wp1} g(\mathbf{c})$, $T_n \xrightarrow{p} g(c)$, or $T_n \xrightarrow{d} g(\mathbf{X})$. However, for situation (c), a different type of theorem is needed. In Section 3.1 we treat the (univariate) case $X_n(\mu, \sigma_n^2)$, $\sigma_n \to 0$, and present theorems which, under additional regularity conditions on g, yield conclusions such as "T_n is $AN(g(\mu), [g'(\mu)]^2\sigma_n^2)$." In Section 3.2 the application of these results, and of Theorem 1.7 as well, is illustrated in connection with the situations (a), (b), and (c). In particular, variance-stabilizing transformations and a device called "Tukey's hanging rootogram" are discussed. Extension of the theorems of Section 3.1 to vector-valued g and vector $\mathbf{X}_n$ is carried out in Section 3.3, followed in Section 3.4 by exemplification for functions of several sample moments and for "best" linear combinations of several estimates.

Section 3.5 treats the application of Theorem 1.7 to the important special case of *quadratic forms* in asymptotically normal random vectors. The asymptotic behavior of the chi-squared statistic, both under the null hypothesis and under local alternatives, is derived.

Finally, in Section 3.6 some statistics which arise naturally as functions of order statistics are discussed.

Although much of the development of this chapter is oriented to the case of functions of asymptotically *normal* vectors, the methods are applicable more widely.

3.1 FUNCTIONS OF ASYMPTOTICALLY NORMAL STATISTICS: UNIVARIATE CASE

Here we present some results apropos to functions g applied to random variables X_n which are asymptotically normal. For convenience and simplicity, we deal with the univariate case separately. Thus here we treat the simple case that g is real-valued and X_n is $AN(\mu, \sigma_n^2)$, with $\sigma_n \to 0$. Multivariate extensions are developed in Section 3.3.

Theorem A. *Suppose that* X_n *is* $AN(\mu, \sigma_n^2)$, *with* $\sigma_n \to 0$. *Let* g *be a real-valued function differentiable at* $x = \mu$, *with* $g'(\mu) \neq 0$. *Then*

$$g(X_n) \quad \textit{is} \quad AN(g(\mu), [g'(\mu)]^2\sigma_n^2).$$

PROOF. We shall show that

$$(*) \qquad \frac{g(X_n) - g(\mu)}{g'(\mu)\sigma_n} - \frac{X_n - \mu}{\sigma_n} \xrightarrow{p} 0.$$

Then, by Theorem 1.5.4 (Slutsky), the random variable $[g(X_n) - g(\mu)]/g'(\mu)\sigma_n$ has the same limit distribution as $(X_n - \mu)/\sigma_n$, namely $N(0, 1)$ by assumption.

Define $h(\mu) = 0$ and

$$h(x) = \frac{g(x) - g(\mu)}{x - \mu} - g'(\mu), \qquad x \neq \mu.$$

Then, by the differentiability of g at μ, $h(x)$ is continuous at μ. Therefore, since $X_n \xrightarrow{p} \mu$ by Problem 1.P.20, it follows by Theorem 1.7 (ii) that $h(X_n) \xrightarrow{p} h(\mu) = 0$ and thus, by Slutsky's Theorem again, that

$$h(X_n)\frac{X_n - \mu}{\sigma_n} \xrightarrow{p} 0,$$

that is, (1) holds. This completes the proof. ∎

Remarks. (i) If, further, g is differentiable in a neighborhood of μ and $g'(x)$ is continuous at μ, then we may replace $g'(\mu)$ by the estimate $g'(X_n)$ and have the modified conclusion

$$\frac{g(X_n) - g(\mu)}{g'(X_n)\sigma_n} \xrightarrow{d} N(0, 1).$$

(ii) If, further, σ_n^2 is given by $\sigma^2(\mu)/n$, where $\sigma(\mu)$ is a continuous function of μ, then we may replace σ_n by the estimate $\sigma(X_n)/n^{1/2}$ and obtain

$$\frac{n^{1/2}[g(X_n) - g(\mu)]}{g'(X_n)\sigma(X_n)} \xrightarrow{d} N(0, 1). \quad \blacksquare$$

Example A. It was seen in **2.2.4** that

$$s^2 \text{ is } AN\left(\sigma^2, \frac{\mu_4 - \sigma^4}{n}\right).$$

It follows that the sample standard deviation s is also asymptotically normal, namely

$$s \text{ is } AN\left(\sigma, \frac{\mu_4 - \sigma^4}{4\sigma^2 n}\right). \quad \blacksquare$$

We now consider the case that g is differentiable at μ but $g'(\mu) = 0$. The following result generalizes Theorem A to include this case.

Theorem B. *Suppose that* X_n *is* $AN(\mu, \sigma_n^2)$, *with* $\sigma_n \to 0$. *Let* g *be a real-valued function differentiable* $m(\geq 1)$ *times at* $x = \mu$, *with* $g^{(m)}(\mu) \neq 0$ *but* $g^{(j)}(\mu) = 0$ *for* $j < m$. *Then*

$$\frac{g(X_n) - g(\mu)}{\frac{1}{m!} g^{(m)}(\mu)\sigma_n^m} \xrightarrow{d} [N(0, 1)]^m.$$

PROOF. The argument is similar to that for Theorem A, this time using the function h defined by $h(\mu) = 0$ and

$$h(x) = \frac{g(x) - g(\mu)}{\frac{1}{m!}(x - \mu)^m} - g^{(m)}(\mu), \qquad x \neq \mu,$$

and applying Young's form of Taylor's Theorem (1.12.1C). $\blacksquare$

Example B. Let X_n be $AN(0, \sigma_n^2)$, $\sigma_n \to 0$. Then

$$\frac{\log^2(1 + X_n)}{\sigma_n^2} \xrightarrow{d} \chi_1^2.$$

(Apply the theorem with $g(x) = \log^2(1 + x)$, $\mu = 0$, $m = 2$.) ■

3.2 EXAMPLES AND APPLICATIONS

Some miscellaneous illustrations of Theorems 1.7, 3.1A and 3.1B are provided in **3.2.1**. Further applications of Theorem 3.1A, in connection with variance-stabilizing transformations and Tukey's hanging rootogram, are provided in **3.2.2** and **3.2.3**.

3.2.1 Miscellaneous Illustrations

In the following, assume that X_n is $AN(\mu, \sigma_n^2)$, $\sigma_n \to 0$. What can be said about the asymptotic behavior of the random variables

$$X_n^2, \frac{1}{X_n}, e^{X_n}, \log|X_n|?$$

Regarding *convergence in probability*, we have $X_n \xrightarrow{p} \mu$ since $\sigma_n \to 0$ and thus, by Theorem 1.7,

$$X_n^2 \xrightarrow{p} \mu^2; \qquad \frac{1}{X_n} \xrightarrow{p} \frac{1}{\mu} \text{ (if } \mu \neq 0); \qquad e^{X_n} \xrightarrow{p} e^{\mu}; \qquad \log|X_n| \xrightarrow{p} \log|\mu|.$$

Moreover, regarding *asymptotic distribution theory*, we have the following results.

(i) For $\mu \neq 0$, X_n^2 is $AN(\mu^2, 4\mu^2\sigma_n^2)$, by Theorem 3.1A. For $\mu = 0$, $X_n^2/\sigma_n^2 \xrightarrow{d} \chi_1^2$, by Theorem 3.1B or Theorem 1.7.

(ii) For $\mu \neq 0$, $1/X_n$ is $AN(1/\mu, \sigma_n^2/\mu^4)$, by Theorem 3.1A. The case $\mu = 0$ is not covered by Theorem 3.1B, but Theorem 1.7 yields $\sigma_n/X_n \xrightarrow{d} 1/N(0, 1)$.

(iii) For any μ, e^{X_n} is $AN(e^{\mu}, e^{2\mu}\sigma_n^2)$.

(iv) For $\mu \neq 0$, $\log|X_n|$ is $AN(\log|\mu|, \sigma_n^2/\mu^2)$. For $\mu = 0$, $\log|X_n/\sigma_n| \xrightarrow{d} \log|N(0, 1)|$.

3.2.2 Variance-Stabilizing Transformations

Sometimes the statistic of interest for inference about a parameter θ is conveniently asymptotically normal, but with an asymptotic variance parameter functionally dependent on θ. That is, we have X_n $AN(\theta, \sigma_n^2(\theta))$. This aspect can pose a difficulty. For example, in testing a hypothesis about θ by using X_n, the rejection region would thus depend upon θ. However, by a suitable transformation $g(\cdot)$, we may equivalently use $Y_n = g(X_n)$ for inference about $g(\theta)$ and achieve the feature that Y_n is $AN(g(\theta), \gamma_n)$, where γ_n does *not* depend upon θ.

In the case that $\sigma_n^2(\theta)$ is the form $\sigma_n^2(\theta) = h^2(\theta)v_n$, where $v_n \to 0$, the appropriate choice of g may be found via Theorem 3.1A. For, if $Y_n = g(X_n)$ and $g'(\theta) \neq 0$, we have

$$Y_n \quad \text{is} \quad AN(g(\theta), [g'(\theta)]^2 h^2(\theta) v_n).$$

Thus, in order to obtain that Y_n is $AN(g(\theta), c^2 v_n)$, where c is a constant independent of θ, we choose g to be the solution of the differential equation

$$\frac{dg}{d\theta} = \frac{c}{h(\theta)}.$$

Example. Let X_n be Poisson with mean θn, where $\theta > 0$. Then (Problem 3.P.1) X_n is $AN(\theta n, \theta n)$, or equivalently,

$$\frac{X_n}{n} \quad \text{is} \quad AN\left(\theta, \frac{\theta}{n}\right).$$

Let g be the solution of

$$\frac{dg(\theta)}{d\theta} = \frac{c}{\theta^{1/2}}.$$

Thus $g(x) = 2cx^{1/2}$. Choose $c = \frac{1}{2}$ for convenience. It follows that $(X_n/n)^{1/2}$ is $AN(\theta^{1/2}, \frac{1}{4}n)$, or equivalently $X_n^{1/2}$ is $AN((\theta n)^{1/2}, \frac{1}{4})$. This result is the basis for the following commonly used approximation: *if X is Poisson with mean μ and μ is large, then $X^{1/2}$ is approximately $N(\mu^{1/2}, \frac{1}{4})$.* ∎

A further illustration of the variance-stabilizing technique arises in the following subsection. Other examples may be found in Rao (1973), Section 6.g.

3.2.3 Tukey's "Hanging Rootogram"

Histograms and other forms of density estimator (recall **2.1.8**) provide popular ways to test a hypothesized distribution. A plot is made depicting both the observed density estimator, say $f_n(x)$, and the hypothesized density, say $f_0(x)$, for $-\infty < x < \infty$ (or $a < x < b$). This enables one to visually assess the disparity between (the population density generating) the observed $f_n(\cdot)$ and the hypothetical $f_0(\cdot)$. Several features are noteworthy, as follows.

(i) Typically, $f_n(x)$ is asymptotically normal. For example, in the case of the simple $f_n(\cdot)$ considered in **2.1.8** and in Problems 2.P.3–5, we have that

$$f_n(x) \quad \text{is} \quad AN(f(x), f(x)/2nb_n),$$

where $nb_n \to \infty$. Thus the observed *discrepancies* $f_n(x) - f_0(x)$ are

$$AN(f(x) - f_0(x), f(x)/2nb_n).$$

(ii) The observed discrepancies fluctuate about the curve traced by $f_0(x)$.

(iii) Under the null hypothesis, all discrepancies are asymptotically normal with mean 0, but nevertheless two observed discrepancies of equal size may have quite different levels of significance since the asymptotic variance in the normal approximation depends on x.

Regarding property (i), we comment that it is quite satisfactory to have a normal approximation available. However, properties (ii) and (iii) make rather difficult a simultaneous visual assessment of the levels of significance of these discrepancies. A solution proposed by Tukey to alleviate this difficulty involves two elements. First, make a variance-stabilizing transformation. From **3.2.2** it is immediately clear that $g(x) = x^{1/2}$ is appropriate, giving

$$f_n^{1/2}(x) \quad \text{is} \quad AN(f^{1/2}(x), \tfrac{1}{8}nb_n)$$

Thus we now compare the curves $f_n^{1/2}(x)$ and $f_0^{1/2}(x)$, and under the null hypothesis the observed discrepancies $f_n^{1/2}(x) - f_0^{1/2}(x)$ are $AN(0, \frac{1}{8}nb_n)$, each x. Secondly, instead of standing the curve $f_n^{1/2}(x)$ on the base line, it is *suspended* from the hypothetical curve $f_0^{1/2}(x)$. This causes the discrepancies $f_n^{1/2}(x) - f_0^{1/2}(x)$ all to fluctuate about a fixed base line, all with a common standard deviation $(8nb_n)^{-1/2}$. The device is picturesquely called a *hanging rootogram*. For an illustrated practical application, see Healy (1968).

3.3 FUNCTIONS OF ASYMPTOTICALLY NORMAL VECTORS

The following theorem extends Theorem 3.1A to the case of a *vector-valued* function g applied to a *vector* $\mathbf{X}_n$ which is $AN(\boldsymbol{\mu}, b_n^2\boldsymbol{\Sigma})$, where $b_n \to 0$.

Theorem A. *Suppose that* $\mathbf{X}_n = (X_{n1}, \ldots, X_{nk})$ *is* $AN(\boldsymbol{\mu}, b_n^2\boldsymbol{\Sigma})$, *with* $\boldsymbol{\Sigma}$ *a covariance matrix and* $b_n \to 0$. *Let* $g(\mathbf{x}) = (g_1(\mathbf{x}), \ldots, g_m(\mathbf{x}))$, $\mathbf{x} = (x_1, \ldots, x_k)$, *be a vector-valued function for which each component function* $g_i(\mathbf{x})$ *is real-valued and has a nonzero differential* $g_i(\boldsymbol{\mu}; \mathbf{t})$, $\mathbf{t} = (t_1, \ldots, t_k)$, *at* $\mathbf{x} = \boldsymbol{\mu}$. *Put*

$$\boldsymbol{D} = \left[\frac{\partial g_i}{\partial x_j}\bigg|_{\mathbf{x}=\boldsymbol{\mu}}\right]_{m\times k}.$$

Then

$$g(\mathbf{X}_n) \quad \textit{is} \quad AN(g(\boldsymbol{\mu}), b_n^2\mathbf{D\Sigma D}').$$

PROOF. Put $\boldsymbol{\Sigma}_n = b_n^2\boldsymbol{\Sigma}$. By the definition of asymptotic multivariate normality (**1.5.5**), we need to show that for every vector $\boldsymbol{\lambda} = (\lambda_1, \ldots, \lambda_m)$ such that $\boldsymbol{\lambda D\Sigma}_n\boldsymbol{D}'\boldsymbol{\lambda}' > 0$ for all sufficiently large n, we have

$$\text{(1)} \qquad \frac{\boldsymbol{\lambda}[g(\mathbf{X}_n) - g(\boldsymbol{\mu})]'}{(\boldsymbol{\lambda D\Sigma}_n\boldsymbol{D}'\boldsymbol{\lambda})^{1/2}} \xrightarrow{d} N(0, 1).$$

Let $\boldsymbol{\lambda}$ satisfy the required condition and suppose that n is already sufficiently large, and put $b_{\boldsymbol{\lambda}n} = (\boldsymbol{\lambda} \boldsymbol{D}\boldsymbol{\Sigma}_n \boldsymbol{D}'\boldsymbol{\lambda}')^{1/2}$. Define functions h_i, $1 \le i \le m$, by $h_i(\boldsymbol{\mu}) = 0$ and

$$h_i(\mathbf{x}) = \frac{g_i(\mathbf{x}) - g_i(\boldsymbol{\mu}) - g_i(\boldsymbol{\mu}; \mathbf{x} - \boldsymbol{\mu})}{\|\mathbf{x} - \boldsymbol{\mu}\|}, \qquad \mathbf{x} \neq \boldsymbol{\mu}.$$

By the definition of g_i having a differential at $\boldsymbol{\mu}$ (**1.12.2**), $h_i(\mathbf{x})$ is continuous at $\boldsymbol{\mu}$.

Now

(2)

$$\begin{aligned}\boldsymbol{\lambda}[g(\mathbf{X}_n) - g(\boldsymbol{\mu})]' b_{\boldsymbol{\lambda}n}^{-1} &= \sum_{i=1}^{m} \lambda_i b_{\boldsymbol{\lambda}n}^{-1}[g_i(\mathbf{X}_n) - g_i(\boldsymbol{\mu})] \\ &= \sum_{i=1}^{m} \lambda_i b_{\boldsymbol{\lambda}n}^{-1} h_i(\mathbf{X}_n)\|\mathbf{X}_n - \boldsymbol{\mu}\| + \sum_{i=1}^{m} \lambda_i b_{\boldsymbol{\lambda}n}^{-1} g_i(\boldsymbol{\mu}; \mathbf{X}_n - \boldsymbol{\mu}).\end{aligned}$$

By the linear form of the differential, we have

(3)
$$\begin{aligned}\sum_{i=1}^{m} \lambda_i b_{\boldsymbol{\lambda}n}^{-1} g_i(\boldsymbol{\mu}; \mathbf{X}_n - \boldsymbol{\mu}) &= \sum_{i=1}^{m} \lambda_i b_{\boldsymbol{\lambda}n}^{-1} \sum_{j=1}^{k} (X_{nj} - \mu_j) \left.\frac{\partial g_i}{\partial x_j}\right|_{\mathbf{x}=\boldsymbol{\mu}} \\ &= b_{\boldsymbol{\lambda}n}^{-1} \sum_{j=1}^{k} (X_{nj} - \mu_j) \sum_{i=1}^{m} \lambda_i \left.\frac{\partial g_i}{\partial x_j}\right|_{\mathbf{x}=\boldsymbol{\mu}} \\ &= \frac{(\boldsymbol{\lambda}\mathbf{D})(\mathbf{X}_n - \boldsymbol{\mu})'}{[(\boldsymbol{\lambda}\mathbf{D})\boldsymbol{\Sigma}_n(\boldsymbol{\lambda}\mathbf{D})']^{1/2}}.\end{aligned}$$

By the assumption on $\boldsymbol{\lambda}$, and by the definition of asymptotic multivariate normality, the right-hand side of (3) converges in distribution to $N(0, 1)$. Thus

(4)
$$\sum_{i=1}^{m} \lambda_i b_{\boldsymbol{\lambda}n}^{-1} g_i(\boldsymbol{\mu}; \mathbf{X}_n - \boldsymbol{\mu}) \xrightarrow{d} N(0, 1).$$

Now write

(5)
$$\sum_{i=1}^{m} \lambda_i b_{\boldsymbol{\lambda}n}^{-1} h_i(\mathbf{X}_n)\|\mathbf{X}_n - \boldsymbol{\mu}\| = b_{\boldsymbol{\lambda}n}^{-1}\|\mathbf{X}_n - \boldsymbol{\mu}\| \sum_{i=1}^{m} \lambda_i h_i(\mathbf{X}_n).$$

By Application C of Corollary 1.7, since $\boldsymbol{\Sigma}_n \to 0$ we have $\mathbf{X}_n \xrightarrow{p} \boldsymbol{\mu}$. Therefore, since each h_i is continuous at $\boldsymbol{\mu}$, Theorem 1.7 yields

$$\sum_{i=1}^{m} \lambda_i h_i(\mathbf{X}_n) \xrightarrow{p} \sum_{i=1}^{m} \lambda_i h_i(\boldsymbol{\mu}) = 0.$$

Also, now utilizing the fact that $\boldsymbol{\Sigma}_n$ is of the form $b_n^2\boldsymbol{\Sigma}$, and applying Application B of Corollary 1.7, we have

$$b_{\boldsymbol{\lambda}n}^{-1}\|\mathbf{X}_n - \boldsymbol{\mu}\| = (\boldsymbol{\lambda}\mathbf{D}\boldsymbol{\Sigma}\mathbf{D}'\boldsymbol{\lambda}')^{-1/2} b_n^{-1}\|\mathbf{X}_n - \boldsymbol{\mu}\| \xrightarrow{d} (\cdot).$$

It follows by Slutsky's Theorem that the right-hand side of (5) converges in probability to 0. Combining this result with (4) and (2), we have (1). ■

Remark A. In the above proof, the limit law of $[g(\mathbf{x}_n) - g(\boldsymbol{\mu})]$, suitably normalized, was found by reducing to the differential, $g(\boldsymbol{\mu}; \mathbf{X}_n - \boldsymbol{\mu})$, likewise normalized, and finding its limit law. The latter determination did not involve the specific form $b_n^2\boldsymbol{\Sigma}$ which was assumed for $\boldsymbol{\Sigma}_n$. Rather, this assumption played a role in the *reduction* step, which had two parts. In one part, only the property $\boldsymbol{\Sigma}_n \to \mathbf{0}$ was needed, to establish $\mathbf{X}_n \xrightarrow{p} \boldsymbol{\mu}$. However, for the other part, to obtain $(\boldsymbol{\lambda}\mathbf{D}\boldsymbol{\Sigma}_n\mathbf{D}'\boldsymbol{\lambda}')^{-1/2}\|\mathbf{X}_n - \boldsymbol{\mu}\| = O_p(1)$, a further restriction is evidently needed. ■

An important special case of the theorem is given by the following result for g real-valued and $b_n = n^{-1/2}$.

Corollary. *Suppose that* $\mathbf{X}_n = (X_{n1}, \ldots, X_{nk})$ *is* $AN(\boldsymbol{\mu}, n^{-1}\boldsymbol{\Sigma})$, *with* $\boldsymbol{\Sigma}$ *a covariance matrix. Let* $g(\mathbf{x})$ *be a real-valued function having a nonzero differential at* $\mathbf{x} = \boldsymbol{\mu}$. *Then*

$$g(\mathbf{X}_n) \quad \text{is} \quad AN\left(g(\boldsymbol{\mu}), \frac{1}{n}\sum_{i=1}^{k}\sum_{j=1}^{k}\sigma_{ij}\left.\frac{\partial g}{\partial x_i}\right|_{\mathbf{x}=\boldsymbol{\mu}} \cdot \left.\frac{\partial g}{\partial x_j}\right|_{\mathbf{x}=\boldsymbol{\mu}}\right).$$

Remarks B. (i) A sufficient condition for g to have a nonzero differential at $\boldsymbol{\mu}$ is that the first partial derivatives $\partial g/\partial x_i$, $1 \le i \le k$, be continuous at $\boldsymbol{\mu}$ and not all zero at $\boldsymbol{\mu}$ (see Lemma 1.12.2).

(ii) Note that in order to obtain the asymptotic normality of $g(X_{n1}, \ldots, X_{nk})$, the asymptotic *joint* normality of $X_{n1}, \ldots, X_{nk}$ is needed. ■

Analogues of Theorem A for the case of a function g having a differential vanishing at $\mathbf{x} = \boldsymbol{\mu}$ may be developed as generalizations of Theorem 3.1B. For simplicity we confine attention to real-valued functions g and state the following.

Theorem B. *Suppose that* $\mathbf{X}_n = (X_{n1}, \ldots, X_{nk})$ *is* $AN(\boldsymbol{\mu}, n^{-1}\boldsymbol{\Sigma})$. *Let* $g(\mathbf{x})$ *be a real-valued function possessing continuous partials of order* m (>1) *in a neighborhood of* $\mathbf{x} = \boldsymbol{\mu}$, *with all the partials of order* j, $1 \le j \le m - 1$, *vanishing at* $\mathbf{x} = \boldsymbol{\mu}$, *but with the* m*th order partials not all vanishing at* $\mathbf{x} = \boldsymbol{\mu}$. *Then*

$$n^{m/2}[g(\mathbf{X}_n) - g(\boldsymbol{\mu})] \xrightarrow{d} \frac{1}{m!}\sum_{i_1=1}^{k}\cdots\sum_{i_m=1}^{k}\left.\frac{\partial^m g}{\partial x_{i_1}\cdots\partial x_{i_m}}\right|_{\mathbf{x}=\boldsymbol{\mu}} \cdot \prod_{j=1}^{m} Z_{i_j},$$

where $\mathbf{Z} = (Z_1, \ldots, Z_k) = N(\mathbf{0}, \boldsymbol{\Sigma})$.

PROOF. In conjunction with the multivariate Taylor expansion (Theorem 1.12.1B), employ arguments similar to those in the proof of Theorem A. ■

Remark C. For the simplest case, $m = 2$, the limit random variable appearing in the preceding result is a *quadratic form* $\mathbf{ZAZ}'$, where

$$A = \left(\frac{1}{2}\frac{\partial^2 g}{\partial x_i\,\partial x_j}\bigg|_{\mathbf{x}=\boldsymbol{\mu}}\right)_{k \times k}.$$

We shall further discuss such random variables in Section 3.5. ■

3.4 FURTHER EXAMPLES AND APPLICATIONS

The behavior of *functions of several sample moments* is discussed in general in **3.4.1** and illustrated for the *sample correlation coefficient* in **3.4.2**. It should be noted that statistics which are functions of several sample quantiles, or of both moments and quantiles, could be treated similarly. In **3.4.3** we consider the problem of forming an "optimal" linear combination of several asymptotically jointly normal statistics.

3.4.1 Functions of Several Sample Moments

Various statistics of interest may be expressed as functions of sample moments. One group of examples consists of the sample "coefficients" of various kinds, such as the sample coefficients of *variation*, of *skewness*, of *kurtosis*, of *regression*, and of *correlation*. By Theorem 2.2.1B, the vector of sample moments $(a_1, \ldots, a_k)$ is $AN((\alpha_1, \ldots, \alpha_k), n^{-1}\boldsymbol{\Sigma})$, for some $\boldsymbol{\Sigma}$. It follows by Corollary 3.3 that statistics which are functions of $(a_1, \ldots, a_k)$ are typically asymptotically normal with means given by the corresponding functions of $(\alpha_1, \ldots, \alpha_k)$ and with variances of the form c/n, c constant. As an example, the correlation coefficient is treated in **3.4.2**. Another example, the sample coefficient of variation $s/\bar{X}$, is assigned as an exercise. Useful further discussion is found in Cramér (1946), Section 28.4. For a treatment of c-sample applications, see Hsu (1945). For Berry–Esséen rates of order $O(n^{-1})$ for the error of approximation in asymptotic normality of functions of sample moments, see Bhattacharya (1977).

3.4.2 Illustration: the Sample Correlation Coefficient

Let $(X_1, Y_1), \ldots, (X_n, Y_n)$ be independent observations on a bivariate distribution. The *correlation* of X_1 and Y_1 is $\rho = \sigma_{xy}/\sigma_x\sigma_y$, where $\sigma_{xy} = E\{(X_1 - \mu_x)(Y_1 - \mu_y)\}$, $\mu_x = E\{X_1\}$, $\mu_y = E\{Y_1\}$, $\sigma_x^2 = \text{Var}\{X_1\}$, $\sigma_y^2 = \text{Var}\{Y_1\}$. The sample analogue,

$$\hat{\rho} = \frac{\dfrac{1}{n}\sum_{i=1}^{n}(X_i - \bar{X})(Y_i - \bar{Y})}{\left[\dfrac{1}{n}\sum_{i=1}^{n}(X_i - \bar{X})^2\right]^{1/2}\left[\dfrac{1}{n}\sum_{i=1}^{n}(Y_i - \bar{Y})^2\right]^{1/2}}$$

may be expressed as $\hat{\rho} = g(\mathbf{V})$, where

$$\mathbf{V} = \left(\bar{X}, \bar{Y}, \frac{1}{n}\sum_{i=1}^{n} X_i^2, \frac{1}{n}\sum_{i=1}^{n} Y_i^2, \frac{1}{n}\sum_{i=1}^{n} X_i Y_i\right)$$

and

$$g(z_1, z_2, z_3, z_4, z_5) = \frac{z_5 - z_1 z_2}{(z_3 - z_1^2)^{1/2}(z_4 - z_2^2)^{1/2}}.$$

The vector $\mathbf{V}$ is $AN(E\{\mathbf{V}\}, n^{-1}\mathbf{\Sigma})$, where $\mathbf{\Sigma}_{5\times 5}$ is the covariance matrix of $(X_1, Y_1, X_1^2, Y_1^2, X_1 Y_1)$. (Compute $\mathbf{\Sigma}$ as an exercise.) It follows from Corollary 3.3 that

$$\hat{\rho} \quad \text{is} \quad AN(\rho, n^{-1}\mathbf{d}\mathbf{\Sigma}\mathbf{d}'),$$

where

$$\mathbf{d} = \left(\left.\frac{\partial g}{\partial z_1}\right|_{\mathbf{z}=E\{\mathbf{V}\}}, \ldots, \left.\frac{\partial g}{\partial z_5}\right|_{\mathbf{z}=E\{\mathbf{V}\}}\right).$$

The elements of $\mathbf{d}$ are readily found. Since

$$\frac{\partial g}{\partial z_1} = \frac{z_1(z_5 - z_1 z_2)}{(z_3 - z_1^2)^{3/2}(z_4 - z_2^2)^{1/2}} - \frac{z_2}{(z_3 - z_1^2)^{1/2}(z_4 - z_2^2)^{1/2}},$$

we obtain

$$d_1 = \left.\frac{\partial g}{\partial z_1}\right|_{\mathbf{z}=E\{\mathbf{V}\}} = \frac{\rho\mu_x}{\sigma_x^2} - \frac{\mu_y}{\sigma_x\sigma_y}.$$

Likewise

$$d_2 = \left.\frac{\partial g}{\partial z_2}\right|_{\mathbf{z}=E\{\mathbf{V}\}} = \frac{\rho\mu_y}{\sigma_y^2} - \frac{\mu_x}{\sigma_x\sigma_y}.$$

Verify that

$$d_3 = -\frac{\rho}{2\sigma_x^2}, \quad d_4 = -\frac{\rho}{2\sigma_y^2}, \quad d_5 = \frac{1}{\sigma_x\sigma_y}.$$

3.4.3 Optimal Linear Combinations

Suppose that we have several estimators $\hat{\theta}_{n1}, \ldots, \hat{\theta}_{nk}$ each having merit as an estimator of the same parameter θ, and suppose that the vector of estimators is asymptotically jointly normal: $\mathbf{X}_n = (\hat{\theta}_{n1}, \ldots, \hat{\theta}_{nk})$ is $AN((\theta, \ldots, \theta), n^{-1}\mathbf{\Sigma})$. Consider estimation of θ by a linear combination of the given estimates, say

$$\hat{\theta}_n = \sum_{i=1}^{k} \beta_i \hat{\theta}_{ni},$$

where $\boldsymbol{\beta} = (\beta_1, \ldots, \beta_k)$ satisfies $\beta_1 + \cdots + \beta_k = 1$. Such an estimator $\hat{\theta}_n$ is $AN(\theta, n^{-1}\boldsymbol{\beta}\boldsymbol{\Sigma}\boldsymbol{\beta}')$. The "**best**" such linear combination may be defined as that which minimizes the asymptotic variance. Thus we seek the choice of $\boldsymbol{\beta}$ which minimizes the quadratic form $\boldsymbol{\beta}\boldsymbol{\Sigma}\boldsymbol{\beta}'$ subject to the restriction $\sum_1^k \beta_i = 1$.

The solution may be obtained as a special case of useful results given by Rao (1973), Section 1.f, on the extreme values attained by quadradic forms under linear and quadratic restrictions on the variables. (Assume, without loss of generality, that $\boldsymbol{\Sigma}$ is nonsingular.) In particular, we have that

$$\inf_{\Sigma_1^k \beta_i = 1} \boldsymbol{\beta}\boldsymbol{\Sigma}\boldsymbol{\beta}' = \frac{1}{\sum_{i=1}^k \sum_{j=1}^k \sigma_{ij}^*},$$

where $\boldsymbol{\Sigma}^* = \boldsymbol{\Sigma}^{-1} = (\sigma_{ij}^*)$, and that this infimum is attained at the point

$$\boldsymbol{\beta}_0 = (\beta_{01}, \ldots, \beta_{0k}) = \left(\frac{\sum_{j=1}^k \sigma_{1j}^*}{\sum_{i=1}^k \sum_{j=1}^k \sigma_{ij}^*}, \ldots, \frac{\sum_{j=1}^k \sigma_{kj}^*}{\sum_{i=1}^k \sum_{j=1}^k \sigma_{ij}^*} \right).$$

For the case $k = 2$, we have

$$\boldsymbol{\Sigma}^* = \begin{bmatrix} \dfrac{\sigma_{22}}{\sigma_{11}\sigma_{22} - \sigma_{12}^2} & \dfrac{-\sigma_{12}}{\sigma_{11}\sigma_{22} - \sigma_{12}^2} \\[2ex] \dfrac{-\sigma_{12}}{\sigma_{11}\sigma_{22} - \sigma_{12}^2} & \dfrac{\sigma_{11}}{\sigma_{11}\sigma_{22} - \sigma_{12}^2} \end{bmatrix}$$

and thus the optimal $\boldsymbol{\beta}$ is

$$\boldsymbol{\beta}_0 = (\beta_{01}, \beta_{02}) = \left(\frac{\sigma_{22} - \sigma_{12}}{\sigma_{11} + \sigma_{22} - 2\sigma_{12}}, \frac{\sigma_{11} - \sigma_{12}}{\sigma_{11} + \sigma_{22} - 2\sigma_{12}} \right),$$

in which case

$$\boldsymbol{\beta}_0\boldsymbol{\Sigma}\boldsymbol{\beta}_0' = \frac{\sigma_{11}\sigma_{22} - \sigma_{12}^2}{\sigma_{11} + \sigma_{22} - 2\sigma_{12}}.$$

Putting $\sigma_1^2 = \sigma_{11}$, $\sigma_2^2 = \sigma_{22}$, $\rho = \sigma_{12}/\sigma_1\sigma_2$, and $\Delta = \sigma_2^2/\sigma_1^2$, we thus have

$$\boldsymbol{\beta}_0\boldsymbol{\Sigma}\boldsymbol{\beta}_0' = \sigma_1^2\Delta \frac{1 - \rho^2}{1 + \Delta - 2\rho\Delta^{1/2}}.$$

Assume, without loss of generality, that $\sigma_1^2 \leq \sigma_2^2$, that is, $\Delta \geq 1$. Then the preceding formula exhibits, in terms of ρ and Δ, the gain due to using the optimal linear combination instead of simply the better of the two given estimators. We have

$$\frac{\boldsymbol{\beta}_0\boldsymbol{\Sigma}\boldsymbol{\beta}_0'}{\sigma_1^2} = \frac{\Delta(1 - \rho^2)}{1 + \Delta - 2\rho\Delta^{1/2}} = 1 - \frac{(1 - \rho\Delta^{1/2})^2}{1 + \Delta - 2\rho\Delta^{1/2}},$$

showing that there is strict reduction of the asymptotic variance if and only if $\rho\Delta^{1/2} \neq 1$. Note also that the "best" linear combination is represented in terms of ρ and Δ as

$$\boldsymbol{\beta}_0 = (\beta_{01}, \beta_{02}) = \left(\frac{\Delta - \rho\Delta^{1/2}}{1 + \Delta - 2\rho\Delta^{1/2}}, \frac{1 - \rho\Delta^{1/2}}{1 + \Delta - 2\rho\Delta^{1/2}}\right).$$

As an exercise, apply these results in connection with estimation of the mean of a symmetric distribution by a linear combination of the sample mean and sample median (Problem 3.P.9).

3.5 QUADRATIC FORMS IN ASYMPTOTICALLY MULTIVARIATE NORMAL VECTORS

In some applications the statistic of interest is a quadratic form, say $T_n = \mathbf{X}_n\mathbf{C}\mathbf{X}_n'$, in a random vector $\mathbf{X}_n$ converging in distribution to $N(\boldsymbol{\mu}, \boldsymbol{\Sigma})$. In this case, we obtain from Corollary 1.7 that $T_n \xrightarrow{d} \mathbf{XCX}'$, where $\mathbf{X}$ is $N(\boldsymbol{\mu}, \boldsymbol{\Sigma})$. In certain other situations, the statistic of interest T_n is such that Theorem 3.3B yields the asymptotic behavior, say $n(T_n - \Delta) \xrightarrow{d} \mathbf{ZAZ}'$, for some Δ and $\mathbf{A}$, where $\mathbf{Z}$ is $N(\mathbf{0}, \boldsymbol{\Sigma})$. (There is a slight overlap of these two situations.)

In both situations just discussed, a (limit) quadratic form in a multivariate normal vector arises for consideration. It is of particular interest to know when the quadratic form has a (possibly noncentral) chi-squared distribution. We give below a basic theorem of use in identifying such distributions, and then we apply the result to examine the behavior of the "chi-squared statistic," a particular quadratic form in a *multinomial vector*. We also investigate other quadratic forms in multinomial vectors.

The theorem we prove will be an extension of the following lemma proved in Rao (1973), Section 3.b.4.

Lemma. *Let* $\mathbf{X} = (X_1, \ldots, X_k)$ *be* $N(\boldsymbol{\mu}, \mathbf{I}_k)$, $\mathbf{I}_k$ *the identity matrix, and let* $\mathbf{C}_{k\times k}$ *be a symmetric matrix. Then the quadratic form* $\mathbf{XCX}'$ *has a (possibly noncentral) chi-squared distribution if and only if* $\mathbf{C}$ *is idempotent, that is,* $\mathbf{C}^2 = \mathbf{C}$, *in which case the degrees of freedom is rank* $(\mathbf{C}) =$ *trace* $(\mathbf{C})$ *and the noncentrality parameter is* $\boldsymbol{\mu}\mathbf{C}\boldsymbol{\mu}'$.

This is a very useful result but yet is seriously limited by the restriction to *independent* $X_1, \ldots, X_k$. For the case $\boldsymbol{\mu} = 0$, an extension to the case of *arbitrary* covariance matrix was given by Ogasawara and Takahashi (1951). A broader generalization is provided by the following result.

Theorem. *Let* $\mathbf{X} = (X_1, \ldots, X_k)$ *be* $N(\boldsymbol{\mu}, \boldsymbol{\Sigma})$, *and let* $\mathbf{C}_{k\times k}$ *be a symmetric matrix. Assume that, for* $\boldsymbol{\eta} = (\eta_1, \ldots, \eta_k)$,

(1) $$\boldsymbol{\eta}\boldsymbol{\Sigma} = \mathbf{0} \Rightarrow \boldsymbol{\eta}\boldsymbol{\mu}' = 0.$$

Then $\mathbf{XCX}'$ *has a (possibly noncentral) chi-squared distribution if and only if*

(2) $$\mathbf{\Sigma C \Sigma C \Sigma} = \mathbf{\Sigma C \Sigma},$$

in which case the degrees of freedom is trace $(\mathbf{C\Sigma})$ *and the noncentrality parameter is* $\boldsymbol{\mu}\mathbf{C}\boldsymbol{\mu}'$.

PROOF. Let $\lambda_1 \geq \cdots \geq \lambda_k \geq 0$ denote the eigenvalues of $\mathbf{\Sigma}$. Since $\mathbf{\Sigma}$ is symmetric, there exists (see Rao (1973), Section 1.C.3(i) and related discussion) an orthogonal matrix $\mathbf{B}$ having rows $\mathbf{b}_1, \ldots, \mathbf{b}_k$ which are eigenvectors corresponding to $\lambda_1, \ldots, \lambda_k$, that is,

$$\mathbf{b}_i \mathbf{\Sigma} = \lambda_i \mathbf{b}_i, \qquad 1 \leq i \leq k.$$

Thus

(*) $$\mathbf{B\Sigma B}' = \mathbf{\Lambda} = \begin{bmatrix} \lambda_1 & & 0 \\ & \ddots & \\ 0 & & \lambda_k \end{bmatrix} = (\lambda_i \delta_{ij}),$$

where $\delta_{ij} = I(i = j)$, or

(**) $$\mathbf{B}'\mathbf{\Lambda B} = \mathbf{\Sigma}.$$

Put

$$\mathbf{V} = \mathbf{XB}'.$$

Since $\mathbf{X}$ is $N(\boldsymbol{\mu}, \mathbf{\Sigma})$, it follows by (*) that $\mathbf{V}$ is $N(\boldsymbol{\mu}\mathbf{B}', \mathbf{\Lambda})$. Also, since $\mathbf{B}$ is orthogonal, $\mathbf{X} = \mathbf{VB}$ and thus

$$\mathbf{XCX}' = \mathbf{VBCB}'\mathbf{V}'.$$

We now seek to represent $\mathbf{V}$ as $\mathbf{V} = \mathbf{W}\mathbf{\Lambda}^{1/2}$, where $\mathbf{W} = N(\boldsymbol{\alpha}, \mathbf{I}_k)$ for some $\boldsymbol{\alpha}$ and $\mathbf{\Lambda}^{1/2} = (\lambda_i^{1/2}\delta_{ij})_{k \times k}$. Since $\boldsymbol{\mu}\mathbf{B}' = (\boldsymbol{\mu}\mathbf{b}_1', \ldots, \boldsymbol{\mu}\mathbf{b}_k')$, we have by (1) that the jth component of $\boldsymbol{\mu}\mathbf{B}'$ is 0 whenever $\lambda_j = 0$. Define

$$\alpha_i = \begin{cases} \boldsymbol{\mu}\mathbf{b}_i'\lambda_i^{-1/2}, & \text{if } \lambda_i \neq 0, \\ 0 & \text{if } \lambda_i = 0. \end{cases}$$

Thus $\boldsymbol{\alpha}\mathbf{\Lambda}^{1/2} = (\alpha_1\lambda_1^{1/2}, \ldots, \alpha_k\lambda_k^{1/2}) = \boldsymbol{\mu}\mathbf{B}'$ and $\mathbf{V}$ has the desired representation for this choice of $\boldsymbol{\alpha}$. Hence we may write

$$\mathbf{XCX}' = \mathbf{W}\mathbf{\Lambda}^{1/2}\mathbf{BCB}'\mathbf{\Lambda}^{1/2}\mathbf{W}' = \mathbf{WDW}',$$

where $\mathbf{D} = \mathbf{\Lambda}^{1/2}\mathbf{BCB}'\mathbf{\Lambda}^{1/2}$. It follows from the lemma that $\mathbf{XCX}'$ has a chi-squared distribution if and only if $\mathbf{D}^2 = \mathbf{D}$. Now

$$\begin{aligned} \mathbf{D}^2 &= (\mathbf{\Lambda}^{1/2}\mathbf{BCB}'\mathbf{\Lambda}^{1/2})(\mathbf{\Lambda}^{1/2}\mathbf{BCB}'\mathbf{\Lambda}^{1/2}) \\ &= \mathbf{\Lambda}^{1/2}\mathbf{BCB}'\mathbf{\Lambda}\mathbf{BCB}'\mathbf{\Lambda}^{1/2} = \mathbf{\Lambda}^{1/2}\mathbf{BC\Sigma CB}'\mathbf{\Lambda}^{1/2}, \end{aligned}$$

making use of (**). Thus $\mathbf{D}^2 = \mathbf{D}$ if and only if

$$\boldsymbol{\Lambda}^{1/2}\mathbf{BC\Sigma CB'}\boldsymbol{\Lambda}^{1/2} = \boldsymbol{\Lambda}^{1/2}\mathbf{BCB'}\boldsymbol{\Lambda}^{1/2}. \tag{3}$$

Now check that

$$\boldsymbol{\Lambda}\mathbf{A}_1 = \boldsymbol{\Lambda}\mathbf{A}_2 \Leftrightarrow \boldsymbol{\Lambda}^{1/2}\mathbf{A}_1 = \boldsymbol{\Lambda}^{1/2}\mathbf{A}_2$$

and

$$\mathbf{A}_1\boldsymbol{\Lambda} = \mathbf{A}_2\boldsymbol{\Lambda} \Leftrightarrow \mathbf{A}_1\boldsymbol{\Lambda}^{1/2} = \mathbf{A}_2\boldsymbol{\Lambda}^{1/2}.$$

Thus (3) is equivalent to

$$\boldsymbol{\Lambda}\mathbf{BC\Sigma CB'}\boldsymbol{\Lambda} = \boldsymbol{\Lambda}\mathbf{BCB'}\boldsymbol{\Lambda}. \tag{4}$$

Now premultiplying by $\mathbf{B}'$ and postmultiplying by $\mathbf{B}$ on each side of (4), we have (2). Thus we have shown that $\mathbf{D}^2 = \mathbf{D}$ if and only if (2) holds. In this case the degrees of freedom is given by rank ($\mathbf{D}$). Since trace $(\boldsymbol{\Lambda}^{1/2}\mathbf{A}\boldsymbol{\Lambda}^{1/2}) =$ trace $(\boldsymbol{\Lambda}\mathbf{A})$ and since $\boldsymbol{\Lambda}\mathbf{B} = \mathbf{B}\boldsymbol{\Sigma}$, we have

$$\text{rank}(\mathbf{D}) = \text{trace}(\boldsymbol{\Lambda}\mathbf{BCB'}) = \text{trace}(\mathbf{B\Sigma CB'}) = \text{trace}(\boldsymbol{\Sigma}\mathbf{C}).$$

It remains to determine the noncentrality parameter, which is given by

$$\boldsymbol{\alpha}\mathbf{D}\boldsymbol{\alpha}' = \boldsymbol{\alpha}\boldsymbol{\Lambda}^{1/2}\mathbf{BCB'}\boldsymbol{\Lambda}^{1/2}\boldsymbol{\alpha}' = \boldsymbol{\mu}\mathbf{B'BCB'B}\boldsymbol{\mu}' = \boldsymbol{\mu}C\boldsymbol{\mu}'. \quad \blacksquare$$

Example A. *The case $\boldsymbol{\Sigma}$ nonsingular and $\mathbf{C} = \boldsymbol{\Sigma}^{-1}$.* In this case conditions (1) and (2) of the theorem are trivially satisfied, and thus $\mathbf{X}\boldsymbol{\Sigma}^{-1}\mathbf{X}'$ is distributed as $\chi_k^2(\boldsymbol{\mu}\boldsymbol{\Sigma}^{-1}\boldsymbol{\mu}')$. $\blacksquare$

Example B. *Multinomial vectors and the "chi-squared statistic."* Let $(n_1, \ldots, n_k)$ be multinomial $(n; p_1, \ldots, p_k)$, with each $p_i > 0$. As seen in Section 2.7, the vector

$$\mathbf{X}_n = n^{1/2}\left(\frac{n_1}{n} - p_1, \ldots, \frac{n_k}{n} - p_k\right) = (X_{n1}, \ldots, X_{nk})$$

converges in distribution to $N(\mathbf{0}, \boldsymbol{\Sigma})$, where

$$\sigma_{ij} = \begin{cases} p_i(1 - p_j), & i = j, \\ -p_i p_j, & i \neq j. \end{cases}$$

A popular statistic for testing hypotheses in various situations is the *chi-squared statistic*

$$T_n = \sum_{i=1}^{k} \frac{(n_i - np_i)^2}{np_i} = n \sum_{i=1}^{k} \frac{1}{p_i}\left(\frac{n_i}{n} - p_i\right)^2.$$

This may be represented as a quadratic form in $\mathbf{X}_n$:

$$T_n = \sum_{i=1}^{n} \frac{1}{p_i} X_{ni}^2 = \mathbf{X}_n \mathbf{C} \mathbf{X}_n',$$

where

$$\mathbf{C} = \begin{bmatrix} 1/p_1 & & 0 \\ & \ddots & \\ 0 & & 1/p_k \end{bmatrix} = \left(\frac{1}{p_i}\delta_{ij}\right).$$

We now apply the theorem to determine that the "chi-squared statistic" is, indeed, asymptotically chi-squared in distribution, with $k - 1$ degrees of freedom. That is,

$$T_n \xrightarrow{d} \chi_{k-1}^2.$$

We apply the theorem with $\mu = 0$, in which case condition (1) is trivially satisfied. Writing $\sigma_{ij} = p_i(\delta_{ij} - p_j)$, we have

$$\mathbf{C\Sigma} = \left(\frac{1}{p_i}\sigma_{ij}\right) = (\delta_{ij} - p_j)$$

and thus

$$\mathbf{C\Sigma C\Sigma} = \left(\sum_{l=1}^{k} (\delta_{il} - p_l)(\delta_{lj} - p_j)\right) = (\delta_{ij} - p_j) = \mathbf{C\Sigma}$$

and hence (2) holds. We also see from the last step that

$$\text{trace}(\mathbf{C\Sigma}) = \sum_{j=1}^{k} (1 - p_j) = k - 1 \;(= \text{rank}(\mathbf{C\Sigma})).$$

Thus we have that, for the given matrices $\mathbf{\Sigma}$ and $\mathbf{C}$,

$$\mathbf{X}_n \xrightarrow{d} N(\mathbf{0}, \mathbf{\Sigma}) \Rightarrow T_n = \mathbf{X}_n \mathbf{C} \mathbf{X}_n' \xrightarrow{d} \chi_{k-1}^2. \quad \blacksquare$$

Example C (*continuation*). It is also of interest to consider the behavior of the statistic T_n when the *actual* distribution of $(n_1, \ldots, n_k)$ is not the hypothesized multinomial $(n; p_1, \ldots, p_k)$ distribution in terms of which T_n is defined, but rather some *other* multinomial distribution, say multinomial $(n; p_{n1}, \ldots, p_{nk})$, where the parameter $(p_{n1}, \ldots, p_{nk})$ converges to $(p_1, \ldots, p_k)$ at a suitable rate. Whereas the foregoing asymptotic result for T_n corresponds to its behavior under the *null hypothesis*, the present consideration concerns the behavior of T_n under a sequence of "*local*" *alternatives* to the null hypothesis. In particular, take

$$p_{ni} = p_i + \Delta_i n^{-1/2}, \qquad 1 \le i \le k, n = 1, 2, \ldots.$$

Then we may express $\mathbf{X}_n$ in the form

$$\mathbf{X}_n = n^{1/2}\left(\frac{n_1}{n} - p_{n1}, \ldots, \frac{n_k}{n} - p_{nk}\right) + (\Delta_1, \ldots, \Delta_k),$$

that is,

$$\mathbf{X}_n = \mathbf{X}_n^* + \mathbf{\Delta},$$

where $\mathbf{\Delta} = (\Delta_1, \ldots, \Delta_k)$ satisfies $\sum_1^k \Delta_i = 0$ and $n^{-1/2}\mathbf{X}_n^*$ is a mean of I.I.D. random vectors, each multinomial $(1; p_{n1}, \ldots, p_{nk})$. By an appropriate multivariate CLT for triangular arrays (Problem 3.P.10), we have $\mathbf{X}_n^* \xrightarrow{d} N(\mathbf{0}, \mathbf{\Sigma})$ and thus $\mathbf{X}_n \xrightarrow{d} N(\mathbf{\Delta}, \mathbf{\Sigma})$. We now apply our theorem to find that in this case T_n converges in distribution to a *noncentral* chi-squared variate. We have already established in Example B that (2) holds and that rank $(\mathbf{C\Sigma}) = k - 1$. This implies that rank $(\mathbf{\Sigma}) = k - 1$ since rank $(\mathbf{C}) = k$. Thus the value 0 occurs with multiplicity 1 as an eigenvalue of $\mathbf{\Sigma}$. Further, note that $\mathbf{1}_k = (1, \ldots, 1)$ is an eigenvector for the eigenvalue 0, that is, $\mathbf{\Sigma 1}_k' = 0$. Finally, $\mathbf{\Delta 1}_k' = \sum_1^k \Delta_i = 0$. It is seen thus that (1) holds. Noting that

$$\mathbf{\Delta C \Delta}' = \sum_{i=1}^k \frac{\Delta_i^2}{p_i},$$

we obtain from the theorem that, for $\mathbf{\Sigma}$ and $\mathbf{C}$ as given,

$$\mathbf{X}_n \xrightarrow{d} N(\mathbf{\Delta}, \mathbf{\Sigma}) \Rightarrow T_n = \mathbf{X}_n \mathbf{C} \mathbf{X}_n' \xrightarrow{d} \chi_{k-1}^2\left(\sum_{i=1}^k \frac{\Delta_i^2}{p_i}\right).$$

Note that this noncentrality parameter may be written as

$$n \sum_{i=1}^k \frac{1}{p_i}\left(E\left\{\frac{n_i}{n}\right\} - p_i\right)^2.$$

An application of the foregoing convergence is to calculate the approximate power of T_n as a test statistic relative to the null hypothesis

$$H_0: (n_1, \ldots, n_k) \quad \text{is multinomial} \quad (n; p_1, \ldots, p_k)$$

against an alternative

$$H_1: (n_1, \ldots, n_k) \quad \text{is multinomial} \quad (n; p_1^*, \ldots, p_k^*).$$

Suppose that the critical region is $\{T_n > t_0\}$, where the choice of t_0 for a level α test would be based upon the null hypothesis asymptotic χ_{k-1}^2 distribution of T_n. Then the approximate power of T_n at the alternative H_1 is given by interpreting $(p_1^*, \ldots, p_k^*)$ as $(p_{n1}, \ldots, p_{nk})$ and calculating the probability that a random variable having the distribution

$$\chi_{k-1}^2\left(n \sum_{i=1}^k \frac{1}{p_i}(p_i^* - p_i)^2\right)$$

exceeds the value t_0. ■

Example D (*continuation*). *Further quadratic forms in multinomial vectors.* Quadratic forms in

$$\mathbf{X}_n = n^{1/2}\left(\frac{n_1}{n} - p_1, \ldots, \frac{n_k}{n} - p_k\right)$$

other than the chi-squared statistic may be of interest. As a general treatment, Rao (1973), Section 6.a.1, considers equivalently the vector

$$\mathbf{V}_n = \left(\frac{n_1 - np_1}{(np_1)^{1/2}}, \ldots, \frac{n_k - np_k}{(np_k)^{1/2}}\right),$$

which is related to $\mathbf{X}_n$ by $\mathbf{V}_n = \mathbf{X}_n\mathbf{D}$, where

$$\mathbf{D} = \begin{bmatrix} p_1^{-1/2} & & 0 \\ & \ddots & \\ 0 & & p_k^{-1/2} \end{bmatrix} = (p_i^{-1/2}\delta_{ij}).$$

Rao puts

$$\boldsymbol{\phi} = (p_1^{1/2}, \ldots, p_k^{1/2})$$

and establishes the proposition: *a sufficient condition for the quadratic form* $\mathbf{V}_n\mathbf{C}\mathbf{V}_n'$, $\mathbf{C}$ *symmetric, to converge in distribution to a chi-squared distribution is*

$$(*) \qquad \mathbf{C}^2 = \mathbf{C} \quad \text{and} \quad \boldsymbol{\phi}\mathbf{C} = \alpha\boldsymbol{\phi},$$

that is, $\mathbf{C}$ *is idempotent and* $\boldsymbol{\phi}$ *is an eigenvector of* $\mathbf{C}$, *in which case the degrees of freedom is rank* ($\mathbf{C}$) *if* $\alpha = 0$ *and rank* ($\mathbf{C}$) $- 1$ *if* $\alpha \neq 0$.

We now show that this result follows from our theorem. By Application A of Corollary 1.7, $\mathbf{V}_n \xrightarrow{d} N(0, \boldsymbol{\Sigma}^*)$, where

$$\boldsymbol{\Sigma}^* = \mathbf{D}\boldsymbol{\Sigma}\mathbf{D}' = \left(\frac{p_i(\delta_{ij} - p_j)}{(p_i p_j)^{1/2}}\right) = (\delta_{ij} - p_i p_j)$$

$$= \mathbf{I}_k - \boldsymbol{\phi}'\boldsymbol{\phi}.$$

Applying (*), we have

$$\mathbf{C}\boldsymbol{\Sigma}^* = \mathbf{C} - \mathbf{C}\boldsymbol{\phi}'\boldsymbol{\phi} = \mathbf{C} - \alpha\boldsymbol{\phi}'\boldsymbol{\phi}$$

and hence (check)

$$\mathbf{C}\boldsymbol{\Sigma}^*\mathbf{C}\boldsymbol{\Sigma}^* = \mathbf{C} - 2\alpha\boldsymbol{\phi}'\boldsymbol{\phi} + \alpha^2\boldsymbol{\phi}'\boldsymbol{\phi}.$$

But (*) implies $\alpha = 0$ or 1, so that

$$\mathbf{C}\boldsymbol{\Sigma}^*\mathbf{C}\boldsymbol{\Sigma}^* = \mathbf{C} - \alpha\boldsymbol{\phi}'\boldsymbol{\phi} = \mathbf{C}\boldsymbol{\Sigma}^*,$$

that is, $\mathbf{C\Sigma^*}$ is idempotent. Further, it is seen that

$$\text{trace}(\mathbf{C\Sigma^*}) = \begin{cases} \text{trace}(\mathbf{C}) = \text{rank}(\mathbf{C}) & \text{if } \alpha = 0 \\ \text{trace}(\mathbf{C}) - 1 = \text{rank}(\mathbf{C}) - 1 & \text{if } \alpha \neq 0. \end{cases}$$

Thus the proposition is established.

In particular, with $\mathbf{C} = I_k$, the quadratic form $\mathbf{V}_n\mathbf{C}\mathbf{V}_n$ is simply the chi-squared statistic and converges in distribution to χ^2_{k-1}, as seen in Example B. ∎

3.6 FUNCTIONS OF ORDER STATISTICS

Order statistics have previously been discussed in some detail in Section 2.4. Here we augment that discussion, giving further attention to statistics which may be expressed as *functions* of order statistics, and giving brief indication of some relevant asymptotic distribution theory. As before, the order statistics of a sample $X_1, \ldots, X_n$ are denoted by $X_{n1} \le \cdots \le X_{nn}$.

A variety of *short-cut* procedures for quick estimates of location or scale parameters, or for quick tests of related hypotheses, are provided in the form of *linear* functions of order statistics, that is statistics of the form

$$\sum_{i=1}^{n} c_{ni} X_{ni}.$$

For example, the *sample range* $X_{nn} - X_{n1}$ belongs to this class. Another example is given by the *α-trimmed mean*,

$$\frac{1}{n - 2[n\alpha]} \sum_{i=[n\alpha]+1}^{n-[n\alpha]} X_{ni},$$

which is a popular competitor of $\bar{X}$ for *robust* estimation of location. A broad treatment of linear functions of order statistics is provided in Chapter 8.

In robustness problems where *outliers* ("contaminated" or "wild" observations) are of concern, a useful statistic for their detection is the studentized range,

$$\frac{X_{nn} - X_{n1}}{\hat{\sigma}_n},$$

where $\hat{\sigma}_n$ is an appropriate estimator of σ. A one-sided version, for detection of excessively large observations, may be based on the so-called *extreme deviate* $X_{nn} - \bar{X}$. Likewise, a *studentized extreme deviate* is given by $(X_{nn} - \bar{X})/s$.

The differences between successive order statistics of the sample are called the *spacings*. These are

$$D_{ni} = X_{ni} - X_{n,i-1}, \qquad 2 \le i \le n.$$

The primary roles of $\mathbf{D}_{(n)} = (D_{n2}, \ldots, D_{nn})$ arise in nonparametric tests of goodness of fit and in tests that F possesses a specified property of interest. As an example of the latter, the hypothesis that F possesses a "monotone failure rate" arises in reliability theory. A general treatment of spacings is given by Pyke (1965), covering the exact distribution theory of spacings, with emphasis on F uniform or exponential, and providing a variety of limit theorems for distributions of spacings and of functions of spacings. Some recent developments on spacings and some open problems in the asymptotic theory are discussed in Pyke (1972).

We conclude with two examples of distribution theory.

Example A. *The sample range.* Suppose that F is symmetric about 0 and that $(X_{nn} - a_n)/b_n$ has the limit distribution

$$G_3(t) = e^{-e^{-t}}, \qquad -\infty < t < \infty.$$

Then, by symmetry, the random variable $-(X_{n1} - a_n)/b_n$ also has limit distribution G_3. Further, these two random variables are asymptotically independent, so that their joint asymptotic distribution has density

$$e^{-s-e^{-s}-t-e^{t}}, \qquad -\infty < s, t < \infty.$$

It follows that the normalized range

$$\frac{(X_{nn} - X_{n1}) - 2a_n}{b_n}$$

has limit distribution with density

$$\int_{-\infty}^{\infty} e^{-z-e^{-z-u}-e^{u}}\, du = 2e^{-z}K_0(2e^{-(1/2)z}),$$

where $K_0(z)$ is a modified Bessel function of the 2nd kind. See David (1970), p. 211. ■

Example B. *The Studentized Extreme Deviate.* Suppose that F has mean 0 and variance 1, that $(X_{nn} - a_n)/b_n$ has limit distribution G, and that

$$\frac{a_n}{n^{1/2}b_n} \to 0.$$

Then it turns out (Berman (1963)) that G is also the limit distribution of the random variable

$$\frac{\dfrac{X_{nn} - \overline{X}}{s} - a_n}{b_n},$$

where $\bar{X} = n^{-1} \sum_1^n X_i$ and $s^2 = (n-1)^{-1} \sum_1^n (X_i - \bar{X})^2$. In particular, for F $N(0, 1)$,

$$\mathscr{L}\left\{(2 \log n)^{1/2}\left(\frac{X_{nn} - \bar{X}}{s} - (2 \log n)^{1/2} + \frac{\log \log n + \log 4\pi}{2(2 \log n)^{1/2}}\right)\right\} \to G_3. \quad \blacksquare$$

3.P PROBLEMS

Sections 3.1, 3.2

1. Let X_n be Poisson with mean λ_n and suppose that $\lambda_n \to \infty$ as $n \to \infty$. Show that X_n is $AN(\lambda_n, \lambda_n)$, $n \to \infty$. (Hint: use characteristic functions.)

2. For the one-sided Kolmogorov–Smirnov distance D_n^+ treated in **2.1.5**, show that $4n(D_n^+)^2 \xrightarrow{d} \chi_2^2$.

3. Let $X_n \xrightarrow{d} N(\mu, \sigma^2)$ and let Y_n be $AN(\mu, \sigma^2/n)$. Let

$$g(x) = \begin{cases} 0, & |x| < 1 \\ ax^2 + b, & |x| \geq 1. \end{cases}$$

Investigate the limiting behavior of $g(X_n)$ and $g(Y_n)$. (By "limiting behavior" is meant both consistency and asymptotic distribution theory.)

4. Let $X_1, \ldots, X_n$ be independent $N(\theta, 1)$ variables, θ unknown. Consider estimation of the parametric function

$$\gamma(\theta) = P_\theta(X_1 \leq c) = \Phi(c - \theta),$$

where c is a specified number. It is well known that the minimum variance unbiased estimator of $\gamma(\theta)$ is

$$\Phi\left(\frac{c - \bar{X}}{\left(\frac{n-1}{n}\right)^{1/2}}\right)$$

Determine the limiting behavior of this estimator.

Sections 3.3, 3.4

5. Provide details of proof for Theorem 3.3B.

6. Complete the details of the sample correlation coefficient illustration in **3.4.2**.

7. Show that the sample correlation coefficient (defined in **3.4.2**) converges with probability 1 to the population correlation coefficient. Show also that it converges in rth mean, each $r > 0$.

8. Let $X_1, X_2, \ldots$ be I.I.D. with mean μ and variance σ^2, and with $\mu_4 < \infty$. The sample coefficient of variation is $s_n/\bar{X}_n$, where $\bar{X}_n = n^{-1} \sum_1^n X_i$

and $s_n^2 = (n-1)^{-1} \sum_1^n (X_i - \bar{X}_n)^2$. Derive the asymptotic behavior of $s_n/\bar{X}_n$. That is, show:

(i) If $\mu \neq 0$, then

(a) $\dfrac{s_n}{\bar{X}_n} \xrightarrow{wp1} \dfrac{\sigma}{\mu}$;

(b) $\dfrac{s_n}{\bar{X}_n}$ is $AN\left(\dfrac{\sigma}{\mu}, \dfrac{1}{n}\left[\dfrac{\sigma^2\mu_2}{\mu^4} - \dfrac{\mu_3}{\mu^3} + \dfrac{\mu_4 - \mu_2^2}{4\mu^2\sigma^2}\right]\right)$.

(ii) If $\mu = 0$, then

$$n^{-1/2}\frac{s_n}{\bar{X}_n} \to \frac{1}{N(0, 1)}.$$

9. Consider independent observations $X_1, X_2, \ldots$ on a distribution F having density $F' = f$ symmetric about μ. Assume that F has finite variance and that F'' exists at μ. Consider estimation of μ by a linear combination of the sample mean $\bar{X}$ and the sample median $\hat{\xi}_{1/2}$.

(a) Derive the asymptotic bivariate normal distribution of $(\bar{X}, \hat{\xi}_{1/2})$. (Hint: use the Bahadur representation.)

(b) Determine the "best" linear combination.

Section 3.5

10. *Multivariate CLT for triangular array.* Let $\mathbf{X}_n = (X_{n1}, \ldots, X_{nk})$ be a mean of n I.I.D. random k-vectors $(\xi_{nj1}, \ldots, \xi_{njk})$, $1 \le j \le n$, each having mean $(0, \ldots, 0)$ and covariance matrix $\mathbf{\Sigma}_n$. Suppose that $\mathbf{\Sigma}_n \to \mathbf{\Sigma}$, $n \to \infty$, where $\mathbf{\Sigma}$ is a covariance matrix. Suppose that all ξ_{nji} satisfy $E|\xi_{nji}|^{2+\varepsilon} < K$ for some fixed $\varepsilon > 0$ and $K < \infty$. Show that $\mathbf{X}_n$ is $AN(\mathbf{0}, n^{-1}\mathbf{\Sigma})$. (Hint: apply Corollary 1.9.3 in conjunction with the Cramér–Wold device.)

11. Discuss the asymptotic distribution theory of $T_n = \mathbf{X}_n \mathbf{C} \mathbf{X}_n'$ when $\mathbf{X}_n \xrightarrow{d} \mathbf{X}$ and $\mathbf{C}_n \xrightarrow{p} \mathbf{C}$, where $\mathbf{C}$ is a constant matrix. In particular, deal with the *modified chi-square statistic*

$$T_n^* = \sum_{i=1}^{k} \frac{(n_i - np_i)^2}{n_i}.$$

CHAPTER 4

Asymptotic Theory in Parametric Inference

This chapter treats statistics which arise in connection with estimation or hypothesis testing relative to a *parametric* family of possible distributions for the data.

Section 4.1 presents a concept of *asymptotic optimality* in the context of estimation on the basis of a random sample from a distribution belonging to the specified family. In particular, Section 4.2 treats estimation by the method of *maximum likelihood*, and Section 4.3 considers some other methods of estimation. Some closely related results concerning hypothesis testing are given in Section 4.4.

We have seen in Section 2.7 how data in the form of a random sample may be reduced to *multinomial form* by grouping the observations into cells. Thus, as an adjunct to the treatment of Sections 4.1–4.4, we deal with "product-multinomial" data in Sections 4.5 (estimation results) and 4.6 (hypothesis testing results). Of course, this methodology is applicable also without reference to a parametric family of distributions.

The concept of asymptotic optimality introduced in Section 4.1 is based on a notion of asymptotic relative efficiency formulated in terms of the *generalized variance* of multidimensional distributions. This generalizes the one-dimensional version given in **1.15.4**. For the *hypothesis testing* context, the treatment of asymptotic relative efficiency is deferred to Chapter 10, which provides several distinctive notions. (These notions may also be recast in the estimation context.)

4.1 ASYMPTOTIC OPTIMALITY IN ESTIMATION

Two notions of asymptotic relative efficiency of estimation procedures were discussed in **1.15.4**, based on the criteria of *variance* and *probability concentration*. The version based on variance has been exemplified in **2.3.5** and **2.6.7**.

Here, in **4.1.1** and **4.1.2**, we further develop the notion based on variance and, in particular, introduce the multidimensional version. On this basis, the classical notion of asymptotic "efficiency" is presented in **4.1.3**. Brief complements are provided in **4.1.4**..

4.1.1 Concentration Ellipsoids and Generalized Variance

The concept of variance as a measure of concentration for a 1-dimensional distribution may be extended to the case of a k-dimensional distribution in two ways—in terms of a *geometrical* entity called the "concentration ellipsoid," and in terms of a *numerical* measure called the "generalized variance." We shall follow Cramér (1946), Section 22.7.

For a distribution in R^k having mean $\boldsymbol{\mu}$ and nonsingular covariance matrix $\boldsymbol{\Sigma}$, the associated *concentration ellipsoid* is defined to be that ellipsoid such that a random vector distributed uniformly throughout the ellipsoid has the same mean $\boldsymbol{\mu}$ and covariance matrix $\boldsymbol{\Sigma}$. This provides a geometrical entity representing the concentration of the distribution about its mean $\boldsymbol{\mu}$. It is found that the concentration ellipsoid is given by the set of points

$$E = \{\mathbf{x}: (\mathbf{x} - \boldsymbol{\mu})\boldsymbol{\Sigma}^{-1}(\mathbf{x} - \boldsymbol{\mu})' \leq k + 2\},$$

or

$$E = \{\mathbf{x}: Q(\mathbf{x}) \leq k + 2\},$$

where

$$Q(\mathbf{x}) = (\mathbf{x} - \boldsymbol{\mu})\boldsymbol{\Sigma}^{-1}(\mathbf{x} - \boldsymbol{\mu})'.$$

In the 1-dimensional case, for a distribution having mean μ and variance σ^2, this ellipsoid is simply the interval $[\mu - 3^{1/2}\sigma, \mu + 3^{1/2}\sigma]$.

The volume of any ellipsoid

$$\{\mathbf{x}: Q(\mathbf{x}) \leq c\},$$

where $c > 0$, is found (see Cramér (1946), Section 11.2) to be

$$\frac{\pi^{\frac{1}{2}k} c^{\frac{1}{2}k} |\boldsymbol{\Sigma}|^{1/2}}{\Gamma(\frac{1}{2}k + 1)}.$$

Thus the determinant $|\boldsymbol{\Sigma}|$ plays in k-dimensions the role played by σ^2 in one dimension and so is called the *generalized variance*.

We may compare two k-dimensional distributions having the same mean $\boldsymbol{\mu}$ by comparing their concentration ellipsoids. If, however, we compare only the *volumes* of these ellipsoids, then it is equivalent to compare the generalized variances.

4.1.2 Application to Estimation: Confidence Ellipsoids and Asymptotic Relative Efficiency

Consider now the context of estimation of a k-dimensional parameter $\boldsymbol{\theta} = (\theta_1, \ldots, \theta_k)$ by $\hat{\boldsymbol{\theta}}_n = (\hat{\theta}_{n1}, \ldots, \hat{\theta}_{nk})$, where $\hat{\boldsymbol{\theta}}_n$ is $AN(\boldsymbol{\theta}, n^{-1}\boldsymbol{\Sigma}_{\boldsymbol{\theta}})$, with $\boldsymbol{\Sigma}_{\boldsymbol{\theta}}$ nonsingular. An ellipsoidal confidence region for $\boldsymbol{\theta}$ is given by

$$\begin{aligned} E_n &= \{\boldsymbol{\theta}: n(\hat{\boldsymbol{\theta}}_n - \boldsymbol{\theta})\boldsymbol{\Sigma}_{\hat{\boldsymbol{\theta}}_n}^{-1}(\hat{\boldsymbol{\theta}}_n - \boldsymbol{\theta})' \le c\} \\ &= \{\boldsymbol{\theta}: Q(n^{1/2}(\hat{\boldsymbol{\theta}}_n - \boldsymbol{\theta}), \boldsymbol{\Sigma}_{\hat{\boldsymbol{\theta}}_n}^{-1}) \le c\}, \end{aligned}$$

where

$$Q(\boldsymbol{\Delta}, \mathbf{C}) = \boldsymbol{\Delta}\mathbf{C}\boldsymbol{\Delta}'$$

and it is assumed that $\boldsymbol{\Sigma}_{\hat{\boldsymbol{\theta}}_n}^{-1}$ is defined. Assuming further that

$$\boldsymbol{\Sigma}_{\hat{\boldsymbol{\theta}}_n}^{-1} \xrightarrow{p_{\boldsymbol{\theta}}} \boldsymbol{\Sigma}_{\boldsymbol{\theta}}^{-1},$$

it follows (why?) that

$$Q(n^{1/2}(\hat{\boldsymbol{\theta}}_n - \boldsymbol{\theta}), \boldsymbol{\Sigma}_{\hat{\boldsymbol{\theta}}_n}^{-1}) - Q(n^{1/2}(\hat{\boldsymbol{\theta}}_n - \boldsymbol{\theta}), \boldsymbol{\Sigma}_{\boldsymbol{\theta}}^{-1}) \xrightarrow{p_{\boldsymbol{\theta}}} 0.$$

Consequently, by Example 3.5A, we have

$$Q(n^{1/2}(\hat{\boldsymbol{\theta}}_n - \boldsymbol{\theta}), \boldsymbol{\Sigma}_{\hat{\boldsymbol{\theta}}_n}^{-1}) \xrightarrow{d_{\boldsymbol{\theta}}} \chi_k^2.$$

Therefore, if $c = c_\alpha$ is chosen so that $P(\chi_k^2 > c_\alpha) = \alpha$, we have

$$P_{\boldsymbol{\theta}}(\boldsymbol{\theta} \in E_n) = P_{\boldsymbol{\theta}}(Q(n^{1/2}(\hat{\boldsymbol{\theta}}_n - \boldsymbol{\theta}), \boldsymbol{\Sigma}_{\hat{\boldsymbol{\theta}}_n}^{-1}) \le c_\alpha) \to P(\chi_k^2 \le c_\alpha) = 1 - \alpha,$$

as $n \to \infty$, so that E_n represents an ellipsoidal confidence region (*confidence ellipsoid*) for $\boldsymbol{\theta}$ having limiting confidence coefficient $1 - \alpha$ as $n \to \infty$.

One approach toward comparison of two such estimation procedures is to compare the volumes of the corresponding confidence ellipsoids, for a specified value of the limiting confidence coefficient. Such a comparison reduces to comparison of the generalized variances of the asymptotic multivariate normal distributions involved and is independent of the choice of confidence coefficient. This is seen as follows. Let us compare the sequences $\{\hat{\boldsymbol{\theta}}_n^{(1)}\}$ and $\{\hat{\boldsymbol{\theta}}_n^{(2)}\}$, where

$$\hat{\boldsymbol{\theta}}_n^{(i)} \quad \text{is} \quad AN(\boldsymbol{\theta}, n^{-1}\boldsymbol{\Sigma}_{\boldsymbol{\theta}}^{(i)}),$$

$$\boldsymbol{\Sigma}_{\hat{\boldsymbol{\theta}}_n^{(i)}}^{(i)} \to \boldsymbol{\Sigma}_{\boldsymbol{\theta}}^{(i)},$$

and

$$(\boldsymbol{\Sigma}_{\hat{\boldsymbol{\theta}}_n^{(i)}}^{(i)})^{-1} \xrightarrow{p_{\boldsymbol{\theta}}} (\boldsymbol{\Sigma}_{\boldsymbol{\theta}}^{(i)})^{-1}$$

for $i = 1, 2$. Then the corresponding confidence ellipsoids

$$E_n^{(i)} = \{\boldsymbol{\theta}: Q(n^{1/2}(\hat{\boldsymbol{\theta}}_n^{(i)} - \boldsymbol{\theta}), (\boldsymbol{\Sigma}_{\hat{\boldsymbol{\theta}}_n^{(i)}}^{(i)})^{-1}) \le c_\alpha\}, \qquad i = 1, 2,$$

each have asymptotic confidence coefficient $1 - \alpha$ and, by **4.1.1**, have volumes

$$\frac{\pi^{(1/2)k}(c_\alpha/n)^{(1/2)k}|\boldsymbol{\Sigma}_{\boldsymbol{\theta}_n^{(i)}}^{(i)}|^{1/2}}{\Gamma(\frac{1}{2}k + 1)}, \qquad i = 1, 2.$$

It follows that the ratio of sample sizes n_2/n_1 at which $\hat{\boldsymbol{\theta}}_{n_1}^{(1)}$ and $\hat{\boldsymbol{\theta}}_{n_2}^{(2)}$ perform "equivalently" (i.e., have confidence ellipsoids whose volumes are asymptotically equivalent "in probability") satisfies

$$\left(\frac{n_2}{n_1}\right)^k \to \frac{|\boldsymbol{\Sigma}_{\boldsymbol{\theta}}^{(2)}|}{|\boldsymbol{\Sigma}_{\boldsymbol{\theta}}^{(1)}|}.$$

Hence a numerical measure of the *asymptotic relative efficiency* of $\{\hat{\boldsymbol{\theta}}_n^{(2)}\}$ with respect to $\{\hat{\boldsymbol{\theta}}_n^{(1)}\}$ is given by

$$\left(\frac{|\boldsymbol{\Sigma}_{\boldsymbol{\theta}}^{(2)}|}{|\boldsymbol{\Sigma}_{\boldsymbol{\theta}}^{(1)}|}\right)^{1/k}.$$

Note that the dimension k is involved in this measure. Note also that we arrive at the same measure if we compare $\{\hat{\boldsymbol{\theta}}_n^{(1)}\}$ and $\{\hat{\boldsymbol{\theta}}_n^{(2)}\}$ on the basis of the *concentration* eillipsoids of the respective asymptotic multivariate normal distributions.

By the preceding approach, we have that $\{\hat{\boldsymbol{\theta}}_n^{(1)}\}$ is *better* than $\{\hat{\boldsymbol{\theta}}_n^{(2)}\}$, in the sense of asymptotically smaller confidence ellipsoids (or concentration ellipsoids), if and only if

(1) $$|\boldsymbol{\Sigma}_{\boldsymbol{\theta}}^{(1)}| \le |\boldsymbol{\Sigma}_{\boldsymbol{\theta}}^{(2)}|.$$

A closely related, but stronger, form of comparison is based on the condition

(2) $$\boldsymbol{\Sigma}_{\boldsymbol{\theta}}^{(2)} - \boldsymbol{\Sigma}_{\boldsymbol{\theta}}^{(1)} \text{ nonnegative definite,}$$

or equivalently (see Rao (1973), p. 70, Problem 9),

(2′) $$(\boldsymbol{\Sigma}_{\boldsymbol{\theta}}^{(1)})^{-1} - (\boldsymbol{\Sigma}_{\boldsymbol{\theta}}^{(2)})^{-1} \text{ nonnegative definite,}$$

or equivalently

(2″) $$\mathbf{x}\boldsymbol{\Sigma}_{\boldsymbol{\theta}}^{(1)}\mathbf{x}' \le \mathbf{x}\boldsymbol{\Sigma}_{\boldsymbol{\theta}}^{(2)}\mathbf{x}', \qquad \text{all } \mathbf{x}.$$

Condition (2) is thus a condition for the asymptotic distribution of $\hat{\boldsymbol{\theta}}_n^{(1)}$ to possess a concentration ellipsoid *contained entirely within* that of the asymptotic distribution of $\hat{\boldsymbol{\theta}}_n^{(2)}$. Note that (2) implies (1).

Under certain regularity conditions, there exists a "best" matrix in the sense of condition (2). This is the topic of **4.1.3**.

4.1.3 The Classical Notion of Asymptotic Efficiency; the Information Inequality

We now introduce a definition of *asymptotic efficiency* which corresponds to the notion of optimal concentration ellipsoid, as discussed in **4.1.2**. Let $X_1, \ldots, X_n$ denote a sample of independent observations from a distribution $F_{\boldsymbol{\theta}}$ belonging to a family $\mathscr{F} = \{F_{\boldsymbol{\theta}}, \boldsymbol{\theta} \in \boldsymbol{\Theta}\}$, where $\boldsymbol{\theta} = (\theta_1, \ldots, \theta_k)$ and $\boldsymbol{\Theta} \subset R^k$. Suppose that the distributions $F_{\boldsymbol{\theta}}$ possess densities or mass functions $f(x; \boldsymbol{\theta})$. Under regularity conditions on $\mathscr{F}$, the matrix

$$\mathbf{I}_{\boldsymbol{\theta}} = \left[E_{\boldsymbol{\theta}}\left\{\frac{\partial \log f(X; \boldsymbol{\theta})}{\partial \theta_i} \cdot \frac{\partial \log f(X; \boldsymbol{\theta})}{\partial \theta_j}\right\}\right]_{k \times k}$$

is defined and is positive definite. Let $\hat{\boldsymbol{\theta}}_n = (\hat{\theta}_{n1}, \ldots, \hat{\theta}_{nk})$ denote an estimator of $\boldsymbol{\theta}$ based on $X_1, \ldots, X_n$. Under regularity conditions on the class of estimators $\hat{\boldsymbol{\theta}}_n$ under consideration, it may be asserted that if $\hat{\boldsymbol{\theta}}_n$ is $AN(\boldsymbol{\theta}, n^{-1}\boldsymbol{\Sigma}_{\boldsymbol{\theta}})$, then the condition

$$(*) \qquad \boldsymbol{\Sigma}_{\boldsymbol{\theta}} - \mathbf{I}_{\boldsymbol{\theta}}^{-1} \text{ is nonnegative definite}$$

must hold. This condition means that the asymptotic distribution of $\hat{\boldsymbol{\theta}}_n$ (suitably normalized) has concentration ellipsoid wholly containing that of the distribution $N(\boldsymbol{\theta}, \mathbf{I}_{\boldsymbol{\theta}}^{-1})$. In this respect, an estimator $\hat{\boldsymbol{\theta}}_n$ which is $AN(\boldsymbol{\theta}, \mathbf{I}_{\boldsymbol{\theta}}^{-1})$ is "optimal." (Such an estimator need not exist.) These considerations are developed in detail in Cramér (1946) and Rao (1973).

The following definition is thus motivated. An estimator $\hat{\boldsymbol{\theta}}_n$ which is $AN(\boldsymbol{\theta}, n^{-1}\mathbf{I}_{\boldsymbol{\theta}}^{-1})$ is called *asymptotically efficient*, or *best asymptotically normal* (*BAN*). Under suitable regularity conditions, an asymptotically efficient estimate exists. One approach toward finding such estimates is the method of maximum likelihood, treated in Section 4.2. Other approaches toward asymptotically efficient estimation are included in the methods considered in Section 4.3.

In the case $k = 1$, the condition (*) asserts that if $\hat{\boldsymbol{\theta}}_n$ is $AN(\theta, n^{-1}\sigma^2)$, then

$$(**) \qquad \sigma^2 \geq \frac{1}{I_\theta} = \frac{1}{E\left\{\left(\dfrac{\partial \log f(X; \theta)}{\partial \theta}\right)^2\right\}}.$$

This lower bound to the parameter σ^2 in the asymptotic normality of $\hat{\theta}_n$ is known as the "*Cramer–Rao lower bound.*" The quantity I_θ is known as the "*Fisher information,*" so that (**) represents a so-called "information inequality." Likewise, for the general k-dimensional case, $\mathbf{I}_{\boldsymbol{\theta}}$ is known as the *information matrix* and (*) is referred to as the *information inequality*.

Example. Consider the family $\mathscr{F} = \{N(\theta, \sigma_0^2), \theta \in R\}$. Writing

$$f(x; \theta) = (2\pi)^{-1/2}\sigma_0^{-1} \exp\left[-\frac{1}{2}\left(\frac{x-\theta}{\sigma_0}\right)^2\right], \qquad -\infty < x < \infty,$$

we have

$$\frac{\partial \log f(x; \theta)}{\partial \theta} = \frac{x - \theta}{\sigma_0^2},$$

so that

$$E_\theta\left\{\left(\frac{\partial \log f(X; \theta)}{\partial \theta}\right)^2\right\} = E_\theta\left\{\left(\frac{X-\theta}{\sigma_0^2}\right)^2\right\} = \frac{\sigma_0^2}{\sigma_0^4} = \frac{1}{\sigma_0^2}.$$

Therefore, for estimation of the mean of a normal distribution with variance σ^2, any "regular" estimator $\hat{\theta}_n$ which is $AN(\theta, n^{-1}v)$ must satisfy $v \geq \sigma^2$. It is thus seen that, in particular, the sample mean $\bar{X}$ is asymptotically efficient whereas the sample median is not. However, $\bar{X}$ is not the only asymptotically efficient estimator in this problem. See Chapters 6, 7, 8 and 9. ■

4.1.4 Complements

(i) *Further discussion of the Cramér–Rao bound.* See Cramér (1946), Sections 32.3, 32.6, 32.7. Also, see Rao (1973), Sections 5a.2–5a.4, for information-theoretic interpretations and references to other results giving different bounds under different assumptions on $\mathscr{F}$ and $\hat{\boldsymbol{\theta}}_n$.

(ii) *Other notions of efficiency.* See Rao (1973), Section 5c.2.

(iii) *Asymptotic effective variance.* To avoid pathologies of "super-efficient" estimates, Bahadur (1967) introduces a quantity, "asymptotic effective variance," to replace asymptotic variance as a criterion.

4.2 ESTIMATION BY THE METHOD OF MAXIMUM LIKELIHOOD

We treat here an approach first suggested by C. F. Gauss, but first developed into a full-fledged methodology by Fisher (1912). Our treatment will be based on Cramér (1946). In **4.2.1** we define the method, and in **4.2.2** we characterize the asymptotic properties of estimates produced by the method.

4.2.1 The Method

Let $X_1, \ldots, X_n$ be I.I.D. with distribution $F_{\boldsymbol{\theta}}$ belonging to a family $\mathscr{F} = \{F_{\boldsymbol{\theta}}, \boldsymbol{\theta} \in \boldsymbol{\Theta}\}$, and suppose that the distributions $F_{\boldsymbol{\theta}}$ possess densities or mass functions $f(x; \boldsymbol{\theta})$. Assume $\boldsymbol{\Theta} \subset R^k$.

The *likelihood function* of the sample $X_1, \ldots, X_n$ is defined as

$$L(\boldsymbol{\theta}; X_1, \ldots, X_n) = \prod_{i=1}^{n} f(X_i; \boldsymbol{\theta}).$$

That is, the joint density (or mass function) of the observations is treated as a function of $\boldsymbol{\theta}$.

The method of *maximum likelihood* provides as estimate of $\boldsymbol{\theta}$ any value $\hat{\boldsymbol{\theta}}$ which maximizes L in $\boldsymbol{\Theta}$. (Equivalently, $\log L$ may be maximized if convenient for computations.)

Often the estimate $\hat{\boldsymbol{\theta}}$ may be obtained by solving the system of *likelihood equations*,

$$\left.\frac{\partial \log L}{\partial \theta_i}\right|_{\boldsymbol{\theta}=\hat{\boldsymbol{\theta}}} = 0, \qquad (i = 1, \ldots, k),$$

and confirming that the solution $\hat{\boldsymbol{\theta}}$ indeed maximizes L.

Remark. Obviously, the method may be formulated analogously without the I.I.D. assumption on $X_1, X_2, \ldots$. However, in our development of the asymptotic behavior of the maximum likelihood estimates, the I.I.D. assumption will be utilized crucially. ■

4.2.2 Consistency, Asymptotic Normality, and Asymptotic Efficiency of Maximum Likelihood Estimates

We shall show that, under regularity conditions on $\mathscr{F}$, the maximum likelihood estimates are *strongly consistent*, *asymptotically normal*, and *asymptotically efficient*. For simplicity, our treatment will be confined to the case of a 1-dimensional parameter. The multivariate extension will be indicated without proof. We also confine attention to the case that $f(x; \theta)$ is a density. The treatment for a mass function is similar.

Regularity Conditions on $\mathscr{F}$. Consider θ to be an open interval (not necessarily finite) in R. We assume:

($R1$) For each $\theta \in \Theta$, the derivatives

$$\frac{\partial \log f(x; \theta)}{\partial \theta}, \frac{\partial^2 \log f(x; \theta)}{\partial \theta^2}, \frac{\partial^3 \log f(x; \theta)}{\partial \theta^3}$$

exist, all x;

($R2$) For each $\theta_0 \in \Theta$, there exist functions $g(x)$, $h(x)$ and $H(x)$ (possibly depending on θ_0) such that for θ in a neighborhood $N(\theta_0)$ the relations

$$\left|\frac{\partial f(x; \theta)}{\partial \theta}\right| \le g(x), \qquad \left|\frac{\partial^2 f(x; \theta)}{\partial \theta^2}\right| \le h(x), \qquad \left|\frac{\partial^3 \log f(x; \theta)}{\partial \theta^3}\right| \le H(x)$$

hold, all x, and

$$\int g(x)dx < \infty, \qquad \int h(x)dx < \infty, \qquad E_\theta\{H(X)\} < \infty \qquad \text{for} \quad \theta \in N(\theta_0)$$

$(R3)$ For each $\theta \in \Theta$,

$$0 < E_\theta\left\{\left(\frac{\partial \log f(X;\theta)}{\partial\theta}\right)^2\right\} < \infty. \quad \blacksquare$$

Some interpretations of these conditions are as follows. Condition $(R1)$ insures that the function $\partial \log f(x;\theta)/\partial\theta$ has, for each x, a Taylor expansion as a function of θ. Condition $(R2)$ insures (*justify*) that $\int f(x;\theta)dx$ and $\int [\partial \log f(x;\theta)/\partial\theta]dx$ may be differentiated with respect to θ under the integral sign. Condition $(R3)$ states that the random variable $\partial \log f(X;\theta)/\partial\theta$ has finite positive variance (we shall see that the mean is 0).

Theorem. *Assume regularity conditions* (R1), (R2) *and* (R3) *on the family* $\mathscr{F}$. *Consider I.I.D. observations on* F_θ, *for* θ *an element of* Θ. *Then, with* P_θ-*probability* 1, *the likelihood equations admit a sequence of solutions* $\{\hat\theta_n\}$ *satisfying*

(i) *strong consistency*: $\hat\theta_n \to \theta$, $n \to \infty$;

(ii) *asymptotic normality and efficiency*:

$$\hat\theta_n \quad \textit{is} \quad \mathrm{AN}\left(\theta, \frac{1}{nE_\theta\{(\partial \log f(X;\theta)/\partial\theta)^2\}}\right).$$

PROOF. (modeled after Cramér (1946)) By (R1) and (R2) we have for λ in the neighborhood $N(\theta)$ a Taylor expansion of $\partial \log f(x;\lambda)/\partial\lambda$ about the point $\lambda = \theta$, as follows;

$$\frac{\partial \log f(x;\lambda)}{\partial\lambda} - \left.\frac{\partial \log f(x;\lambda)}{\partial\lambda}\right|_{\lambda=\theta} = (\lambda - \theta)\left.\frac{\partial^2 \log f(x;\lambda)}{\partial\lambda^2}\right|_{\lambda=\theta} + \tfrac{1}{2}\xi(\lambda-\theta)^2 H(x),$$

where $|\xi| < 1$. Therefore, putting

$$A_n = \frac{1}{n}\sum_{i=1}^n \left.\frac{\partial \log f(X_i;\lambda)}{\partial\lambda}\right|_{\lambda=\theta},$$

$$B_n = \frac{1}{n}\sum_{i=1}^n \left.\frac{\partial^2 \log f(X_i;\lambda)}{\partial\lambda^2}\right|_{\lambda=\theta},$$

and

$$C_n = \frac{1}{n}\sum_{i=1}^n H(X_i),$$

we have

$$(*) \qquad \frac{1}{n}\frac{\partial \log L(\lambda)}{\partial\lambda} = A_n + B_n(\lambda-\theta) + \tfrac{1}{2}\xi^* C_n(\lambda-\theta)^2,$$

where $|\xi^*| < 1$. (Note that the left-hand side of the likelihood equation, which is an average of I.I.D.'s depending on λ, thus becomes represented by an expression involving λ and averages of I.I.D.'s not depending on λ.)

By $(R1)$ and $(R2)$

$$\int \frac{\partial f(x;\lambda)}{\partial \lambda}dx = \frac{\partial}{\partial \lambda}\int f(x;\lambda)dx = \frac{\partial}{\partial \lambda}(1) = 0$$

and thus also

$$\int \frac{\partial^2 f(x;\lambda)}{\partial \lambda^2}dx = 0.$$

It follows that

$$E_\theta\left\{\frac{\partial \log f(X;\theta)}{\partial \theta}\right\} = \int \frac{1}{f(x;\theta)}\frac{\partial f(x;\theta)}{\partial \theta} f(x;\theta)dx = 0$$

and

$$\begin{aligned} E_\theta&\left\{\frac{\partial^2 \log f(x;\theta)}{\partial \theta^2}\right\} \\ &= \int\left[\frac{1}{f(x;\theta)}\frac{\partial^2 f(x;\theta)}{\partial \theta^2} - \left(\frac{1}{f(x;\theta)}\frac{\partial f(x;\theta)}{\partial \theta}\right)^2\right] f(x;\theta)dx \\ &= -E_\theta\left\{\left(\frac{\partial \log f(X;\theta)}{\partial \theta}\right)^2\right\}. \end{aligned}$$

By $(R3)$, the quantity

$$v_\theta = E_\theta\left\{\left(\frac{\partial \log f(X;\theta)}{\partial \theta}\right)^2\right\}$$

satisfies $0 < v_\theta < \infty$.

It follows that

(a) A_n is a mean of I.I.D.'s with mean 0 and variance v_θ;
(b) B_n is a mean of I.I.D.'s with mean $-v_\theta$;
(c) C_n is a mean of I.I.D.'s with mean $E_\theta\{H(X)\}$.

Therefore, by the SLLN (Theorem 1.8B),

$$A_n \xrightarrow{wp1} 0,\ B_n \xrightarrow{wp1} -v_\theta,\ C_n \xrightarrow{wp1} E_\theta\{H(X)\},$$

and, by the CLT (Theorem 1.9.1A),

$$A_n \quad \text{is} \quad AN(0, n^{-1}v_\theta).$$

Now let $\varepsilon > 0$ be given, such that $\varepsilon < v_\theta / E_\theta\{H(X)\}$ and such that the points $\lambda_1 = \theta - \varepsilon$ and $\lambda_2 = \theta + \varepsilon$ lie in $N(\theta)$, the neighborhood specified in condition $(R2)$. Then, by (*),

$$\left| \frac{1}{n} \frac{\partial \log L(\lambda)}{\partial \lambda}\bigg|_{\lambda_1} - v_\theta \varepsilon \right| \leq |A_n| + \varepsilon |B_n + v_\theta| + \tfrac{1}{2}\varepsilon^2 |C_n|$$

and

$$\left| \frac{1}{n} \frac{\partial \log L(\lambda)}{\partial \lambda}\bigg|_{\lambda_2} + v_\theta \varepsilon \right| \leq |A_n| + \varepsilon |B_n + v_\theta| + \tfrac{1}{2}\varepsilon^2 |C_n|.$$

By the strong convergences of A_n, B_n and C_n noted above, we have that with P_θ-probability 1 the right-hand side of each of the above inequalities becomes $\leq (\frac{3}{4}) v_\theta \varepsilon$ for all n sufficiently large. *For such n*, the interval

$$\left[\frac{1}{n} \frac{\partial \log L(\lambda)}{\partial \lambda}\bigg|_{\lambda_2}, \quad \frac{1}{n} \frac{\partial \log L(\lambda)}{\partial \lambda}\bigg|_{\lambda_1} \right]$$

thus contains the point 0 and hence, by the continuity of $\partial \log L(\lambda)/\partial \lambda$, the interval

$$[\theta - \varepsilon, \theta + \varepsilon] = [\lambda_1, \lambda_2]$$

contains a solution of the likelihood equation. In particular, it contains the solution

$$\hat{\theta}_{n\varepsilon} = \inf\left\{ \lambda \colon \theta - \varepsilon \leq \lambda \leq \theta + \varepsilon \text{ and } \frac{\partial \log L(\lambda)}{\partial \lambda} = 0 \right\}.$$

Before going further, let us verify that $\hat{\theta}_{n\varepsilon}$ is a proper random variable, that is, is measurable. Note that, for all $t \geq \theta - \varepsilon$,

$$\{\hat{\theta}_{n\varepsilon} > t\} = \left\{ \inf_{\theta - \varepsilon \leq \lambda \leq t} \frac{\partial \log L(\lambda)}{\partial \lambda} > 0 \right\} \cup \left\{ \sup_{\theta - \varepsilon \leq \lambda \leq t} \frac{\partial \log L(\lambda)}{\partial \lambda} < 0 \right\}.$$

Also, by continuity of $\partial \log L(\lambda)/\partial \lambda$ in $[\theta - \varepsilon, \theta + \varepsilon]$,

$$\inf_{\theta - \varepsilon \leq \lambda \leq t} \frac{\partial \log L(\lambda)}{\partial \lambda} = \inf_{\substack{\theta - \varepsilon \leq \lambda \leq t \\ \lambda \text{ rational}}} \frac{\partial \log L(\lambda)}{\partial \lambda}$$

and

$$\sup_{\theta - \varepsilon \leq \lambda \leq t} \frac{\partial \log L(\lambda)}{\partial \lambda} = \sup_{\substack{\theta - \varepsilon \leq \lambda \leq t \\ \lambda \text{ rational}}} \frac{\partial \log L(\lambda)}{\partial \lambda}.$$

Thus $\{\hat{\theta}_{n\varepsilon} > t\}$ is a measurable set.

Next let us obtain a sequence of solutions $\{\hat{\theta}_n\}$ not depending upon the choice of ε. For this, let us denote by $(\Omega, \mathscr{A}, P_\theta)$ the underlying probability space and let us express $\hat{\theta}_{n\varepsilon}$ explicitly as $\hat{\theta}_{n\varepsilon}(\omega)$. Our definition of $\hat{\theta}_{n\varepsilon}(\omega)$ required that n be *sufficiently large*, $n \geq N_\varepsilon(\omega)$, say, and that ω belong to a set Ω_ε having P_θ-probability 1. Let us now define

$$\Omega_0 = \bigcap_{k=1}^{\infty} \Omega_{1/k}.$$

Then $P_\theta(\Omega_0) = 1$ also. For the moment, confine attention to $\omega \in \Omega_0$. Here, without loss of generality, we may require that

$$N_1(\omega) \leq N_{1/2}(\omega) \leq N_{1/3}(\omega) \leq \cdots.$$

Hence, for $N_{1/k}(\omega) \leq n < N_{1/(k+1)}(\omega)$, we may define

$$\hat{\theta}_n(\omega) = \hat{\theta}_{n,\,1/k}(\omega),$$

for $k = 1, 2, \ldots$. And for $n < N_1(\omega)$, we set $\hat{\theta}_n(\omega) = 0$. Finally, for $\omega \notin \Omega_0$, we set $\hat{\theta}_n(\omega) \equiv 0$, all n. It is readily seen that $\{\hat{\theta}_n\}$ is a sequence of random variables which with P_θ-probability 1 satisfies:

(1) $\hat{\theta}_n$ is a solution of the likelihood equation for all n sufficiently large, and

(2) $\hat{\theta}_n \to \theta$, $n \to \infty$.

We have thus established strong consistency, statement (i) of the theorem. To obtain statement (ii), write

$$0 = \frac{1}{n} \frac{\partial \log L(\lambda)}{\partial \lambda}\bigg|_{\lambda = \hat{\theta}_n} = A_n + B_n(\hat{\theta}_n - \theta) + \tfrac{1}{2}\xi^* C_n(\hat{\theta}_n - \theta)^2,$$

which with P_θ-probability 1 is valid for all n sufficiently large. Therefore,

$$n^{1/2}(\hat{\theta}_n - \theta) - \frac{-n^{1/2} A_n}{B_n + \frac{1}{2}\xi^* C_n(\hat{\theta}_n - \theta)} \xrightarrow{wp1} 0.$$

Also, since $\hat{\theta}_n \xrightarrow{wp1} \theta$, we have $B_n + \frac{1}{2}\xi^* C_n(\hat{\theta}_n - \theta) \xrightarrow{wp1} -v_\theta$. Further, $n^{1/2} A_n \xrightarrow{d} N(0, v_\theta)$. Consequently, by Slutsky's Theorem,

$$n^{1/2}(\hat{\theta}_n - \theta) \xrightarrow{d} N(0, v_\theta^{-1}),$$

establishing statement (ii) of the theorem. ■

Multidimensional Generalization. For the case of several unknown parameters $\boldsymbol{\theta} = (\theta_1, \ldots, \theta_k)$, and under appropriate generalizations of the regularity conditions $(R1)$–$(R3)$, there exists a sequence $\{\hat{\boldsymbol{\theta}}_n\}$ of solutions to the likelihood equations such that $\hat{\boldsymbol{\theta}}_n \xrightarrow{wp1} \boldsymbol{\theta}$ and $\hat{\boldsymbol{\theta}}_n$ is $AN(\boldsymbol{\theta}, n^{-1}\mathbf{I}_{\boldsymbol{\theta}}^{-1})$, where $\mathbf{I}_{\boldsymbol{\theta}}$ is the information matrix defined in **4.1.3**. ■

Remarks. (i) *Other sequences of solutions.* The argument leading to statement (ii) of the theorem may be modified to handle any sequence $\{\hat{\theta}_n^*\}$ of solutions which are weakly consistent for θ. Therefore, if $\hat{\theta}_n^*$ is any solution of the likelihood equations satisfying $\hat{\theta}_n^* \xrightarrow{p_\theta} \theta$, then $\hat{\theta}_n^*$ is $AN(\theta, n^{-1}v_\theta^{-1})$ (Problem 4.P.3).

(ii) *Transformation of parameters.* It is readily seen that if we transform to new parameters $\boldsymbol{\beta} = (\beta_1, \ldots, \beta_r)$, where $\beta_i = g_i(\theta_1, \ldots, \theta_k)$, then the maximum likelihood estimate of $\boldsymbol{\beta}$ is given by the corresponding transformation of the maximum likelihood estimate $\hat{\boldsymbol{\theta}}_n$ of $\boldsymbol{\theta}$. Thus, under mild regularity conditions on the transformation, the consistency and asymptotic normality properties survive under the transformation.

(iii) "*Likelihood processes*" *associated with a sample.* See Rubin (1961).

(iv) *Regularity assumptions not involving differentiability.* See Wald (1949) for other assumptions yielding consistency of $\hat{\theta}_n$.

(v) *Iterative Solution of the Likelihood Equations.* The Taylor expansion appearing in the proof of the theorem is the basis for the following iterative approach. For an initial guess $\hat{\theta}_{n0}$, we have

$$0 = \frac{\partial \log L(\theta)}{\partial \theta} \doteq \frac{\partial \log L(\lambda)}{\partial \theta}\bigg|_{\lambda=\hat{\theta}_{n0}} - (\hat{\theta}_{n0} - \theta)\,\frac{\partial^2 \log L(\lambda)}{\partial \lambda^2}\bigg|_{\lambda=\hat{\theta}_{n0}}.$$

This yields the next iterate

$$\hat{\theta}_{n1} = \hat{\theta}_{n0} - \frac{\left.\dfrac{\partial \log L(\lambda)}{\partial \lambda}\right|_{\lambda=\hat{\theta}_{n0}}}{\left.\dfrac{\partial^2 \log L(\lambda)}{\partial \lambda^2}\right|_{\lambda=\hat{\theta}_{n0}}}.$$

The process is continued until the sequence $\hat{\theta}_{n0}, \hat{\theta}_{n1}, \hat{\theta}_{n2}, \ldots$ has converged to a solution $\hat{\theta}_n$. A modification of this procedure is to replace

$$\frac{\partial^2 \log L(\lambda)}{\partial \lambda^2}\bigg|_{\lambda=\hat{\theta}_{ni}}$$

by its expected value, in order to simplify computations. This version is called *scoring*, and the quantity

$$\frac{\partial \log L(\lambda)}{\partial \lambda}\bigg|_{\lambda=\hat{\theta}_{ni}}$$

is called the "*efficient score.*"

(vi) *Further reading.* For techniques of application, see Rao (1973), Sections 5f, 5g and 8a. ■

4.3 OTHER APPROACHES TOWARD ESTIMATION

Here we discuss the method of moments (**4.3.1**), minimization methods (**4.3.2**), and statistics of special form (**4.3.3**).

4.3.1 The Method of Moments

Consider a sample $X_1, \ldots, X_n$ from a distribution $F_{\boldsymbol{\theta}}$ of known form but with unknown parameter $\boldsymbol{\theta} = (\theta_1, \ldots, \theta_k)$ to be estimated. The *method of moments* consists of producing estimates of $\theta_1, \ldots, \theta_k$ by first estimating the distribution $F_{\boldsymbol{\theta}}$ by estimating its moments. This is carried out by equating an appropriate number of sample moments to the corresponding population moments, the latter being expressed as functions of $\boldsymbol{\theta}$. The estimates of $\theta_1, \ldots, \theta_k$ are then obtained by inverting the relationships with the moments.

For example, a $N(\mu, \sigma^2)$ distribution may be estimated by writing $\sigma^2 = \alpha_2 - \mu^2$ and estimating μ by $\bar{X}$ and α_2 by $a_2 = n^{-1} \sum_1^n X_i^2$. This leads to estimation of the $N(\mu, \sigma^2)$ distribution by $N(\bar{X}, s^2)$, where $s^2 = a_2 - \bar{X}^2$.

Of course, in general, the parameters $\theta_1, \ldots, \theta_k$ need not be such simple functions of the moments of $F_{\boldsymbol{\theta}}$ as in the preceding example.

The method of moments, introduced by Pearson (1894), has enjoyed wide appeal because of its naturalness and expediency. Further, typically the parameters $\theta_1, \ldots, \theta_k$ are well-behaved functions of the population moments, so that the estimates given by the corresponding functions of the sample moments are *consistent* and *asymptotically normal.* Indeed, as discussed in **3.4.1**, the asymptotic variances are of the form c/n.

On the other hand, typically the method-of-moments estimators are *not* asymptotically efficient (an exception being the example considered above). Thus various authors have introduced schemes for *modified* method-of-moments estimators possessing enhanced efficiency. For example, a relatively simple approach is advanced by Soong (1969), whose "combined moment estimators" for parameters $\theta_1, \ldots, \theta_k$ are optimal linear combinations (recall **3.4.3**) of simple moment estimators. Soong also discusses related earlier work of other investigators and provides for various examples the asymptotic efficiency curves of several estimators.

Further reading on the method of moments is available in Cramér (1946), Section 33.1.

4.3.2 Minimization Methods; *M*-Estimation

A variety of estimation methods are based on *minimization* of some function of the observations $\{X_i\}$ and the unknown parameter $\boldsymbol{\theta}$. For example, if θ is a location parameter for the observations $X_1, \ldots, X_n$, the "*least-squares estimator*" of θ is found by minimizing

$$d(\theta; X_1, \ldots, X_n) = \sum_{i=1}^{n} (X_i - \theta)^2,$$

considered as a function of θ. Similarly, the "*least-absolute-values estimator*" of θ is given by minimizing $\sum_1^n |X_i - \theta|$. (These solutions are found to be the sample mean and sample median, respectively.) Likewise, the *maximum likelihood method* of Section 4.2 may be regarded as an approach of this type.

In Section 4.5 we shall consider approaches of this type in connection with product-multinomial data. There the function to be minimized will be a *distance function* $d(g(\boldsymbol{\theta}), \hat{g})$ between a parametric function $g(\boldsymbol{\theta})$ and an estimator $\hat{g}$ of $g(\boldsymbol{\theta})$ based on the data. Several distance functions will be considered.

Typically, the problem of minimizing a function of data and parameter reduces to a problem involving solution of a system of equations for an estimator $\hat{\boldsymbol{\theta}}$. In Chapter 7 we treat in general the properties of statistics given as solutions of equations. Such statistics are termed "*M-statistics.*"

A related approach toward estimation is to consider a particular class of estimators, for example those obtained as solutions of equations, and, within this class, to select the estimator for which a *nonrandom* function of $\boldsymbol{\theta}$ and $\hat{\boldsymbol{\theta}}$ is minimized. For example, the mean square error $E(\hat{\theta} - \theta)^2$ might be minimized. The method of maximum likelihood may also be derived by this approach. See also **4.3.3** below.

4.3.3 Statistics of Special Form; *L*-Estimation and *R*-Estimation

As mentioned above, the principle of minimization typically leads to the class of M-estimates (having the special form of being given as solutions of equations). On the other hand, it is sometimes of interest to restrict attention to some class of statistics quite different (perhaps more appealing, or simpler) in form, and within the given class to select an estimator which optimizes some specfied criterion. The criterion might be to minimize $E(\hat{\theta} - \theta)^2$, or $E|\hat{\theta} - \theta|$, for example.

A case of special interest consists of *linear functions of order statistics*, which we have considered already in Sections 2.4 and 3.6. A general treatment of these "*L-statistics*" is provided in Chapter 8, including discussion of *efficient* estimation via L-estimates.

Another case of special interest concerns estimators which are expressed as *functions of the ranks* of the observations. These "*R-statistics*" are treated in Chapter 9, and again the question of *efficient* estimation is considered.

4.4 HYPOTHESIS TESTING BY LIKELIHOOD METHODS

Here we shall consider hypothesis testing and shall treat three special test statistics, each based on the maximum likelihood method. A reason for involving the maximum likelihood method is to exploit the asymptotic efficiency. Thus other asymptotically efficient estimates, where applicable, could be used in the role of the maximum likelihood estimates.

We formulate the hypothesis testing problem in **4.4.1** and develop certain preliminaries in **4.4.2**. For the case of a *simple* null hypothesis, the relevant test statistics are formulated in **4.4.3** and their null-hypothesis asymptotic distributions are derived. Also, extension to "local" alternatives is considered. The case of a *composite* null hypothesis is treated in **4.4.4**.

4.4.1 Formulation of the Problem

Let $X_1, \ldots, X_n$ be I.I.D. with distribution $F_{\boldsymbol{\theta}}$ belonging to a family $\mathscr{F} = \{F_{\boldsymbol{\theta}}, \boldsymbol{\theta} \in \boldsymbol{\Theta}\}$, where $\boldsymbol{\Theta} \subset R^k$. Let the distributions $F_{\boldsymbol{\theta}}$ possess densities or mass functions $f(x; \boldsymbol{\theta})$. Assume that the information matrix

$$\mathbf{I}_{\boldsymbol{\theta}} = \left[E_{\boldsymbol{\theta}}\left\{ \frac{\partial \log f(X; \boldsymbol{\theta})}{\partial \theta_i} \frac{\partial \log f(X; \boldsymbol{\theta})}{\partial \theta_j} \right\} \right]_{k \times k}$$

exists and is positive definite.

A null hypothesis H_0 (to be tested) will be specified as a subset $\boldsymbol{\Theta}_0$ of $\boldsymbol{\Theta}$, where $\boldsymbol{\Theta}_0$ is determined by a set of $r(\leq k)$ restrictions given by equations

$$R_i(\boldsymbol{\theta}) = 0, \qquad 1 \leq i \leq r.$$

In the case of a *simple* hypothesis $H_0: \boldsymbol{\theta} = \boldsymbol{\theta}_0$, we have $\boldsymbol{\Theta}_0 = \{\boldsymbol{\theta}_0\}$, and the functions $R_i(\boldsymbol{\theta})$ may be taken to be

$$R_i(\boldsymbol{\theta}) = \theta_i - \theta_{0i}, \qquad 1 \leq i \leq k.$$

In the case of a *composite* hypothesis, the set $\boldsymbol{\Theta}_0$ contains more than one element and we necessarily have $r < k$. For example, for $k = 3$, we might have $H_0: \boldsymbol{\theta} \in \boldsymbol{\Theta}_0 = \{\boldsymbol{\theta} = (\theta_1, \theta_2, \theta_3): \theta_1 = \theta_{01}\}$. In this case $r = 1$ and the function $R_1(\boldsymbol{\theta})$ may be taken to be

$$R_1(\boldsymbol{\theta}) = \theta_1 - \theta_{01}.$$

4.4.2 Preliminaries

Throughout we assume the *regularity conditions* and *results* given in **4.2.2**, explicitly in connection with Theorem 4.2.2 and implicitly in connection with its multidimensional extension. Define for $\boldsymbol{\theta} = (\theta_1, \ldots, \theta_k)$, the vectors

$$\mathbf{a}_{n\boldsymbol{\theta}} = \left(\frac{1}{n} \sum_{i=1}^{n} \frac{\partial \log f(X_i; \boldsymbol{\theta})}{\partial \theta_1}, \ldots, \frac{1}{n} \sum_{i=1}^{n} \frac{\partial \log f(X_i; \boldsymbol{\theta})}{\partial \theta_k} \right)$$

and

$$\mathbf{d}_{n\theta} = \hat{\boldsymbol{\theta}}_n - \boldsymbol{\theta} = (\hat{\theta}_{n1} - \theta_1, \ldots, \hat{\theta}_{nk} - \theta_k),$$

where $\hat{\boldsymbol{\theta}}_n = (\hat{\theta}_{n1}, \ldots, \hat{\theta}_{nk})$ denotes a consistent, asymptotically normal, and asymptotically efficient sequence of solutions of the likelihood equations, as given by Theorem 4.2.2 (multidimensional extension).

Lemma A. *Let* $X_1, X_2, \ldots$ *be I.I.D. with distribution* $F_{\boldsymbol\theta}$. *Then (under appropriate regularity conditions)*

(i) $n^{1/2}\mathbf{a}_{n\boldsymbol\theta} \xrightarrow{d} N(\mathbf{0}, \mathbf{I}_{\boldsymbol\theta})$;

(ii) $n^{1/2}\mathbf{d}_{n\boldsymbol\theta} \xrightarrow{d} N(\mathbf{0}, \mathbf{I}_{\boldsymbol\theta}^{-1})$;

(iii) $n\mathbf{a}_{n\boldsymbol\theta}\mathbf{I}_{\boldsymbol\theta}^{-1}\mathbf{a}'_{n\boldsymbol\theta} \xrightarrow{d} \chi^2_k$;

(iv) $n\mathbf{d}_{\boldsymbol\theta}\mathbf{I}_{\boldsymbol\theta}\mathbf{d}'_{n\boldsymbol\theta} \xrightarrow{d} \chi^2_k$.

PROOF. (i) follows directly from the multivariate Lindeberg–Levy CLT; (ii) is simply the multidimensional version of Theorem 4.2.2; (iii) and (iv) follow from (i) and (ii), respectively, by means of Example 3.5A. ■

It is seen from (i) and (ii) that the vectors

$$n^{1/2}\mathbf{a}_{n\boldsymbol\theta},\ n^{1/2}\mathbf{d}_{n\boldsymbol\theta}\mathbf{I}_{\boldsymbol\theta}$$

have the same limit distribution namely $N(\mathbf{0}, \mathbf{I}_{\boldsymbol\theta})$. In fact, there holds the following stronger relationship.

Lemma B. *Let* $X_1, X_2, \ldots$ *be I.I.D. with distribution* $F_{\boldsymbol\theta}$. *Then (under appropriate regularity conditions)*

$$n^{1/2}(\mathbf{a}_{n\boldsymbol\theta} - \mathbf{d}_{n\boldsymbol\theta}\mathbf{I}_{\boldsymbol\theta}) \xrightarrow{p} \mathbf{0}.$$

PROOF. Noting that

$$\frac{1}{n}\sum_{m=1}^{n} \frac{\partial \log f(X_m; \boldsymbol\theta)}{\partial \theta_i}\bigg|_{\boldsymbol\theta = \hat{\boldsymbol\theta}_n} = 0,$$

we obtain by Theorem 1.12B the Taylor expansion

$$0 - \frac{1}{n}\sum_{m=1}^{n} \frac{\partial \log f(X_m; \boldsymbol\theta)}{\partial \theta_i} = \sum_{j=1}^{k}\frac{1}{n}\sum_{m=1}^{n}\frac{\partial^2 \log f(X_m; \boldsymbol\theta)}{\partial\theta_i\,\partial\theta_j}(\hat\theta_{nj} - \theta_j)$$
$$+ \frac{1}{2}\sum_{k=1}^{k}\sum_{l=1}^{k}\frac{1}{n}\sum_{m=1}^{n}\frac{\partial^3 \log f(X_m; \boldsymbol\theta)}{\partial\theta_i\,\partial\theta_j\,\partial\theta_l}\bigg|_{\boldsymbol\theta=\boldsymbol\theta_n^*}\cdot(\hat\theta_{nj} - \theta_j)(\hat\theta_{nl} - \theta_l),$$

where $\boldsymbol\theta_n^*$ lies on the line joining $\boldsymbol\theta$ and $\hat{\boldsymbol\theta}_n$. From the regularity conditions (extended to the multidimensional parameter case), and from the convergence in distribution of the normalized maximum likelihood estimates, we see that the second term on the right-hand side may be characterized as $O_p(n^{-1/2})$. Thus we have, for each $i = 1, \ldots, k$,

$$n^{1/2}\left[-\frac{1}{n}\sum_{m=1}^{n}\frac{\partial \log f(X_m; \boldsymbol\theta)}{\partial\theta_i} - \sum_{j=1}^{k}\frac{1}{n}\sum_{m=1}^{n}\frac{\partial^2 \log f(X_m; \boldsymbol\theta)}{\partial\theta_i\,\partial\theta_j}(\hat\theta_{nj} - \theta_j)\right]$$
$$= o_p(1).$$

That is,

$$n^{1/2}(-\mathbf{a}_{n\theta} - \mathbf{d}_{n\theta}\mathbf{J}_{n\theta}) \xrightarrow{p} 0,$$

where

$$\mathbf{J}_{n\theta} = \left[\frac{1}{n}\sum_{m=1}^{n} \frac{\partial^2 \log f(X_m;\boldsymbol{\theta})}{\partial\theta_i\,\partial\theta_j}\right]_{k\times k}.$$

Thus

$$n^{1/2}(\mathbf{a}_{n\theta} - \mathbf{d}_{n\theta}\mathbf{I}_{\theta}) = n^{1/2}\mathbf{d}_{n\theta}(-\mathbf{I}_{\theta} - \mathbf{J}_{n\theta}) + o_p(1).$$

As an exercise, show that the convergence and equality

$$\mathbf{J}_{n\theta} \xrightarrow{p} E_{\theta}\left\{\frac{\partial^2 \log f(X;\boldsymbol{\theta})}{\partial\theta_i\,\partial\theta_j}\right\}$$

$$= -E_{\theta}\left\{\frac{\partial \log f(X;\boldsymbol{\theta})}{\partial\theta_i}\frac{\partial \log f(X;\boldsymbol{\theta})}{\partial\theta_j}\right\} = -\mathbf{I}_{\theta}$$

hold. We thus have

$$n^{1/2}(\mathbf{a}_{n\theta} - \mathbf{d}_{n\theta}\mathbf{I}_{\theta}) = n^{1/2}\mathbf{d}_{n\theta}\,o_p(1) + o_p(1) = o_p(1),$$

since $n^{1/2}\mathbf{d}_{n\theta}$ converges in distribution. ■

We further define

$$l_n(\boldsymbol{\theta}) = \log L(\boldsymbol{\theta}; X_1, \ldots, X_n) = \sum_{i=1}^{n} \log f(X_i;\boldsymbol{\theta}).$$

Lemma C. *Let* $X_1, X_2, \ldots$ *be I.I.D. with distribution* F_{θ}. *Then (under appropriate regularity conditions)*

(i) $[l_n(\hat{\boldsymbol{\theta}}_n) - l_n(\boldsymbol{\theta})] - \frac{1}{2}n\mathbf{d}_{n\theta}\mathbf{I}_{\theta}\mathbf{d}'_{n\theta} \xrightarrow{p} 0;$

(ii) $2[l_n(\hat{\boldsymbol{\theta}}_n) - l_n(\boldsymbol{\theta})] \xrightarrow{d} \chi^2_k.$

PROOF. (ii) is a direct consequence of (i) and Lemma A(iv). It remains to prove (i). By an argument similar to that of Lemma B, we have

$$l_n(\boldsymbol{\theta}) - l_n(\hat{\boldsymbol{\theta}}_n) = \frac{1}{2}\sum_{i=1}^{k}\sum_{j=1}^{k} \left.\frac{\partial^2 l_n(\boldsymbol{\theta})}{\partial\theta_i\,\partial\theta_j}\right|_{\boldsymbol{\theta}=\hat{\boldsymbol{\theta}}_n}(\theta_i - \hat{\theta}_{ni})(\theta_j - \hat{\theta}_{nj}) + o_p(1)$$

$$= \tfrac{1}{2}n\mathbf{d}_{n\theta}\mathbf{I}_{\theta}\mathbf{d}'_{n\theta}$$

$$+ \tfrac{1}{2}n\mathbf{d}_{n\theta}\left[\frac{1}{n}\sum_{m=1}^{n}\left.\frac{\partial^2 \log f(X_m;\boldsymbol{\theta})}{\partial\theta_i\,\partial\theta_j}\right|_{\boldsymbol{\theta}=\hat{\boldsymbol{\theta}}_n} - \mathbf{I}_{\theta}\right]\mathbf{d}'_{n\theta} + o_p(1)$$

$$= \tfrac{1}{2}n\mathbf{d}_{n\theta}\mathbf{I}_{\theta}\mathbf{d}'_{n\theta} + o_p(1).$$ ■

4.4.3 Test Statistics for a Simple Null Hypothesis

Consider testing $H_0: \boldsymbol{\theta} = \boldsymbol{\theta}_0$.

A "likelihood ratio" statistic,

$$\Lambda_n = \frac{L(\boldsymbol{\theta}_0)}{\sup_{\boldsymbol{\theta} \in \Theta} L(\boldsymbol{\theta})},$$

was introduced by Neyman and Pearson (1928). Clearly, Λ_n takes values in the interval [0, 1] and H_0 is to be rejected for sufficiently small values of Λ_n. Equivalently, the test may be carried out in terms of the statistic

$$\lambda_n = -2 \log \Lambda_n,$$

which turns out to be more convenient for asymptotic considerations.

A second statistic,

$$W_n = n\mathbf{d}_{n\boldsymbol{\theta}_0} \mathbf{I}_{\hat{\boldsymbol{\theta}}_n} \mathbf{d}'_{n\boldsymbol{\theta}_0},$$

was introduced by Wald (1943).

A third statistic,

$$V_n = n\mathbf{a}_{n\boldsymbol{\theta}_0} \mathbf{I}_{\boldsymbol{\theta}_0}^{-1} \mathbf{a}_{n\boldsymbol{\theta}_0},$$

was introduced by Rao (1947).

The three statistics differ somewhat in computational features. Note that Rao's statistic does not require explicit computation of the maximum likelihood estimates. Nevertheless all three statistics have the same limit chi-squared distribution under the null hypothesis:

Theorem. *Under* H_0, *the statistics* λ_n, W_n, *and* V_n *each converge in distribution to* χ_k^2.

PROOF. The result for λ_n follows by observing that

$$\lambda_n = 2[l_n(\hat{\boldsymbol{\theta}}_n) - l_n(\boldsymbol{\theta}_0)]$$

and applying Lemma 4.4.2C (ii). (It is assumed that the solution $\hat{\boldsymbol{\theta}}_n$ of the likelihood equations indeed maximizes the likelihood function.) The result for W_n follows from Lemma 4.4.2A (iv) and the fact that $\mathbf{I}_{\hat{\boldsymbol{\theta}}_n} \xrightarrow{wp1} \mathbf{I}_{\boldsymbol{\theta}}$. The result for V_n is given by Lemma 4.4.2A (iii). ∎

Let us now consider the behavior of λ_n, W_n and V_n under "local" alternatives, that is, for a sequence $\{\boldsymbol{\theta}_n\}$ of the form

$$\boldsymbol{\theta}_n = \boldsymbol{\theta}_0 + n^{-1/2}\boldsymbol{\Delta},$$

where $\mathbf{\Delta} = (\Delta_1, \ldots, \Delta_k)$. Let us suppose that the convergences expressed in Lemmas 4.4.2A (ii), B, and C (i) may be established *uniformly* in $\boldsymbol{\theta}$ for $\boldsymbol{\theta}$ in a neighborhood of $\boldsymbol{\theta}_0$. It *then* would follow that

(1) $$n^{1/2}\mathbf{d}_{n\boldsymbol{\theta}_0} = n^{1/2}\mathbf{d}_{n\boldsymbol{\theta}_n} + \mathbf{\Delta} \xrightarrow{d} N(\mathbf{\Delta}, \mathbf{I}_{\boldsymbol{\theta}_0}^{-1}),$$

(2) $$n^{1/2}\mathbf{a}_{n\boldsymbol{\theta}_0} = n^{1/2}\mathbf{d}_{n\boldsymbol{\theta}_0}\mathbf{I}_{\boldsymbol{\theta}_0} + o_{p_{\boldsymbol{\theta}_n}}(1) \xrightarrow{d} N(\mathbf{\Delta}\mathbf{I}_{\boldsymbol{\theta}_0}, \mathbf{I}_{\boldsymbol{\theta}_0}),$$

and

(3) $$\lambda_n - W_n \xrightarrow{p_{\boldsymbol{\theta}n}} 0,$$

where by (3) is meant that $P_{\boldsymbol{\theta}_n}(|\lambda_n - W_n| > \varepsilon) \to 0$, $n \to \infty$, for each $\varepsilon > 0$. By (1), (2), (3) and Lemma 3.5B, since $\mathbf{I}_{\boldsymbol{\theta}}$ is nonsingular, it *then* would follow that the statistics λ_n, W_n and V_n each converge in distribution to $\chi_k^2(\mathbf{\Delta}\mathbf{I}_{\boldsymbol{\theta}}\mathbf{\Delta}')$.

Therefore, under appropriate regularity conditions, the statistics λ_n, W_n and V_n are asymptotically *equivalent* in distribution, both under the null hypothesis and under local alternatives converging sufficiently fast. However, at *fixed* alternatives these equivalences are not anticipated to hold.

The technique of application of the limit distribution $\chi_k^2(\mathbf{\Delta}\mathbf{I}_{\boldsymbol{\theta}}\mathbf{\Delta}')$ to calculate the power of the test statistics λ_n, W_n or V_n is as for the chi-squared statistic discussed in Example 3.5C.

Regarding the *uniformity* assumed above, see the references cited at the end of **4.4.4**.

4.4.4 Test Statistics for a Composite Null Hypothesis

We adopt the formulation given in **4.4.1**, and we assume also that the specification of $\boldsymbol{\Theta}_0$ may equivalently be given as a transformation

$$\theta_1 = g_1(v_1, \ldots, v_{k-r}),$$

$$\ldots,$$

$$\theta_k = g_k(v_1, \ldots, v_{k-r}),$$

where $\mathbf{v} = (v_1, \ldots, v_{k-r})$ ranges through an open subset $\mathbf{N} \subset R^{k-r}$. For example, if $k = 3$ and $\boldsymbol{\Theta}_0 = \{\boldsymbol{\theta}: \theta_1 = \theta_{01}\}$, then we may take $\mathbf{N} = \{(v_1, v_2): (\theta_{01}, v_1, v_2) \in \boldsymbol{\Theta}_0\}$ and the functions g_1, g_2, g_3 to be $g_1(v_1, v_2) = \theta_{01}$, $g_2(v_1, v_2) = v_1$, and $g_3(v_1, v_2) = v_2$.

Assume that R_i and g_i possess continuous first order partial derivatives and that

$$\mathbf{C}_{\boldsymbol{\theta}} = \left[\frac{\partial R_i}{\partial \theta_j}\right]_{r \times k}$$

is of rank r and

$$\mathbf{D}_{\mathbf{v}} = \left[\frac{\partial g_i}{\partial v_j}\right]_{k \times (k-r)}$$

is of rank $k - r$.

In the present context the three test statistics considered in **4.4.3** have the following more general formulations. The likelihood ratio statistic is given by

$$\Lambda_n = \frac{\sup_{\boldsymbol{\theta} \in \boldsymbol{\Theta}_0} L(\boldsymbol{\theta})}{\sup_{\boldsymbol{\theta} \in \boldsymbol{\Theta}} L(\boldsymbol{\theta})} = \frac{\sup_{R_1(\boldsymbol{\theta}) = \cdots = R_r(\boldsymbol{\theta}) = 0} L(\boldsymbol{\theta})}{\sup_{\boldsymbol{\theta} \in \boldsymbol{\Theta}} L(\boldsymbol{\theta})}.$$

Equivalently, we use

$$\lambda_n = -2 \log \Lambda_n.$$

The Wald statistic will be based on the vector

$$\mathbf{b}_{\boldsymbol{\theta}} = (R_1(\boldsymbol{\theta}), \ldots, R_r(\boldsymbol{\theta})).$$

Concerning this vector, we have by Theorem 3.3A the following result. (Here $\hat{\boldsymbol{\theta}}_n$ is as in **4.4.2** and **4.4.3**.)

Lemma A. *Let* $X_1, X_2, \ldots$ *be I.I.D. with distribution* $F_{\boldsymbol{\theta}}$. *Then*

$$\mathbf{b}_{\hat{\boldsymbol{\theta}}_n} \quad \textit{is} \quad \mathrm{AN}(\mathbf{b}_{\boldsymbol{\theta}}, n^{-1}\mathbf{C}_{\boldsymbol{\theta}}\mathbf{I}_{\boldsymbol{\theta}}^{-1}\mathbf{C}'_{\boldsymbol{\theta}}).$$

The Wald statistic is defined as

$$W_n = n\mathbf{b}_{\hat{\boldsymbol{\theta}}_n}(\mathbf{C}_{\hat{\boldsymbol{\theta}}_n}\mathbf{I}_{\hat{\boldsymbol{\theta}}_n}^{-1}\mathbf{C}'_{\hat{\boldsymbol{\theta}}_n})^{-1}\mathbf{b}_{\hat{\boldsymbol{\theta}}_n}.$$

The Rao statistic is based on the estimate $\boldsymbol{\theta}_n^*$ which maximizes $L(\boldsymbol{\theta})$ subject to the restrictions $R_i(\boldsymbol{\theta}) = 0$, $1 \le i \le r$. Equivalently, $\boldsymbol{\theta}_n^*$ may be represented as

$$\boldsymbol{\theta}_n^* = g(\hat{\mathbf{v}}_n) = (g_1(\hat{\mathbf{v}}_n), \ldots, g_k(\hat{\mathbf{v}}_n)),$$

where $\hat{\mathbf{v}}_n$ is the maximum likelihood estimate of $\mathbf{v}$ in the reparametrization specified by the null hypothesis. Denoting by $\mathbf{J}_{\mathbf{v}}$ the information matrix for the $\mathbf{v}$-formulation of the model, we have by Theorems 4.2.2 and 3.3A the following result.

Lemma B. *Under* H_0, *that is, if* $X_1, X_2, \ldots$ *have a distribution* $F_{\boldsymbol{\theta}}$ *for* $\boldsymbol{\theta} \in \boldsymbol{\Theta}_0$, *and thus* $\boldsymbol{\theta} = g(\mathbf{v})$ *for some* $\mathbf{v} \in \mathbf{N}$, *we have*

(i) $\hat{\mathbf{v}}_n$ *is* $\mathrm{AN}(\mathbf{v}, n^{-1}\mathbf{J}_{\mathbf{v}}^{-1})$

and

(ii) $\hat{\boldsymbol{\theta}}_n$ *is* $\mathrm{AN}(\boldsymbol{\theta}, n^{-1}\mathbf{D}_{\mathbf{v}}\mathbf{J}_{\mathbf{v}}^{-1}\mathbf{D}'_{\mathbf{v}})$.

Noting that for $\boldsymbol{\theta} \in \boldsymbol{\Theta}_0$, that is, for $\boldsymbol{\theta} = g(\mathbf{v})$,

$$\frac{\partial \log f(x; \boldsymbol{\theta})}{\partial v_j} = \sum_{i=1}^{k} \frac{\partial \log f(x; \boldsymbol{\theta})}{\partial \theta_i} \cdot \frac{\partial g_i}{\partial v_j},$$

we have

$$\mathbf{t}_{n\mathbf{v}} = \mathbf{a}_{n\boldsymbol{\theta}} \mathbf{D}_{\mathbf{v}},$$

where

$$\mathbf{t}_{n\mathbf{v}} = \left(\frac{1}{n} \sum_{m=1}^{n} \frac{\partial \log f(X_m; g(\mathbf{v}))}{\partial v_1}, \ldots, \frac{1}{n} \sum_{m=1}^{n} \frac{\partial \log f(X_m; g(\mathbf{v}))}{\partial v_{k-r}}\right),$$

which is the analogue in the $\mathbf{v}$-formulation of $\mathbf{a}_{n\boldsymbol{\theta}}$ in the unrestricted model.

An immediate application of Lemma 4.4.2A(i), but in the $\mathbf{v}$-formulation, yields

Lemma C. *Under* H_0,

$$\mathbf{t}_{n\mathbf{v}} \quad \textit{is} \quad \mathrm{AN}(\mathbf{0}, \mathrm{n}^{-1}\mathbf{J}_{\mathbf{v}}).$$

On the other hand, application of Lemma 4.4.2A (i) to $\mathbf{a}_{n\boldsymbol{\theta}}$, with the use of the relation $\mathbf{t}_{n\mathbf{v}} = \mathbf{a}_{n\boldsymbol{\theta}} \mathbf{D}_{\mathbf{v}}$, yields that

$$\mathbf{t}_{n\mathbf{v}} \quad \text{is} \quad AN(\mathbf{0}, n^{-1}\mathbf{D}_{\mathbf{v}}' \mathbf{I}_{\boldsymbol{\theta}} \mathbf{D}_{\mathbf{v}}).$$

Hence

Lemma D. *For* $\boldsymbol{\theta} = \mathrm{g}(\mathbf{v})$, $\mathbf{J}_{\mathbf{v}} = \mathbf{D}_{\mathbf{v}}' \mathbf{I}_{\boldsymbol{\theta}} \mathbf{D}_{\mathbf{v}}$.

Thus the analogue of the Rao statistic given in **4.4.3** is

$$V_n = n \mathbf{t}_{n\hat{\mathbf{v}}_n} \mathbf{J}_{\hat{\mathbf{v}}_n}^{-1} \mathbf{t}_{n\hat{\mathbf{v}}_n}',$$

which may be expressed in terms of the statistic $\boldsymbol{\theta}_n^*$ as

$$V_n = n \mathbf{a}_{n\boldsymbol{\theta}_n^*} \mathbf{D}_{\hat{\mathbf{v}}_n} (\mathbf{D}_{\hat{\mathbf{v}}_n}' \mathbf{I}_{\boldsymbol{\theta}_n^*} \mathbf{D}_{\hat{\mathbf{v}}_n})^{-1} \mathbf{D}_{\hat{\mathbf{v}}_n}' \mathbf{a}_{n\boldsymbol{\theta}_n^*}'.$$

The asymptotic distribution theory of λ_n, W_n and V_n under the null hypothesis is given by

Theorem. *Under* H_0, *each of the statistics* λ_n, W_n *and* V_n *converges in distribution to* χ_r^2.

PROOF. We first deal with W_n, which presents the least difficulty. Under H_0, we have $\mathbf{b}_{\boldsymbol{\theta}} = \mathbf{0}$ and thus, by Lemma A,

$$n^{1/2}\mathbf{b}_{\hat{\boldsymbol{\theta}}_n} \xrightarrow{d} N(\mathbf{0}, \mathbf{C}_{\boldsymbol{\theta}} \mathbf{I}_{\boldsymbol{\theta}}^{-1} \mathbf{C}_{\boldsymbol{\theta}}').$$

Hence Theorem 3.5 immediately yields

$$n\mathbf{b}_{\hat{\boldsymbol{\theta}}_n}(\mathbf{C}_{\boldsymbol{\theta}}\mathbf{I}_{\boldsymbol{\theta}}^{-1}\mathbf{C}_{\boldsymbol{\theta}}')^{-1}\mathbf{b}_{\hat{\boldsymbol{\theta}}_n} \xrightarrow{d} \chi_r^2.$$

Since

$$(\mathbf{C}_{\hat{\boldsymbol{\theta}}_n}\mathbf{I}_{\hat{\boldsymbol{\theta}}_n}^{-1}\mathbf{C}_{\hat{\boldsymbol{\theta}}_n}')^{-1} \xrightarrow{p} (\mathbf{C}_{\boldsymbol{\theta}}\mathbf{I}_{\boldsymbol{\theta}}^{-1}\mathbf{C}_{\boldsymbol{\theta}}')^{-1},$$

we thus have

$$W_n \xrightarrow{d} \chi_r^2.$$

Next we deal with λ_n. By an argument similar to the proof of Lemma 4.4.2C, it is established that

$$\text{(1)} \qquad \lambda_n = -2[l_n(\hat{\boldsymbol{\theta}}_n) - l_n(\boldsymbol{\theta}_n^*)] = n(\hat{\boldsymbol{\theta}}_n - \boldsymbol{\theta}_n^*)\mathbf{I}_{\boldsymbol{\theta}}(\hat{\boldsymbol{\theta}}_n - \boldsymbol{\theta}_n^*)' + o_p(1)$$

and that

$$\mathbf{b}_{\hat{\boldsymbol{\theta}}_n} = \mathbf{b}_{\hat{\boldsymbol{\theta}}_n} - \mathbf{b}_{\hat{\boldsymbol{\theta}}_n^*} = (\hat{\boldsymbol{\theta}}_n - \boldsymbol{\theta}_n^*)\mathbf{C}_{\boldsymbol{\theta}}' + o_p(|\hat{\boldsymbol{\theta}}_n - \boldsymbol{\theta}_n^*|)$$

and

$$n^{1/2}(\hat{\boldsymbol{\theta}}_n - \boldsymbol{\theta}_n^*) = O_p(1),$$

whence

$$\text{(2)} \qquad W_n = n(\hat{\boldsymbol{\theta}}_n - \boldsymbol{\theta}^*)\mathbf{C}_{\boldsymbol{\theta}}'(\mathbf{C}_{\boldsymbol{\theta}}\mathbf{I}_{\boldsymbol{\theta}}^{-1}\mathbf{C}_{\boldsymbol{\theta}}')^{-1}\mathbf{C}_{\boldsymbol{\theta}}(\hat{\boldsymbol{\theta}}_n - \boldsymbol{\theta}_n^*)' + o_p(1).$$

Writing

$$\mathbf{C}_{\boldsymbol{\theta}}'(\mathbf{C}_{\boldsymbol{\theta}}\mathbf{I}_{\boldsymbol{\theta}}^{-1}\mathbf{C}_{\boldsymbol{\theta}}')\mathbf{C}_{\boldsymbol{\theta}} = \mathbf{K}_{\boldsymbol{\theta}}$$

and defining $\mathbf{B}_{\boldsymbol{\theta}}$ by

$$\mathbf{B}_{\boldsymbol{\theta}}'\mathbf{B}_{\boldsymbol{\theta}} = \mathbf{I}_{\boldsymbol{\theta}}^{-1},$$

we have (check) that

$$\mathbf{B}_{\boldsymbol{\theta}}'\mathbf{K}_{\boldsymbol{\theta}}\mathbf{B}_{\boldsymbol{\theta}} \text{ is indempotent}$$

and hence

$$\begin{aligned}
\text{rank } \mathbf{B}_{\boldsymbol{\theta}}'\mathbf{K}_{\boldsymbol{\theta}}\mathbf{B}_{\boldsymbol{\theta}} &= \text{trace } \mathbf{B}_{\boldsymbol{\theta}}'\mathbf{K}_{\boldsymbol{\theta}}\mathbf{B}_{\boldsymbol{\theta}} \\
&= \text{trace } \mathbf{B}_{\boldsymbol{\theta}}'\mathbf{C}_{\boldsymbol{\theta}}'(\mathbf{C}_{\boldsymbol{\theta}}\mathbf{B}_{\boldsymbol{\theta}}\mathbf{B}_{\boldsymbol{\theta}}'\mathbf{C}_{\boldsymbol{\theta}}')^{-1}\mathbf{C}_{\boldsymbol{\theta}}\mathbf{B}_{\boldsymbol{\theta}} \\
&= \text{trace } (\mathbf{C}_{\boldsymbol{\theta}}\mathbf{B}_{\boldsymbol{\theta}}\mathbf{B}_{\boldsymbol{\theta}}'\mathbf{C}_{\boldsymbol{\theta}}')(\mathbf{C}_{\boldsymbol{\theta}}\mathbf{B}_{\boldsymbol{\theta}}\mathbf{B}_{\boldsymbol{\theta}}'\mathbf{C}_{\boldsymbol{\theta}}')^{-1} \\
&= \text{trace } \mathbf{I}_{k \times k} \\
&= k.
\end{aligned}$$

Since $\mathbf{B}_{\boldsymbol{\theta}}'\mathbf{K}_{\boldsymbol{\theta}}\mathbf{B}_{\boldsymbol{\theta}}$ is idempotent, symmetric, of order k and rank k,

$$\mathbf{B}_{\boldsymbol{\theta}}'\mathbf{K}_{\boldsymbol{\theta}}\mathbf{B}_{\boldsymbol{\theta}} = \mathbf{I}_{k \times k}.$$

Hence

$$\mathbf{K}_{\boldsymbol{\theta}} = (\mathbf{B}'_{\boldsymbol{\theta}})^{-1}\mathbf{B}_{\boldsymbol{\theta}}^{-1} = (\mathbf{I}_{\boldsymbol{\theta}}^{-1})^{-1} = \mathbf{I}_{\boldsymbol{\theta}}.$$

Therefore, combining (1) and (2), we see that

$$\lambda_n - W_n \xrightarrow{p} 0.$$

Hence

$$\lambda_n \xrightarrow{d} \chi_r^2.$$

For V_n, see Rao (1973), Section 6e. ■

The null hypothesis asymptotic distribution of λ_n was originally obtained by Wilks (1938). The limit theory of λ_n under local alternatives and of W_n under both null hypothesis and local alternatives was initially explored by Wald (1943). For further development, see Chernoff (1954, 1956), Feder (1968), and Davidson and Lever (1970).

4.5 ESTIMATION VIA PRODUCT-MULTINOMIAL DATA

In this section, and in Section 4.6, we consider data corresponding to a *product-multinomial* model. In **4.5.1** the model is formulated and the business of estimating parameters is characterized. Methods of obtaining asymptotically *efficient* estimates are presented in **4.5.2**. A simplifying computational device is given in **4.5.3**, and brief complements in **4.5.4**. In Section 4.6 we consider the closely related matter of testing hypotheses.

4.5.1 The Model, the Parameters, and the Maximum Likelihood Estimates

Multinomial models and "cell frequency vectors" have been discussed in Section 2.7. The "*product-multinomial*" model is simply an extension of the scheme to the case of c populations.

Let the ith population have r_i "categories" or "cells," $1 \le i \le c$. Let p_{ij} denote the probability that an observation taken on the ith population falls in the jth cell. Let n_i denote the (nonrandom) sample size taken in the ith population and n_{ij} the (random) observed frequency in the jth cell of the ith population. Let $N = n_1 + \cdots + n_c$ denote the total sample size. We have the following constraints on the p_{ij}'s:

(1)
$$\sum_{j=1}^{r_i} p_{ij} - 1 = 0, \qquad 1 \le i \le c.$$

Likewise

$$\sum_{j=1}^{r_i} n_{ij} = n_i, \qquad 1 \le i \le c.$$

Finally, the probability of the observed frequency matrix

$$\{n_{ij}: 1 \le j \le r_i, 1 \le i \le c\}$$

is

$$\prod_{i=1}^{c} \frac{n_i!}{\prod_{j=1}^{r_i} n_{ij}!} \prod_{j=1}^{r_i} p_{ij}^{n_{ij}}.$$

Regarding estimation, let us first note (Problem 4.P.6) that the *maximum likelihood estimates* of the p_{ij}'s are given by their sample analogues,

$$\hat{p}_{ij} = \frac{n_{ij}}{n_i}, \qquad 1 \le j \le r_i, 1 \le i \le c.$$

(This is found by maximizing the likelihood function subject to the constraints (1).) We shall employ the notation

$$\mathbf{p} = (p_{11}, \ldots, p_{1r_1}; \ldots; p_{c1}, \ldots, p_{cr_c})$$

for the vector of parameters, and

$$\hat{\mathbf{p}} = (\hat{p}_{11}, \ldots, \hat{p}_{1r_1}; \ldots; \hat{p}_{c1}, \ldots, \hat{p}_{cr_c}).$$

for the vector of maximum likelihood estimates.

More generally, we shall suppose that the p_{ij}'s are given as specified functions of a set of parameters $\theta_1, \ldots, \theta_k$, and that the problem is to estimate $\boldsymbol{\theta} = (\theta_1, \ldots, \theta_k)$. An example of such a problem was seen in Section 2.7. Another example follows.

Example A. Suppose that the c populations of the product-multinomial model represent *different levels of a treatment*, and that the r_i cells of the ith population represent *response categories*. Let us take $r_1 = \cdots = r_c = r$. Further, suppose that the response and factor are each "structured." That is, attached to the response categories are certain known weights $a_1, \ldots, a_r$, and attached to the treatment levels are known weights $b_1, \ldots, b_c$. Finally, suppose that the expected response weights at the various treatment levels have a linear regression on the treatment level weights. This latter supposition is expressed as a set of relations

$$(*) \qquad \sum_{j=1}^{r} a_j p_{ij} = \lambda + \mu b_i, \qquad 1 \le i \le c,$$

where λ and μ are unknown parameters. We now identify the relevant parameter vector $\boldsymbol{\theta}$. First, suppose (without loss of generality) that $a_1 \ne a_r$. Now note that, by the constraints (1), we may write

$$\text{(i)} \qquad p_{i1} = 1 - \sum_{j=2}^{r} p_{ij}, \qquad 1 \le i \le c.$$

Also, after eliminating each p_{i1} by (i), we have by (*) that

$$\text{(ii)} \qquad p_{ir} = \frac{\lambda + \mu b_i - a_1 - \sum_{j=2}^{r-1} (a_j - a_1)p_{ij}}{a_r - a_1}, \qquad 1 \le i \le c.$$

Finally, we also write

$$\text{(iii)} \qquad p_{ij} = p_{ij}, \qquad 2 \le j \le r - 1, \qquad 1 \le i \le c.$$

It thus follows from (i), (ii) and (iii) that the components of **p** may be expressed entirely in terms of the parameters

$$\theta_1 = \lambda; \qquad \theta_2 = \mu; \qquad \theta_{ij} = p_{ij}, \qquad 2 \le j \le r - 1, \qquad 1 \le i \le c,$$

that is, in terms of $\boldsymbol{\theta}$ containing $k = (r - 2)c + 2$ components. We shall consider this example further below, as well as in **4.6.3**. ■

The condition that the p_{ij}'s are specified functions of $\boldsymbol{\theta}$,

$$p_{ij} = p_{ij}(\boldsymbol{\theta}), \qquad 1 \le j \le r_i, \qquad 1 \le i \le c,$$

is equivalent to a set of $m = \sum_{i=1}^{c} r_i - c - k$ constraints, say

$$\text{(2)} \qquad H_l(\mathbf{p}) = 0, \qquad 1 \le l \le m,$$

obtained by eliminating the parameters $\theta_1, \ldots, \theta_k$. These equations are independent of the c constraints given by (1). ■

Example B (continuation). For the preceding example, we have $m = cr - c - [(r - 2)c + 2] = c - 2$. These $c - 2$ constraints are obtained from (*) by eliminating λ and μ. (Problem 4.P.7). ■

Example C. The problem of estimation of **p** may be represented as estimation of $\boldsymbol{\theta}$, where the θ_i's consist of the $k = \sum_{i=1}^{c} r_i - c$ p_{ij}'s remaining after elimination of $p_{1r_1}, \ldots, p_{cr_c}$ by the use of (1). In this case $m = 0$, that is, there are no additional constraint equations (2). ■

The problem of estimation of $\boldsymbol{\theta}$ thus becomes equivalent to that of estimation of the original vector **p** subject to the combined set of $m + c$ constraint equations (1) and (2). If the representation of $\boldsymbol{\theta}$ in terms of p_{ij}'s is given by

$$\boldsymbol{\theta} = g(\mathbf{p}) = (g_1(\mathbf{p}), \ldots, g_k(\mathbf{p})),$$

then an estimator of $\boldsymbol{\theta}$ is given by

$$\hat{\boldsymbol{\theta}} = g(\hat{\mathbf{p}}) = (g_1(\hat{\mathbf{p}}), \ldots, g_k(\hat{\mathbf{p}})),$$

where $\hat{\hat{\mathbf{p}}} = (\hat{\hat{p}}_{11}, \ldots ; \ldots ; \ldots, \hat{\hat{p}}_{cr_c})$ denotes a vector estimate of $\mathbf{p}$ under the constraints (1) and (2). In particular, if $\hat{\hat{\mathbf{p}}}$ denotes the *maximum likelihood estimate* of $\mathbf{p}$ subject to these constraints, then (under appropriate regularity conditions on g) the maximum likelihood estimate of $\boldsymbol{\theta}$ is given by $\hat{\boldsymbol{\theta}} = g(\hat{\hat{\mathbf{p}}})$. Therefore, asymptotically efficient estimates of $\boldsymbol{\theta}$ are provided by $g(\hat{\mathbf{p}}^*)$ for *any* BAN estimate $\hat{\mathbf{p}}^*$ of $\mathbf{p}$ subject to (1) and (2).

There are two principal advantages to the formulation entirely in terms of constraints on the p_{ij}'s:

(a) in testing, it is sometimes *convenient* to express the null hypothesis in the form of a set of constraint equations on the p_{ij}'s, rather than by a statement naming further parameters $\theta_1, \ldots, \theta_k$ (see **4.6.2** and **4.6.3**);

(b) this formulation is suitable for making a computational simplification of the problem by a linearization technique (**4.5.3**).

4.5.2 Methods of Asymptotically Efficient Estimation

Regarding estimation of the θ_i's, several approaches will be considered, following Neyman (1949). Neyman's objective was to provide estimators possessing the same *large sample efficiency* as the maximum likelihood estimates but possibly superior *computational ease* or *small sample efficiency*. Although the issue of computational ease is now of less concern after great advances in computer technology, the small sample efficiency remains an important consideration.

The "maximum likelihood" approach consists of maximizing

$$\prod_{i=1}^{c} \prod_{j=1}^{r_i} p_{ij}(\boldsymbol{\theta})^{n_{ij}}$$

with respect to $\theta_1, \ldots, \theta_k$, subject to the constraints (1) (of **4.5.1**).

The "minimum χ^2" approach consists of minimizing

$$d_1(\mathbf{p}(\boldsymbol{\theta}), \hat{\mathbf{p}}) = \sum_{i=1}^{c} n_i \sum_{j=1}^{r_i} \frac{[\hat{p}_{ij} - p_{ij}(\boldsymbol{\theta})]^2}{p_{ij}(\boldsymbol{\theta})}$$

with respect to $\theta_1, \ldots, \theta_k$, subject to the constraints (1).

Finally, the "modified minimum χ^2" approach consists of minimizing

$$d_2(\mathbf{p}(\boldsymbol{\theta}), \hat{\mathbf{p}}) = \sum_{i=1}^{c} n_i \sum_{j=1}^{r_i} \frac{[\hat{p}_{ij} - p_{ij}(\boldsymbol{\theta})]^2}{\hat{p}_{ij}}$$

with respect to $\theta_1, \ldots, \theta_k$, subject to the constraints (1).

Noting that d_1 and d_2 are measures of discrepancy between $\mathbf{p}$ and $\hat{\mathbf{p}}$, we may characterize the maximum likelihood approach in this fashion in terms of

$$d_0(\mathbf{p}(\boldsymbol{\theta}), \hat{\mathbf{p}}) = -2 \log \lambda(\mathbf{p}(\boldsymbol{\theta}), \hat{\mathbf{p}}),$$

where

$$\lambda(\mathbf{p}(\boldsymbol{\theta}), \hat{\mathbf{p}}) = \prod_{i=1}^{c} \prod_{j=1}^{r_i} \left[\frac{p_{ij}(\boldsymbol{\theta})}{\hat{p}_{ij}}\right]^{n_{ij}}.$$

Each approach leads to a system of equations. However, the relative convenience of the three systems of equations depends on the nature of the functions $p_{ij}(\boldsymbol{\theta})$. In the case that these are *linear* in $\theta_1, \ldots, \theta_k$, the modified minimum χ^2 approach yields a *linear* system of equations for $\theta_1, \ldots, \theta_k$.

In any case, the three systems of equations are *asymptotically equivalent in probability*, in the sense that the estimates produced differ only by $o_p(N^{-1/2})$, as $N \to \infty$ in such fashion that each n_i/N has a limit l_i, $0 < l_i < 1$, $1 \le i \le c$. For these details, see Cramér (1946), Sections 30.3 and 33.4, and Neyman (1949).

For appropriate regularity conditions on the parameter space Θ and the functions $p_{ij}(\boldsymbol{\theta})$, in order for the maximum likelihood estimates to be asymptotically efficient, see Rao (1973), Section 5e.2.

4.5.3 Linearization Technique

Corresponding to the set of (possibly nonlinear) constraint equations (2) (of **4.5.1**), we associate the set of *linear* constraint equations

$$H_l^*(\mathbf{p}) = 0, \qquad 1 \le l \le m, \tag{2*}$$

where

$$H_l^*(\mathbf{p}) = H_l(\hat{\mathbf{p}}) + \sum_{i=1}^{c} \sum_{j=1}^{r_i} \left.\frac{\partial H_l(\mathbf{p})}{\partial p_{ij}}\right|_{\mathbf{p}=\hat{\mathbf{p}}} \cdot (p_{ij} - \hat{p}_{ij}),$$

which is the linear part of the Taylor expansion of $H_l(\mathbf{p})$ about the point $\mathbf{p} = \hat{\mathbf{p}}$, the maximum likelihood estimate in the model unrestricted by the constraints (2).

Neyman (1949) proves that minimization of $d_0(\mathbf{p}, \hat{\mathbf{p}})$, $d_1(\mathbf{p}, \hat{\mathbf{p}})$, or $d_2(\mathbf{p}, \hat{\mathbf{p}})$ with respect to the p_{ij}'s, subject to the constraints (1) and (2), and minimization alternatively subject to the constraints (1) and (2*), yields estimates $\hat{\hat{\mathbf{p}}}$ and $\hat{\hat{\mathbf{p}}}^*$, respectively, which satisfy

$$\hat{\hat{\mathbf{p}}} - \hat{\hat{\mathbf{p}}}^* = o_p(N^{-1/2}).$$

Further, regarding estimation of the parameters θ_i, Neyman establishes analogous results for estimates $\hat{\hat{\boldsymbol{\theta}}}$ and $\hat{\hat{\boldsymbol{\theta}}}^*$ based on (2) and (2*), respectively.

As shown in the following example, the application of the linearization technique in conjunction with the modified minimum χ^2 approach produces a *linear* system of equations for asymptotically efficient estimates.

Example. *Linearized constraints with modified minimum χ^2 approach.* In order to minimize $d_2(\mathbf{p}, \hat{\mathbf{p}})$ with respect to the p_{ij}'s subject to the constraints (1) and (2*), we introduce Lagrangian multipliers $\lambda_i(1 \le i \le c)$ and $\mu_l(1 \le l \le m)$ and minimize the function

$$D_2(\mathbf{p}, \hat{\mathbf{p}}, \boldsymbol{\lambda}, \boldsymbol{\mu}) = d_2(\mathbf{p}, \hat{\mathbf{p}}) + \sum_{i=1}^{c} \lambda_i\left(\sum_{j=1}^{r_i} p_{ij} - 1\right) + \sum_{l=1}^{m} \mu_l H_l^*(\mathbf{p})$$

with respect to the p_{ij}'s, λ_j's and μ_l's. The system of equations obtained by equating to 0 the partials of D_2 with respect to the p_{ij}'s, λ_i's and μ_l's is a *linear* system. Thus one may obtain asymptotically efficient estimates of the p_{ij}'s under the constraints (1) and (2), and thus of the θ_i's likewise, by solving a certain *linear* system of equations, that is, by inverting a matrix. ■

4.5.4 Complements

(i) *Further "minimum χ^2 type" approaches.* For a review of such approaches and of work subsequent to Neyman (1949), see Ferguson (1958).

(ii) *Distance measures.* The three approaches in **4.5.2** may be regarded as methods of estimation of $\boldsymbol{\theta}$ by minimization of a *distance* measure between the *observed* **p** vector (i.e., $\hat{\mathbf{p}}$) and the *hypothetical* **p** vector (i.e., $\mathbf{p}(\boldsymbol{\theta})$). (Recall **4.3.2**.) For further distance measures, see Rao (1973), Section 5d.2.

4.6 HYPOTHESIS TESTING VIA PRODUCT-MULTINOMIAL DATA

Continuing the set-up introduced in **4.5.1**, we consider in **4.6.1** three test statistics, each having asymptotic chi-squared distribution under the null hypothesis. Simplified schemes for computing the test statistics are described in **4.6.2**. Application to the analysis of variance of product-multinomial data is described in **4.6.3**.

4.6.1 Three Test Statistics

For the product-multinomial of **4.5.1**, the constraints (1) are an inherent part of the "unrestricted" model. In this setting, a null hypothesis H_0 may be formulated as

$$H_0: p_{ij} = p_{ij}(\theta_1, \ldots, \theta_k),$$

where the p_{ij}'s are given as specified functions of unknown parameters $\boldsymbol{\theta} = (\theta_1, \ldots, \theta_k)$, or equivalently as

$$H_0: H_l(\mathbf{p}) = 0, \qquad 1 \le l \le m.$$

As in **4.5.2**, denote by $\hat{\mathbf{p}}$ the *maximum likelihood estimate* of $\mathbf{p}$ in the unrestricted model, and let $\hat{\mathbf{p}}^*$ denote an *asymptotically efficient* estimate of $\mathbf{p}$ under H_0 or under the corresponding linearized hypothesis (**4.5.3**). Each of the three distance measures considered in **4.5.2** serves as a *test statistic* when evaluated at $\hat{\mathbf{p}}^*$ and $\hat{\mathbf{p}}$. That is, each of

$$d_i(\hat{\mathbf{p}}^*, \hat{\mathbf{p}}), \qquad i = 0, 1, 2,$$

is considered as a test statistic for H_0, with H_0 to be rejected for large values of the statistic. Thus the null hypothesis becomes rejected if $\hat{\mathbf{p}}$ and $\hat{\mathbf{p}}^*$ are sufficiently "far apart."

Theorem (Neyman (1949)). *Under* H_0, *each of* $\mathrm{d}_i(\hat{\mathbf{p}}^*, \hat{\mathbf{p}})$, $\mathrm{i} = 0, 1, 2$, *converges in distribution to* χ^2.

4.6.2 Simplified Computational Schemes

Consider the statistic $d_2(\hat{\mathbf{p}}^*, \hat{\mathbf{p}})$ in the case that $\hat{\mathbf{p}}^*$ denotes the estimate obtained by minimizing $d_2(\mathbf{p}, \hat{\mathbf{p}})$ with respect to $\mathbf{p}$ under (1) and the constraints specified by H_0. For some types of hypothesis H_0, the statistic $d_2(\hat{\mathbf{p}}^*, \hat{\mathbf{p}})$ can actually be computed *without* first computing $\hat{\mathbf{p}}^*$. These computational schemes are due to Bhapkar (1961, 1966).

Bhapkar confines attention to *linear* hypotheses, on the grounds that nonlinear hypotheses may be reduced to linear ones if desired, by Neyman's linearization technique (**4.5.3**). Also, we shall now confine attention to the case of an equal number of cells in each population: $r_1 = \cdots = r_c = r$.

Two forms of linear hypothesis H_0 will be considered. *Firstly*, let H_0 be defined by m linearly independent constraints (also independent of (1) of **4.5.1**),

$$H_0\colon H_l(\mathbf{p}) = \sum_{i=1}^{c} \sum_{j=1}^{r} h_{lij} p_{ij} + h_l = 0, \qquad 1 \le l \le m,$$

where h_{lij} and h_l are known constants such that the hypothesis equations together with (1) have at least one solution for which the p_{ij}'s are positive. For this hypothesis, Bhapkar shows that

$$d_2(\hat{\mathbf{p}}^*, \mathbf{p}) = [H_1(\hat{\mathbf{p}}), \ldots, H_m(\hat{\mathbf{p}})]\mathbf{C}_N^{-1}[H_1(\hat{\mathbf{p}}), \ldots, H_m(\hat{\mathbf{p}})]',$$

where $\mathbf{C}_N$ denotes the sample estimate of the covariance matrix of the vector $[H_1(\hat{\mathbf{p}}), \ldots, H_m(\hat{\mathbf{p}})]$. Check that this vector has covariance matrix $[c_{lk}]_{m \times m}$, where

$$\begin{aligned} c_{lk} &= \sum_{i=1}^{c} \sum_{j=1}^{r} \frac{h_{lij} h_{kij} p_{ij}}{n_i} - \sum_{i=1}^{c} \frac{1}{n_i} \left(\sum_{j=1}^{r} h_{lij} p_{ij} \right) \left(\sum_{j=1}^{r} h_{kij} p_{ij} \right) \\ &= \sum_{i=1}^{c} \frac{1}{n_i} \left[\sum_{j=1}^{r} h_{lij} h_{kij} p_{ij} - \left(\sum_{j=1}^{r} h_{lij} p_{ij} \right) \left(\sum_{j=1}^{r} h_{kij} p_{ij} \right) \right]. \end{aligned}$$

Thus the matrix $\mathbf{C}_N$ is $[c_{Nlk}]_{m \times m}$, where c_{Nlk} is obtained by putting $\hat{p}_{ij}$ for p_{ij} in c_{lk}.

Note that the use of $d_2(\hat{\mathbf{p}}^*, \hat{\mathbf{p}})$ for testing H_0 is thus *exactly equivalent* to the "natural" test based on the asymptotic normality of the unbiased estimate $[H_1(\hat{\mathbf{p}}), \ldots, H_m(\hat{\mathbf{p}})]$ of $[H_1(\mathbf{p}), \ldots, H_m(\mathbf{p})]$, with the covariance matrix of this estimate estimated by its sample analogue. Note also that, in this situation, $d_2(\hat{\mathbf{p}}^*, \hat{\mathbf{p}})$ represents the Wald-type statistic of **4.4.4**.

Secondly, consider a hypothesis of the form

$$H_0\colon \sum_{j=1}^{r} a_j p_{ij} = \sum_{t=1}^{k} b_{it}\theta_t, \qquad 1 \le i \le c,$$

where the a_j's and b_{it}'s are known constants and the θ_t's are unknown parameters, and rank $[b_{it}]_{c \times k} = u \le c - 1$. This is a *linear* hypothesis, defined by linear functions of unknown parameters, and so it may be reduced to the form of H_0 considered previously. (In this case we would have $m = c - u$.) For example, recall Example 4.5.1 A, B. However, in many cases the reduction would be tedious to carry out and not of intrinsic interest. Instead, the problem may be viewed as a standard problem in "least squares analysis," Bhapkar shows. That is,

$$d_2(\hat{\mathbf{p}}^*, \hat{\mathbf{p}}) = \text{"Residual Sum of Squares,"}$$

corresponding to application of the general least squares technique on the variables $\sum_{j=1}^{r} a_j p_{ij}$ with the variances estimated by sample variances. Thus $d_2(\hat{\mathbf{p}}^*, \hat{\mathbf{p}})$ may be obtained as the residual sum of squares corresponding to minimization of

$$\sum_{i=1}^{c} \lambda_i^{-1}\left(\alpha_i - \sum_{t=1}^{k} b_{it}\theta_t\right)^2,$$

where

$$\alpha_i = \sum_{j=1}^{r} a_j \hat{p}_{ij}, \qquad \lambda_i = n_i^{-1} \sum_{j=1}^{r} (a_j - \alpha_i)^2 \hat{p}_{ij}.$$

4.6.3 Applications: Analysis of Variance of Product-Multinomial Data

For a product-multinomial model as in **4.5.1**, let "i" correspond to *factor* and "j" to *response*. Thus factor categories are indexed by $i = 1, \ldots, c$ and response categories by $j = 1, \ldots, r$. (For simplicity, assume $r_1 = \cdots = r_c = r$.) A response or factor is said to be *structured* if weights are attached to its categories, as illustrated earlier in Example 4.5.1A. We now examine some typical hypotheses and apply the results of **4.6.1** and **4.6.2**.

Hypothesis of homogeneity. (Neither response nor factor is structured.) The null hypothesis is

$$H_0\colon p_{ij} \text{ does not depend on } i.$$

In terms of constraint functions, this is written

$$H_0: H_{ij}(\mathbf{p}) = p_{ij} - p_{cj} = 0, \qquad (i = 1, \ldots, c - 1; j = 1, \ldots, r - 1).$$

The hypothesis thus specifies $m = (r - 1)(c - 1)$ constraints in addition to the constraints (1).

Under H_0, the product-multinomial model reduces to a single multinomial model, and corresponding BAN estimates of the p_{ij}'s are

$$\hat{p}_{ij}^* = \frac{n_{1j} + \cdots + n_{cj}}{N}, \qquad i = 1, \ldots, c; j = 1, \ldots, r.$$

Therefore, by Theorem 4.6.1, each of the statistics $d_i(\hat{\mathbf{p}}^*, \hat{\mathbf{p}})$, $i = 0, 1, 2$, is asymptotically $\chi^2_{(r-1)(c-1)}$. ■

Hypothesis of mean homogeneity. (The response is structured, and the hypothesis is "no treatment effects,")

$$H_0: \sum_{j=1}^{r} a_j p_{ij} \text{ does not depend on } j.$$

In terms of constraint functions, this is written

$$H_0: H_i(\mathbf{p}) = \sum_{j=1}^{r} a_j p_{ij} - \sum_{j=1}^{r} a_j p_{1j} = 0, \qquad (i = 1, \ldots, c - 1).$$

In terms of further parameters θ_t, this is written

$$H_0: \sum_{j=1}^{r} a_j p_{ij} = \theta, \qquad (i = 1, \ldots, c).$$

Instead of estimating the p_{ij}'s under H_0 (as we did in the previous illustration), we may apply either of Bhapkar's devices to evaluate $d_2(\hat{\mathbf{p}}^*, \hat{\mathbf{p}})$. The least-squares representation enables us to write immediately

$$d_2(\hat{\mathbf{p}}^*, \hat{\mathbf{p}}) = \sum_{i=1}^{c} \gamma_i \alpha_i^2 - \frac{(\sum_{i=1}^{c} \gamma_i \alpha_i)^2}{\sum_{i=1}^{c} \gamma_i},$$

where

$$\alpha_i = \sum_{j=1}^{r} a_j \hat{p}_{ij}, \; \gamma_i = \frac{n_i}{\sum_{j=1}^{r} (a_j - \alpha_i)^2 \hat{p}_{ij}}.$$

(As an exercise, check that this is the proper identification with standard least-squares formulas.) By Theorem 4.6.1, $d_2(\hat{\mathbf{p}}^*, \hat{\mathbf{p}})$ is asymptotically χ^2_{c-1}. ■

Hypothesis of linearity of regression. (Both response and factor are structured. The hypothesis of linearity of the regression of response on "treatment level" is to be tested.)

$$H_0: \sum_{j=1}^{r} a_j p_{ij} = \lambda + \mu b_i, \qquad (i = 1, \ldots, c).$$

By the least-squares analogy,

$$d_2(\hat{\mathbf{p}}^*, \hat{\mathbf{p}}) = \sum_{i=1}^{c} \gamma_i \alpha_i^2 - \frac{(\varepsilon s - \alpha\delta sd + \gamma d^2)}{(\gamma\varepsilon - \delta)^2}$$

where α_i and γ_i are as in the preceding illustration, and

$$\gamma = \sum_{i=1}^{c} \gamma_i, \qquad \delta = \sum_{i=1}^{c} b_i \gamma_i, \qquad \varepsilon = \sum_{i=1}^{c} b_i^2 \gamma_i,$$

$$s = \sum_{i=1}^{c} \alpha_i \gamma_i, \qquad d = \sum_{i=1}^{c} \alpha_i b_i \gamma_i.$$

Estimates of λ and μ are

$$\lambda = \frac{\varepsilon s - \delta d}{\gamma\varepsilon - \sigma^2}, \qquad \hat{\mu} = \frac{\gamma d - \delta s}{\gamma\varepsilon - \delta^2}.$$

The statistic $d_2(\hat{\mathbf{p}}^*, \hat{\mathbf{p}})$ is asymptotically χ^2_{c-2}.

If linearity is not sustained by the test, then the method may be extended to test for quadratic regression, etc. ■

Further examples and discussion. See, for example, Wilks (1962), Problems 13.7–13.9 and 13.11–13.12. ■

4.P PROBLEMS

Miscellaneous

1. Suppose that

(a) $\mathbf{X}_n = (X_{n1}, \ldots, X_{nk}) \xrightarrow{d} \mathbf{X}_0 = (X_{01}, \ldots, X_{0k})$

and

(b) $\mathbf{Y}_n = (Y_{n1}, \ldots, Y_{nk}) \xrightarrow{p} \mathbf{c} = (c_1, \ldots, c_k)$.

(i) Show that $(\mathbf{X}_n, \mathbf{Y}_n) \xrightarrow{d} (\mathbf{X}_0, \mathbf{c})$.

(ii) Apply (i) to obtain that $\mathbf{X}_n + \mathbf{Y}_n \xrightarrow{d} \mathbf{X}_0 + \mathbf{c}$ and $\mathbf{X}_n \mathbf{Y}_n' \xrightarrow{d} \mathbf{X}_0 \mathbf{c}'$.

(iii) What are your conclusions in (i) if (b) is replaced by (b') $\mathbf{Y}_n \xrightarrow{d} \mathbf{c}$?

(iv) What are your conclusions in (i) if (b) is replaced by (b'') $\mathbf{Y}_n \xrightarrow{p} \mathbf{Y}_0 = (Y_{01}, \ldots, Y_{0k})$?

(Justify all answers.)

Section 4.2

2. Justify the interpretations of regularity conditions (R1)–(R3) in **4.2.2**.
3. Prove Remark 4.2.2 (i).

Section 4.4

4. Do the exercise assigned in the proof of Lemma 4.4.2B.
5. Check that $\mathbf{B}_\theta' \mathbf{K}_\theta \mathbf{B}_\theta$ is idempotent (in the proof of Theorem 4.4.4).

Section 4.5

6. Verify for the product-multinomial model of **4.5.1** that the maximum likelihood estimates of the p_{ij}'s are $\hat{p}_{ij} = n_{ij}/n_i$, respectively.
7. Provide the details for Example 4.5.1B.

Section 4.6

8. Verify the covariance matrix $[c_{lk}]$ asserted in **4.6.2**.
9. Do the exercises assigned in **4.6.3**.

CHAPTER 5

U-Statistics

From a purely mathematical standpoint, it is desirable and appropriate to view any given statistic as but a single member of some general class of statistics having certain important features in common. In such fashion, several interesting and useful collections of statistics have been formulated as generalizations of particular statistics that have arisen for consideration as special cases.

In this and the following four chapters, five such classes will be introduced. For each class, key features and propositions will be examined, with emphasis on results pertaining to consistency and asymptotic distribution theory. As a by-product, new ways of looking at some familiar statistics will be discovered.

The class of statistics to be considered in the present chapter was introduced in a fundamental paper by Hoeffding (1948). In part, the development rested upon a paper of Halmos (1946). The class arises as a generalization of the sample mean, that is, as a generalization of the notion of forming an *average*. Typically, although not without important exceptions, the members of the class are asymptotically normal statistics. They also have good consistency properties.

The so-called "*U-statistics*" are closely connected with a class of statistics introduced by von Mises (1947), which we shall examine in Chapter 6. Many statistics of interest fall within these two classes, and many other statistics may be approximated by a member of one of these classes.

The basic description of U-statistics is provided in Section 5.1. This includes relevant definitions, examples, connections with certain other statistics, martingale structure and other representations, and an optimality property of U-statistics among unbiased estimators. Section 5.2 deals with the moments, especially the variance, of U-statistics. An important tool in deriving the asymptotic theory of U-statistics, the "*projection*" of a U-statistic on the basic observations of the sample, is introduced in Section 5.3. Sections 5.4 and 5.5 treat, respectively, the almost sure behavior and asymptotic distribution theory of U-statistics. Section 5.6 provides some further probability bounds and limit theorems. Several complements are provided in Section 5.7, including a look at stochastic processes associated with a sequence of U-statistics, and an examination of the Wilcoxon one-sample statistic as a U-statistic in connection with the problem of confidence intervals for quantiles (recall **2.6.5**).

The method of "projection" introduced in Section 5.3 is of quite general scope and will be utilized again with other types of statistic in Chapters 8 and 9.

5.1 BASIC DESCRIPTION OF U-STATISTICS

Basic definitions and examples are given in **5.1.1**, and a class of closely related statistics is noted in **5.1.2**. These considerations apply to *one-sample U-statistics*. Generalization to *several samples* is given in **5.1.3**, and to *weighted versions* in **5.1.7**. An important *optimality* property of U-statistics in unbiased estimation is shown in **5.1.4**. The representation of a U-statistic as a *martingale* is provided in **5.1.5**, and as an *average of I.I.D. averages* in **5.1.6**.

Additional general discussion of U-statistics may be found in Fraser (1957), Section 4.2, and in Puri and Sen (1971), Section 3.3.

5.1.1 First Definitions and Examples

Let $X_1, X_2, \ldots$ be independent observations on a distribution F. (They may be vector-valued, but usually for simplicity we shall confine attention to the real-valued case.) Consider a "parametric function" $\theta = \theta(F)$ for which there is an unbiased estimator. That is, $\theta(F)$ may be represented as

$$\theta(F) = E_F\{h(X_1, \ldots, X_m)\} = \int \cdots \int h(x_1, \ldots, x_m)dF(x_1) \cdots dF(x_m),$$

for some function $h = h(x_1, \ldots, x_m)$, called a "kernel." Without loss of generality, we may assume that h is *symmetric*. For, if not, it may be replaced by the symmetric kernel

$$\frac{1}{m!} \sum_p h(x_{i_1}, \ldots, x_{i_m}),$$

where $\sum_p$ denotes summation over the $m!$ permutations $(i_1, \ldots, i_m)$ of $(1, \ldots, m)$.

For any kernel h, the corresponding *U-statistic* for estimation of θ on the basis of a sample $X_1, \ldots, X_n$ of size $n \geq m$ is obtained by averaging the kernel h symmetrically over the observations:

$$U_n = U(X_1, \ldots, X_n) = \frac{1}{\binom{n}{m}} \sum_c h(X_{i_1}, \ldots, X_{i_m}),$$

where $\sum_c$ denotes summation over the $\binom{n}{m}$ combinations of m distinct elements $\{i_1, \ldots, i_m\}$ from $\{1, \ldots, n\}$. Clearly, U_n is an *unbiased* estimate of θ.

Examples. (i) $\theta(F) =$ mean of $F = \mu(F) = \int x\,dF(x)$. For the kernel $h(x) = x$, the corresponding U-statistic is

$$U(X_1, \ldots, X_n) = \frac{1}{n}\sum_{i=1}^{n} X_i = \bar{X},$$

the *sample mean.*

(ii) $\theta(F) = \mu^2(F) = [\int x\,dF(x)]^2$. For the kernel $h(x_1, x_2) = x_1 x_2$, the corresponding U-statistic is

$$U(X_1, \ldots, X_n) = \frac{2}{n(n-1)} \sum_{1 \le i < j \le n} X_i X_j.$$

(iii) $\theta(F) =$ variance of $F = \sigma^2(F) = \int (x - \mu)^2\,dF(x)$. For the kernel

$$h(x_1, x_2) = \frac{x_1^2 + x_2^2 - 2x_1x_2}{2} = \frac{1}{2}(x_1 - x_2)^2.$$

the corresponding U-statistic is

$$\begin{aligned} U(X_1, \ldots, X_n) &= \frac{2}{n(n-1)} \sum_{1 \le i < j \le n} h(X_i, X_j) \\ &= \frac{1}{n-1}\left(\sum_{i=1}^{n} X_i^2 - n\bar{X}^2\right) \\ &= s^2, \end{aligned}$$

the *sample variance.*

(iv) $\theta(F) = F(t_0) = \int_{-\infty}^{t_0} dF(x) = P_F(X_1 \le t_0)$. For the kernel $h(x) = I(x \le t_0)$, the corresponding U-statistic is

$$U(X_1, \ldots, X_n) = \frac{1}{n}\sum_{i=1}^{n} I(X_i \le t_0) = F_n(t_0),$$

where F_n denotes the sample distribution function.

(v) $\theta(F) = \alpha_k(F) = \int x^k\,dF(x) = k$th moment of F. For the kernel $h(x) = x^k$, the corresponding U-statistic is

$$U(X_1, \ldots, X_n) = \frac{1}{n}\sum_{i=1}^{n} X_i^k = a_k,$$

the *sample* kth *moment.*

(vi) $\theta(F) = E_F|X_1 - X_2|$, a measure of concentration. For the kernel $h(x_1, x_2) = |x_1 - x_2|$, the corresponding U-statistic is

$$U(X_1, \ldots, X_n) = \frac{2}{n(n-1)} \sum_{1 \le i < j \le n} |X_i - X_j|,$$

the statistic known as "Gini's mean difference."

(vii) Fisher's k-statistics for estimation of cumulants are U-statistics (see Wilks (1962), p. 200).

(viii) $\theta(F) = E_F\gamma(X_1) = \int \gamma(x)dF(x)$; $U_n = n^{-1}\sum_1^n \gamma(X_i)$.

(ix) *The Wilcoxon one-sample statistic.* For estimation of $\theta(F) = P_F(X_1 + X_2 \le 0)$, a kernel is given by $h(x_1, x_2) = I(x_1 + x_2 \le 0)$ and the corresponding U-statistic is

$$U(X_1, \ldots, X_n) = \frac{2}{n(n-1)} \sum_{1 \le i < j \le n} I(X_i + X_j \le 0).$$

(x) $\theta(F) = \iint [F(x, y) - F(x, \infty)F(\infty, y)]^2\, dF(x, y)$, a measure of dependence for a bivariate distribution F. Putting

$$\psi(z_1, z_2, z_3) = I(z_2 \le z_1) - I(z_3 \le z_1)$$

and

$$\begin{aligned} h((x_1, y_1), \ldots, (x_5, y_5)) &= \tfrac{1}{4}\psi(x_1, x_2, x_3)\psi(x_1, x_4, x_5) \\ &\quad \times \psi(y_1, y_2, y_3)\psi(y_1, y_4, y_5), \end{aligned}$$

we have $E_F\{h\} = \theta(F)$, and the corresponding U-statistic is

$$U_n = \frac{5!}{n(n-1)(n-2)(n-3)(n-4)} \sum_c h((X_{i_1}, Y_{i_1}), \ldots, (X_{i_5}, Y_{i_5})). \quad \blacksquare$$

5.1.2 Some Closely Related Statistics: V-Statistics

Corresponding to a U-statistic

$$U_n = \frac{1}{\binom{n}{m}} \sum_c h(X_{i_1}, \ldots, X_{i_m})$$

for estimation of $\theta(F) = E_F\{h\}$, the *associated von Mises statistic* is

$$\begin{aligned} V_n &= \frac{1}{n^m} \sum_{i_1=1}^{n} \cdots \sum_{i_m=1}^{n} h(X_{i_1}, \ldots, X_{i_m}) \\ &= \theta(F_n), \end{aligned}$$

where F_n denotes the sample distribution function. Let us term this statistic, in connection with a kernel h, the associated *V-statistic*. The connection between U_n and V_n will be examined closely in **5.7.3** and pursued further in Chapter 6.

Certain other statistics, too, may be treated as approximately a U-statistic, the gap being bridged via Slutsky's Theorem and the like. Thus the domain of application of the asymptotic theory of U-statistics is considerably wider than the context of unbiased estimation.

5.1.3 Generalized U-Statistics

The extension to the case of several samples is straightforward. Consider k independent collections of independent observations $\{X_1^{(1)}, X_2^{(1)}, \ldots\}, \ldots, \{X_1^{(k)}, X_2^{(k)}, \ldots\}$ taken from distributions $F^{(1)}, \ldots, F^{(k)}$, respectively. Let $\theta = \theta(F^{(1)}, \ldots, F^{(k)})$ denote a parametric function for which there is an unbiased estimator. That is,

$$\theta = E\{h(X_1^{(1)}, \ldots, X_{m_1}^{(1)}; \ldots; X_1^{(k)}, \ldots, X_{m_k}^{(k)})\},$$

where h is assumed, without loss of generality, to be symmetric within each of its k blocks of arguments. Corresponding to the "kernel" h and assuming $n_1 \geq m_1, \ldots, n_k \geq m_k$, the U-statistic for estimation of θ is defined as

$$U_n = \frac{1}{\prod_{j=1}^{k} \binom{n_j}{m_j}} \sum_c h(X_{i_{11}}^{(1)}, \ldots, X_{i_{1m_1}}^{(1)}; \ldots; X_{i_{k1}}^{(k)}, \ldots, X_{i_{km_k}}^{(k)}).$$

Here $\{i_{j1}, \ldots, i_{jm_j}\}$ denotes a set of m_j distinct elements of the set $\{1, 2, \ldots, n_j\}$, $1 \leq j \leq k$, and $\sum_c$ denotes summation over all such combinations.

The extension of Hoeffding's treatment of one-sample U-statistics to the k-sample case is due to Lehmann (1951) and Dwass (1956). Many statistics of interest are of the k-sample U-statistic type.

Example. *The Wilcoxon 2-sample statistic.* Let $\{X_1, \ldots, X_{n_1}\}$ and $\{Y_1, \ldots, Y_{n_2}\}$ be independent observations from continuous distributions F and G, respectively. Then, for

$$\theta(F, G) = \int G\, dF = P(X \leq Y),$$

an unbiased estimator is

$$U = \frac{1}{n_1 n_2} \sum_{i=1}^{n_1} \sum_{j=1}^{n_2} I(X_i \leq Y_j). \quad \blacksquare$$

5.1.4 An Optimality Property of U-Statistics

A U-statistic may be represented as the result of conditioning the kernel on the order statistic. That is, for a kernel $h(x_1, \ldots, x_m)$ and a sample $X_1, \ldots, X_n$, $n \geq m$, the corresponding U-statistic may be expressed as

$$U_n = E\{h(X_1, \ldots, X_m) | \mathbf{X}_{(n)}\},$$

where $\mathbf{X}_{(n)}$ denotes the order statistic $(X_{n1}, \ldots, X_{nn})$.

One implication of this representation is that any statistic $S = S(X_1, \ldots, X_n)$ for unbiased estimation of $\theta = \theta(F)$ may be "improved" by the corresponding U-statistic. That is, we have

Theorem. *Let* $S = S(X_1, \ldots, X_n)$ *be an unbiased estimator of* $\theta(F)$ *based on a sample* $X_1, \ldots, X_n$ *from the distribution* F. *Then the corresponding* U*-statistic is also unbiased and*

$$\mathrm{Var}_F\{U\} \leq \mathrm{Var}_F\{S\},$$

with equality if and only if $P_F(U = S) = 1$.

PROOF. The "kernel" associated with S is

$$\frac{1}{n!} \sum_p S(x_{i_1}, \ldots, x_{i_n}),$$

which in this case $(m = n)$ is the U-statistic associated with itself. That is, the U-statistic associated with S may be expressed as

$$U = E\{S | \mathbf{X}_{(n)}\}.$$

Therefore,

$$E_F\{U^2\} = E_F\{E^2\{S | \mathbf{X}_{(n)}\}\} \leq E_F\{E\{S^2 | \mathbf{X}_{(n)}\}\} = E_F\{S^2\},$$

with equality if and only if $E\{S | \mathbf{X}_{(n)}\}$ is degenerate and equals S with P_F-probability 1. Since $E_F\{U\} = E_F\{S\}$, the proof is complete. ■

Since the order statistic $\mathbf{X}_{(n)}$ is sufficient (in the usual technical sense) for any family $\mathscr{F}$ of distributions containing F, the U-statistic is the result of conditioning on a sufficient statistic. Thus the preceding result is simply a special case of the Rao–Blackwell theorem (see Rao (1973), §5a.2). In the case that $\mathscr{F}$ is rich enough that $\mathbf{X}_{(n)}$ is *complete* sufficient (e.g., if $\mathscr{F}$ contains all absolutely continuous F), then U_n is the minimum variance *unbiased* estimator of θ.

5.1.5 Martingale Structure of U-Statistics

Some important properties of U-statistics (see **5.2.1**, **5.3.3**, **5.3.4**, Section 5.4) flow from their martingale structure and a related representation.

Definitions. Consider a probability space $(\Omega, \mathscr{A}, P)$, a sequence of random variables $\{Y_n\}$, and a sequence of σ-fields $\{\mathscr{F}_n\}$ contained in $\mathscr{A}$, such that Y_n is $\mathscr{F}_n$-measurable and $E|Y_n| < \infty$. Then the sequence $\{Y_n, \mathscr{F}_n\}$ is called a *forward martingale* if

(a) $\mathscr{F}_1 \subset \mathscr{F}_2 \subset \ldots,$

(b) $E\{Y_{n+1}|\mathscr{F}_n\} = Y_n \ wp1$, all n,

and a *reverse martingale* if

(a′) $\mathscr{F}_1 \supset \mathscr{F}_2 \supset \ldots,$

(b′) $E\{Y_n|\mathscr{F}_{n+1}\} = Y_{n+1} \ wp1$, all n. ∎

The following lemmas, due to Hoeffding (1961) and Berk (1966), respectively, provide both forward and reverse martingale characterizations for U-statistics. For the first lemma, some preliminary notation is needed. Consider a symmetric kernel $h(x_1, \ldots, x_m)$ satisfying $E_F|h(X_1, \ldots, X_m)| < \infty$. We define the associated functions

$$h_c(x_1, \ldots, x_c) = E_F\{h(x_1, \ldots, x_c, X_{c+1}, \ldots, X_m)\}$$

for each $c = 1, \ldots, m - 1$ and put $h_m \equiv h$. Since

$$\int_A h_c(x_1, \ldots, x_c)dF(x_1)\cdots dF(x_c) = \int_{A \times R^{m-c}} h(x_1, \ldots, x_m)dF(x_1)\cdots dF(x_m)$$

for every Borel set A in R^c, h_c is (a version of) the conditional expectation of $h(X_1, \ldots, X_m)$ given $X_1, \ldots, X_c$:

$$h_c(x_1, \ldots, x_c) = E_F\{h(X_1, \ldots, X_m)|X_1 = x_1, \ldots, X_c = x_c\}.$$

Further, note that for $1 \le c \le m - 1$

$$h_c(x_1, \ldots, x_c) = E_F\{h_{c+1}(x_1, \ldots, x_c, X_{c+1})\}.$$

It is convenient to center at expectations, by defining

$$\theta(F) = E_F\{h(X_1, \ldots, X_m)\},$$
$$\tilde{h} = h - \theta(F),$$

and

$$\tilde{h}_c = h_c - \theta(F), \qquad 1 \le c \le m.$$

We now define

$$g_1(x_1) = \tilde{h}_1(x_1),$$
$$g_2(x_1, x_2) = \tilde{h}_2(x_1, x_2) - g_1(x_1) - g_1(x_2),$$
$$g_3(x_1, x_2, x_3) = \tilde{h}_3(x_1, x_2, x_3) - \sum_{i=1}^{3} g_1(x_i) - \sum_{1 \le i < j \le 3} g_2(x_i, x_j),$$
$$\ldots,$$
$$(*) \quad g_m(x_1, \ldots, x_m) = \tilde{h}(x_1, \ldots, x_m) - \sum_{i=1}^{m} g_1(x_i) - \sum_{1 \le i_1 < i_2 \le m} g_2(x_{i_1}, x_{i_2})$$
$$- \cdots - \sum_{1 \le i_1 < \cdots < i_{m-1} \le m} g_{m-1}(x_{i_1}, \ldots, x_{i_{m-1}}).$$

Clearly, the g_c's are symmetric in their arguments. Also, it is readily seen (check) that

$$E_F\{g_1(X_1)\} = 0,$$
$$E_F\{g_2(x_1, X_2)\} = 0,$$
$$\ldots,$$
$$E_F\{g_m(x_1, \ldots, x_{m-1}, X_m)\} = 0.$$

Now consider a sample $X_1, \ldots, X_n (n \ge m)$ and note that the U-statistic U_n corresponding to the kernel h satisfies

$$U_n - \theta(F) = \binom{n}{m}^{-1} S_n,$$

where

$$(1) \qquad S_n = \sum_{1 \le i_1 < \cdots < i_m \le n} \tilde{h}(X_{i_1}, \ldots, X_{i_m}).$$

Finally, for $1 \le c \le m$, put

$$S_{cn} = \sum_{1 \le i_1 < \cdots < i_c \le n} g_c(X_{i_1}, \ldots, X_{i_c}).$$

Hoeffding's lemma, which we now state, asserts a martingale property for the sequence $\{S_{cn}\}_{n \ge c}$ for each $c = 1, \ldots, m$, and gives a representation for U_n in terms of $S_{1n}, \ldots, S_{mn}$.

Lemma A (Hoeffding). *Let* $h = h(x_1, \ldots, x_m)$ *be a symmetric kernel for* $\theta = \theta(F)$, *with* $E_F|h| < \infty$. *Then*

$$(2) \qquad U_n - \theta = \sum_{c=1}^{m} \binom{m}{c} \binom{n}{c}^{-1} S_{cn}.$$

Further, for each c = 1, ..., m,

(3) $$E_F\{S_{cn}|X_1, \ldots, X_k\} = S_{ck}, \qquad c \le k \le n.$$

Thus, with $\mathscr{F}_k = \sigma\{X_1, \ldots, X_k\}$, *the sequence* $\{S_{cn}, \mathscr{F}_n\}_{n \ge c}$ *is a forward martingale.*

PROOF. The definition of g_m in (*) expresses $\tilde{h}$ in terms of $g_1, \ldots, g_m$. Substitution in (1) then yields

$$S_n = S_{mn} + \sum_{c=1}^{m-1} \sum_{1 \le i_1 < \cdots < i_m \le n} \sum_{1 \le j_1 < \cdots < j_c \le m} g_c(X_{i_{j_1}}, \ldots, X_{i_{j_c}}).$$

On the right-hand side, the term for $c = 1$ may be written

$$\sum_{1 \le i_1 < \cdots < i_m \le n} \sum_{j=1}^{m} g(X_{i_j}).$$

In this sum, each $g(X_i)$, $1 \le i \le n$, is represented the same number of times. Since the sum contains $\binom{n}{m} \cdot m$ terms, each $g(X_i)$ appears $n^{-1}\binom{n}{m}m$ times. That is, the sum $S_{1n} = \sum_1^n g(X_i)$ appears $\binom{n}{1}^{-1}\binom{n}{m}\binom{m}{1}$ times. In this fashion we obtain

$$S_n = \sum_{c=1}^{m} \binom{n}{c}^{-1} \binom{n}{m} \binom{m}{c} S_{cn},$$

which yields (2). To see the martingale property (3), observe that

$$E_F\{g_c(X_{i_1}, \ldots, X_{i_c})|X_1, \ldots, X_k\} = 0$$

if one of $i_1, \ldots, i_c$ is not contained in $\{1, \ldots, k\}$. For example, if $i_1 \notin \{1, \ldots, k\}$, then

$$\begin{aligned} &E_F\{g_c(X_{i_1}, \ldots, X_{i_c})|X_1, \ldots, X_k\} \\ &\quad = E_F\{E_F[g_c(X_{i_1}, \ldots, X_{i_c})|X_1, \ldots, X_k, X_{i_2}, \ldots, X_{i_c}]|X_1, \ldots, X_k\} \\ &\quad = E_F\{E_F[g_c(X_{i_1}, \ldots, X_{i_c})|X_{i_2}, \ldots, X_{i_c}]|X_1, \ldots, X_k\} \\ &\quad = E_F\{0|X_1, \ldots, X_k\} = 0. \end{aligned}$$

Thus

$$E_F\{S_{cn}|X_1, \ldots, X_k\} = \sum_{1 \le i_1 < \cdots < i_c \le k} g_c(X_{i_1}, \ldots, X_{i_c}) = S_{ck}. \quad ■$$

Example A. For the case $m = 1$ and $h(x) = x$, Lemma A states simply that

$$U_n - \theta = \frac{1}{n} \sum_{i=1}^{n} (X_i - \theta)$$

and that $\{\sum_1^n (X_i - \theta), \sigma(X_1, \ldots, X_n)\}$ is a forward martingale. ■

The other martingale representation for U_n is much simpler:

Lemma B (Berk). *Let* $h = h(x_1, \ldots, x_m)$ *be a symmetric kernel for* $\theta = \theta(F)$, *with* $E_F|h| < \infty$. *Then, with* $\mathscr{F}_n = \sigma\{\mathbf{X}_{(n)}, X_{n+1}, X_{n+2}, \ldots\}$, *the sequence* $\{U_n, \mathscr{F}_n\}_{n \geq m}$ *is a reverse martingale.*

PROOF. (exercise) Apply the representation

$$U_n = E\{h(X_1, \ldots, X_m)|\mathbf{X}_{(n)}\}$$

considered in **5.1.4**. ■

Example B (*continuation*). For the case $m = 1$ and $h(x) = x$, Lemma B asserts that $\bar{X}$ is a reverse martingale. ■

5.1.6 Representation of a *U*-Statistic as an Average of (Dependent) Averages of I.I.D. Random Variables

Consider a symmetric kernel $h(x_1, \ldots, x_m)$ and a sample $X_1, \ldots, X_n$ of size $n \geq m$. Define $k = [n/m]$, the greatest integer $\leq n/m$, and define

$$W(x_1, \ldots, x_n) = \frac{h(x_1, \ldots, x_m) + h(x_{m+1}, \ldots, x_{2m}) + \cdots + h(x_{km-m+1}, \ldots, x_{km})}{k}.$$

Letting $\sum_p$ denote summation over all $n!$ permutations $(i_1, \ldots, i_n)$ of $(1, \ldots, n)$ and $\sum_c$ denote summation over all $\binom{n}{m}$ combinations $\{i_1, \ldots, i_m\}$ from $\{1, \ldots, n\}$, we have

$$k \sum_p W(x_{i_1}, \ldots, x_{i_n}) = km!(n-m)! \sum_c h(x_{i_1}, \ldots, x_{i_m}),$$

and thus

$$\sum_p W(X_{i_1}, \ldots, X_{i_n}) = m!(n-m)!\binom{n}{m} U_n,$$

or

$$U_n = \frac{1}{n!} \sum_p W(X_{i_1}, \ldots, X_{i_n}).$$

This expresses U_n as an average of $n!$ terms, each of which is itself an average of k I.I.D. random variables. This type of representation was introduced and utilized by Hoffding (1963). We shall apply it in Section 5.6.

5.1.7 Weighted *U*-Statistics

Consider now an arbitrary kernel $h(x_1, \ldots, x_n)$, not necessarily symmetric, to be applied as usual to observations $X_1, \ldots, X_n$ taken m at a time. Suppose also that each term $h(X_{i_1}, \ldots, X_{i_m})$ becomes weighted by a factor $w(i_1, \ldots, i_m)$

depending only on the indices $i_1, \ldots, i_m$. In this case the U-statistic sum takes the more general form

$$T_n = \sum_c w(i_1, \ldots, i_m) h(X_{i_1}, \ldots, X_{i_m}).$$

In the case that h is symmetric and the weights $w(i_1, \ldots, i_m)$ take only 0 or 1 as values, a statistic of this form represents an "incomplete" or "reduced" U-statistic sum, designed to be computationally simpler than the usual sum. This is based on the notion that, on account of the dependence among the $\binom{n}{m}$ terms of the complete sum, it should be possible to use less terms without losing much information. Such statistics have been investigated by Blom (1976) and Brown and Kildea (1978).

Certain "permutation statistics" arising in nonparametric inference are asymptotically equivalent to statistics of the above form, with weights not necessarily 0- and 1-valued. For these and other applications, the statistics of form T_n with h symmetric and $m = 2$ have been studied by Shapiro and Hubert (1979).

Finally, certain "weighted rank statistics" for simple linear regression take the form T_n. Following Sievers (1978), consider the simple linear regression model

$$y_i = \alpha + \beta x_i + e_i, \qquad 1 \le i \le n,$$

where α and β are unknown parameters, $x_1, \ldots, x_n$ are known regression scores, and $e_1, \ldots, e_n$ are I.I.D. with distribution F. Sievers considers inferences for β based on the random variables

$$T_\beta = \sum_{i=1}^{n-1} \sum_{j=i+1}^{n} a_{ij} \phi(Y_i - \alpha - \beta x_i, Y_j - \alpha - \beta x_j),$$

where $\phi(u, v) = I(u \le v)$, the weights $a_{ij} \ge 0$ are arbitrary, and it is assumed that $x_1 \le \cdots \le x_n$ with at least one strict inequality. For example, a test of $H_0\colon \beta = \beta_0$ against $H_1\colon \beta > \beta_0$ may be based on the statistic T_{β_0}. Under the null hypothesis, the distribution of T_{β_0} is the same as that of T_0 when $\beta = 0$. That is, it is the same as

$$\sum_{i=1}^{n} \sum_{j=i+1}^{n} a_{ij} \phi(e_i, e_j),$$

which is of the form T_n above. The a_{ij}'s here are selected to achieve high asymptotic efficiency. Recommended weights are $a_{ij} = x_j - x_i$.

5.2 THE VARIANCE AND OTHER MOMENTS OF A U-STATISTIC

Exact formulas for the variance of a U-statistic are derived in **5.2.1**. The higher moments are difficult to deal with exactly, but useful bounds are obtained in **5.2.2**.

5.2.1 The Variance of a U-Statistic

Consider a symmetric kernel $h(x_1, \ldots, x_m)$ satisfying

$$E_F\{h^2(X_1, \ldots, X_m)\} < \infty.$$

We shall again make use of the functions h_c and $\tilde{h}_c$ introduced in **5.1.5**. Recall that $h_m = h$ and, for $1 \leq c \leq m - 1$,

$$h_c(x_1, \ldots, x_c) = E_F\{h(x_1, \ldots, x_c, X_{c+1}, \ldots, X_m)\},$$

that $\tilde{h} = h - \theta$, $\tilde{h}_c = h_c - \theta (1 \leq c \leq m)$, where

$$\theta = \theta(F) = E_F\{h(X_1, \ldots, X_m)\},$$

and that, for $1 \leq c \leq m - 1$,

$$h_c(x_1, \ldots, x_c) = E_F\{h_{c+1}(x_1, \ldots, x_c, X_{c+1})\}.$$

Note that

$$E_F\tilde{h}_c(X_1, \ldots, X_c) = 0, \qquad 1 \leq c \leq m.$$

Define $\zeta_0 = 0$ and, for $1 \leq c \leq m$,

$$\zeta_c = \text{Var}_F\{h_c(X_1, \ldots, X_c)\} = E_F\{\tilde{h}_c^2(X_1, \ldots, X_c)\}.$$

We have (Problem 5.P.3(i))

$$0 = \zeta_0 \leq \zeta_1 \leq \cdots \leq \zeta_m = \text{Var}_F\{h\} < \infty.$$

Before proceeding further, let us exemplify these definitions. Note from the following example that the functions h_c and $\tilde{h}_c$ depend on F for $c \leq m - 1$. The role of these functions is technical.

Example A. $\theta(F) = \sigma^2(F)$. Writing $\mu = \mu(F)$, $\sigma^2 = \sigma^2(F)$ and $\mu_4 = \mu_4(F)$, we have

$$\begin{aligned}
h(x_1, x_2) &= \tfrac{1}{2}(x_1^2 + x_2^2 - 2x_1x_2) = \tfrac{1}{2}(x_1 - x_2)^2, \\
\tilde{h}(x_1, x_2) &= h(x_1, x_2) - \sigma^2, \\
h_1(x) &= \tfrac{1}{2}(x^2 + \sigma^2 + \mu^2 - 2x\mu), \\
\tilde{h}_1(x) &= \tfrac{1}{2}(x^2 - \sigma^2 + \mu^2 - 2x\mu) = \tfrac{1}{2}[(x - \mu)^2 - \sigma^2], \\
E\{h^2\} &= \tfrac{1}{4}E\{[(X_1 - \mu) - (X_2 - \mu)]^4\} \\
&= \frac{1}{4}\sum_{j=0}^{4}\binom{4}{j}(-1)^{4-j}E\{(X_1 - \mu)^j\}E\{(X_2 - \mu)^{4-j}\} \\
&= \tfrac{1}{4}(2\mu_4 + 6\sigma^4), \\
\zeta_2 &= E\{h^2\} - \sigma^4 = \tfrac{1}{2}(\mu_4 + \sigma^4), \\
\zeta_1 &= E\{\tilde{h}_1^2\} = \tfrac{1}{4}\,\text{Var}_F\{(X_1 - \mu)^2\} = \tfrac{1}{4}(\mu_4 - \sigma^4). \quad \blacksquare
\end{aligned}$$

Next let us consider two sets $\{a_1, \ldots, a_m\}$ and $\{b_1, \ldots, b_m\}$ of m distinct integers from $\{1, \ldots, n\}$ and let c be the number of integers common to the two sets. It follows (Problem 5.P.4) by symmetry of $\tilde{h}$ and by independence of $\{X_1, \ldots, X_n\}$ that

$$E_F\{\tilde{h}(X_{a_1}, \ldots, X_{a_n})\tilde{h}(X_{b_1}, \ldots, X_{b_m})\} = \zeta_c.$$

Note also that the number of distinct choices for two such sets having exactly c elements in common is $\binom{n}{m}\binom{m}{c}\binom{n-m}{m-c}$.

With these preliminaries completed, we may now obtain the variance of a U-statistic. Writing

$$U_n - \theta = \binom{n}{m}^{-1} \sum_c \tilde{h}(X_{i_1}, \ldots, X_{i_m}),$$

we have

$$\begin{aligned} \mathrm{Var}_F\{U_n\} &= E_F\{(U_n - \theta)^2\} \\ &= \binom{n}{m}^{-2} \sum_c \sum_c E_F\{\tilde{h}(X_{a_1}, \ldots, X_{a_m})\tilde{h}(X_{b_1}, \ldots, X_{b_m})\} \\ &= \binom{n}{m}^{-2} \sum_{c=0}^{n} \binom{n}{m}\binom{m}{c}\binom{n-m}{m-c}\zeta_c. \end{aligned}$$

This result and other useful relations from Hoeffding (1948) may be stated as follows.

Lemma A. *The variance of* $\mathrm{U_n}$ *is given by*

(*) $$\mathrm{Var_F\{U_n\}} = \binom{\mathrm{n}}{\mathrm{m}}^{-1} \sum_{\mathrm{c}=1}^{\mathrm{m}} \binom{\mathrm{m}}{\mathrm{c}}\binom{\mathrm{n-m}}{\mathrm{m-c}}\zeta_{\mathrm{c}}$$

and satisfies

(i) $\dfrac{\mathrm{m}^2}{\mathrm{n}}\zeta_1 \le \mathrm{Var_F\{U_n\}} \le \dfrac{\mathrm{m}}{\mathrm{n}}\zeta_{\mathrm{m}}$;

(ii) $(\mathrm{n}+1)\mathrm{Var_F\{U_{n+1}\}} \le \mathrm{n}\,\mathrm{Var_F\{U_n\}}$;

(iii) $\mathrm{Var_F\{U_n\}} = \dfrac{\mathrm{m}^2\zeta_1}{\mathrm{n}} + \mathrm{O(n^{-2})}, \qquad \mathrm{n} \to \infty.$

Note that (*) is a fixed sample size formula. Derive (i), (ii), and (iii) from (*) as an exercise.

Example B (*Continuation*).

$$\begin{aligned}\mathrm{Var}_F\{s^2\} &= \binom{n}{2}^{-1}[2(n-2)\zeta_1 + \zeta_2] \\ &= \frac{4\zeta_1}{n} + \frac{2\zeta_2}{n(n-1)} - \frac{4\zeta_1}{n(n-1)} \\ &= \frac{\mu_4 - \sigma^4}{n} + \frac{2\sigma^4}{n(n-1)} \\ &= \frac{\mu_4 - \sigma^4}{n} + O(n^{-2}). \quad \blacksquare\end{aligned}$$

The extension of (*) to the case of a *generalized* U-statistic is straightforward (Problem 5.P.6).

An alternative formula for $\mathrm{Var}_F\{U_n\}$ is obtained by using, instead of h_c and $\tilde{h}_c$, the functions g_c introduced in **5.1.5** and the representation given by Lemma 5.1.5A.

Consider a set $\{i_1, \ldots, i_c\}$ of c distinct integers from $\{1, \ldots, n\}$ and a set $\{j_1, \ldots, j_d\}$ of d distinct integers from $\{1, \ldots, n\}$, where $1 \le c, d \le m$. It is evident from the proof of Lemma 5.1.5A that if one of $\{i_1, \ldots, i_c\}$ is not contained in $\{j_1, \ldots, j_d\}$, then

$$E_F\{g_c(X_{i_1}, \ldots, X_{i_c}) | X_{j_1}, \ldots, X_{j_d}\} = 0.$$

From this it follows that $E_F\{g_c(X_{i_1}, \ldots, X_{i_c})g_d(X_{j_1}, \ldots, X_{j_d})\} = 0$ *unless* $c = d$ and $\{i_1, \ldots, i_c\} = \{j_1, \ldots, j_d\}$. Therefore, for the functions

$$S_{cn} = \sum_{1 \le i_1 < \cdots < i_c \le n} g_c(X_{i_1}, \ldots, X_{i_c}),$$

we have

$$E\{S_c S_d\} = \begin{cases} \binom{n}{c} E\{g_c^2\}, & c = d, \\ 0, & c \ne d. \end{cases}$$

Hence

Lemma B. *The variance of* U_n *is given by*

(**) $$\mathrm{Var}_F\{U_n\} = \sum_{c=1}^{m} \binom{m}{c}^{-2} \binom{n}{c}^{-1} E_F\{g_c^2\}.$$

The result (iii) of Lemma A follows slightly more readily from (**) than from (*).

Example C (Continuation). We have

$$g_1(x) = \tilde{h}_1(x) = \tfrac{1}{2}[(x-\mu)^2 - \sigma^2],$$
$$g_2(x_1, x_2) = \tilde{h}_2(x_1, x_2) - \tilde{h}_1(x_1) - \tilde{h}_1(x_2) = \mu^2 + x_1\mu + x_2\mu - x_1x_2,$$
$$E\{g_1^2\} = \zeta_1 = \tfrac{1}{4}(\mu_4 - \sigma^4), \text{ as before,}$$
$$E\{g_2^2\} = \sigma^4,$$

and thus

$$\begin{aligned}\operatorname{Var}_F\{s^2\} &= \frac{4}{n}E\{g_1^2\} + \frac{2}{n(n-1)}E\{g_2^2\}\\ &= \frac{\mu_4 - \sigma^4}{n} + \frac{2\sigma^4}{n(n-1)}, \text{ as before. } \blacksquare\end{aligned}$$

The rate of convergence of $\operatorname{Var}\{U_n\}$ to 0 depends upon the least c for which ζ_c is nonvanishing. From either Lemma A or Lemma B, we obtain

Corollary. *Let* $c \geq 1$ *and suppose that* $\zeta_0 = \cdots = \zeta_{c-1} = 0 < \zeta_c$. *Then*

$$E(U_n - \theta)^2 = O(n^{-c}), \qquad n \to \infty.$$

Note that the condition $\zeta_d = 0$, $d < c$, is equivalent to the condition $E\{\tilde{h}_d^2\} = 0$, $d < c$, and also to the condition $E\{g_d^2\} = 0$, $d < c$.

5.2.2 Other Moments of *U*-Statistics

Exact generalizations of Lemmas 5.2.1 A, B for moments of order other than 2 are difficult to work out and complicated even to state. However, for most purposes, suitable bounds suffice. Fortunately, these are rather easily obtained.

Lemma A. *Let* r *be a real number* ≥ 2. *Suppose that* $E_F|h|^r < \infty$. *Then*

$$(*) \qquad E|U_n - \theta|^r = O(n^{-(1/2)r}), \qquad n \to \infty.$$

PROOF. We utilize the representation of U_n as an average of averages of I.I.D.'s (**5.1.6**),

$$U_n - \theta = (n!)^{-1} \sum_p \tilde{W}(X_{i_1}, \ldots, X_{i_n}),$$

where $\tilde{W}(X_{i_1}, \ldots, X_{i_n}) = W(X_{i_1}, \ldots, X_{i_n}) - \theta$ is an average of $k = [n/m]$ I.I.D. terms of the form $\tilde{h}(X_{i_1}, \ldots, X_{i_m})$. By Minkowski's inequality,

$$E|U_n - \theta|^r \leq E|\tilde{W}(X_1, \ldots, X_n)|^r.$$

By Lemma 2.2.2B, $E|\tilde{W}(X_1, \ldots, X_n)|^r = O(k^{-(1/2)r})$, $k \to \infty$. $\blacksquare$

Lemma B. *Let* $c \geq 1$ *and suppose that* $\zeta_0 = \cdots = \zeta_{c-1} = 0 < \zeta_c$. *Let* r *be an integer* ≥ 2 *and suppose that* $E_F|h|^r < \infty$. *Then*

$$(**) \qquad E(U_n < \theta)^r = O(n^{-[(1/2)(rc+1)]}), \qquad n \to \infty,$$

where $[\cdot]$ *denotes integer part.*

PROOF. Write

$$(1) \qquad E(U_n - \theta)^r = \binom{n}{m}^{-r} \sum E\left\{\prod_{j=1}^{r} \tilde{h}(X_{i_{j1}}, \ldots, X_{i_{jm}})\right\},$$

where "j" identifies the factor within the product, and $\sum$ denotes summation over all $\binom{n}{m}^r$ of the indicated terms. Consider a typical term. For the jth factor, let p_j denote the number of indices repeated in other factors. If $p_j \leq c - 1$, then (justify)

$$E\{\tilde{h}(X_{i_{j1}}, \ldots, X_{i_{jm}}) | \text{the } p_j \text{ repeated } X_{i_{jk}}\text{'s}\} = 0.$$

Thus a term in (1) can have nonzero expectation only if each factor in the product contains at least c indices which appear in other factors in the product. Denote by q the number of distinct elements among the repeated indices in the r factors of a given product. Then (justify)

$$(2) \qquad 2q \leq \sum_{j=1}^{r} p_j.$$

For fixed values of $q, p_1, \ldots, p_r$, the number of ways to select the indices in the r factors of a product is of order

$$(3) \qquad O(n^{q+(m-p_1)+\cdots+(m-p_r)}),$$

where the implicit constants depend upon r and m, but not upon n. Now, by (2), $q \leq [\frac{1}{2}\sum_{j=1}^{r} p_j]$. Thus

$$q + \sum_{j=1}^{r}(m - p_j) \leq rm + \left[\frac{1}{2}\sum_{j=1}^{m} p_j\right] - \sum_{j=1}^{m} p_j = rm - \left[\frac{1}{2}\left(\sum_{j=1}^{r} p_j + 1\right)\right],$$

since (verify), for any integer x, $x - [\frac{1}{2}x] = [\frac{1}{2}(x+1)]$. Confining attention to the case that $p_1 \geq c, \ldots, p_r \geq c$, we have $\sum_{j=1}^{r} p_j \geq rc$, so that

$$(4) \qquad q + \sum_{j=1}^{r}(m - p_j) \leq rm - [\tfrac{1}{2}(rc + 1)].$$

The number of ways to select the values $q, p_1, \ldots, p_r$ depends on r and m, but not upon n. Thus, by (3) and (4), it follows that the number of terms in the sum in (1) for which the expectation is possibly nonzero is of order

$$O(n^{rm-[(1/2)(rc+1)]}), \qquad n \to \infty.$$

Since $\binom{n}{m}^{-1} = O(n^{-m})$, the relation (*) is proved. ∎

Remarks. (i) Lemma A generalizes to rth order the relation $E(U_n - \theta)^2 = O(n^{-1})$ expressed in Lemma 5.2.1A.

(ii) Lemma B generalizes to rth order the relation $E(U_n - \theta)^2 = O(n^{-c})$, given $\zeta_{c-1} = 0$, expressed in Corollary 5.2.1.

(iii) In the proof of Theorem 2.3.3, it was seen that

$$E(\bar{X} - \mu)^3 = \mu_3 n^{-2} = O(n^{-2}).$$

This corresponds to (**) in the case $m = 1, c = 1, r = 3$ of Lemma B.

(iv) For a *generalized* U-statistic based on k samples, (**) holds with n given by $n = \min\{n_1, \ldots, n_k\}$. The extension of the preceding proof is straightforward (Problem 5.P.8).

(v) An application of Lemma B in the case $c \geq 2$ arises in connection with the approximation of a U-statistic by its *projection*, as discussed in **5.3.2** below. (Indeed, the proof of Lemma B is based on the method used by Grams and Serfling (1973) to prove Theorem 5.3.2.) ■

5.3 THE PROJECTION OF A U-STATISTIC ON THE BASIC OBSERVATIONS

An appealing feature of a U-statistic is its simple structure as a sum of identically distributed random variables. However, except in the case of a kernel of dimension $m = 1$, the summands are *not* all independent, so that a direct application of the abundant theory for sums of *independent* random variables is not possible. On the other hand, by the special device of "projection," a U-statistic may be approximated within a sufficient degree of accuracy by a sum of I.I.D. random variables. In this way, classical limit theory for sums does carry over to U-statistics and yields the relevant asymptotic distribution theory and almost sure behavior.

Throughout we consider as usual a U-statistic U_n based on a symmetric kernel $h = h(x_1, \ldots, x_m)$ and a sample $X_1, \ldots, X_n$ $(n \geq m)$ from a distribution F, with $\theta = E_F\{h(X_1, \ldots, X_m)\}$.

In **5.3.1** we define and evaluate the projection $\hat{U}_n$ of a U-statistic U_n. In **5.3.2** the moments of $U_n - \hat{U}_n$ are characterized, thus providing the basis for negligibility of $U_n - U_n$ in appropriate senses. As an application, a representation for U_n as a mean of I.I.D.'s plus a negligible random variable is obtained in **5.3.3**. Further applications are made in Sections 5.4 and 5.5.

In the case $\zeta_1 = 0$, the projection $\hat{U}_n$ serves no purpose. Thus, in **5.3.4**, we consider an extended notion of projection for the general case $\zeta_0 = \cdots = \zeta_{c-1} = 0 < \zeta_c$.

In Chapter 9 we shall further treat the concept of projection, considering it in general for an arbitrary statistic S_n in place of the U-statistic U_n.

5.3.1 The Projection of U_n

Assume $E_F|h| < \infty$. The *projection* of the U-statistic U_n is defined as

$$\hat{U}_n = \sum_{i=1}^{n} E_F\{U_n|X_i\} - (n-1)\theta. \tag{1}$$

Note that it is exactly a sum of I.I.D. random variables. In terms of the function $\tilde{h}_1$ considered in Section 5.2. we have

$$\hat{U}_n - \theta = \frac{m}{n}\sum_{i=1}^{n} \tilde{h}_1(X_1). \tag{2}$$

Verify (Problem 5.P.9) this in the wider context of a *generalized* U-statistic. From (2) it is evident that $\hat{U}_n$ is of no interest in the case $\zeta_1 = 0$. However, in this case we pass to a certain analogue (**5.3.4**).

5.3.2 The Moments of $U_n - \hat{U}_n$

Here we treat the difference $U_n - \hat{U}_n$. It is useful that $U_n - \hat{U}_n$ may itself be expressed as a U-statistic, namely (Problem 5.P.10).

$$U_n - \hat{U}_n = \frac{1}{\binom{n}{m}} \sum_c H(X_{i_1}, \ldots, X_{i_m}),$$

based on the symmetric kernel

$$H(x_1, \ldots, x_m) = h(x_1, \ldots, x_m) - \tilde{h}_1(x_1) - \cdots - \tilde{h}_1(x_m) - \theta.$$

Note that $E_F\{H\} = E_F\{H|X_1\} = 0$. That is, in an obvious notation, $\zeta_1^{(H)} = 0$. An application of Lemma 5.2.2B with $c = 2$ thus yields

Theorem. *Let* v *be an even integer. If* $E_F H^v < \infty$ *(implied by* $E_F h^v < \infty$*), then*

$$E_F(U_n - \hat{U}_n)^v = O(n^{-v}), \qquad n \to \infty. \tag{*}$$

For $v = 2$, relation (*) was established by Hoeffding (1948) and applied to obtain the CLT for U-statistics, as will be seen in Section 5.5. It also yields the LIL for U-statistics (Section 5.4). Indeed, as seen below in **5.3.3**, it leads to an almost sure representation of U_n as a mean of I.I.D.'s. However, for information on the *rates* of convergence in such results as the CLT and SLLN for U-statistics, the case $v > 2$ in (*) is apropos. This extension was obtained by Grams and Serfling (1973). Sections 5.4 and 5.5 exhibit some relevant rates of convergence.

5.3.3 Almost Sure Representation of a *U*-Statistic as a Mean of I.I.D.'s

Theorem. *Let* ν *be an even integer. Suppose that* $E_F h^\nu < \infty$. *Put*

$$U_n = \hat{U}_n + R_n.$$

Then, for any $\delta > 1/\nu$, *with probability* 1

$$R_n = o(n^{-1}(\log n)^\delta), \qquad n \to \infty,$$

PROOF. Let $\delta > 1/\nu$. Put $\lambda_n = n(\log n)^{-\delta}$. It suffices to show that, for any $\varepsilon > 0$, *wp*1 $\lambda_n|R_n| < \varepsilon$ for all n sufficiently large, that is,

(1) $$P(\lambda_n|R_n| > \varepsilon \text{ for infinitely many } n) = 0.$$

Let $\varepsilon > 0$ be given. By the Borel–Cantelli lemma, and since λ_n is nondecreasing for large n, it suffices for (1) to show that

(2) $$\sum_{k=0}^{\infty} P\left(\lambda_{2^{k+1}} \max_{2^k \le n \le 2^{k+1}} |R_n| > \varepsilon\right) < \infty.$$

Since $R_n = U_n - \hat{U}_n$ is itself a U-statistic as noted in **5.3.2** and hence a reverse martingale as noted in Lemma 5.1.5B, we may apply a standard result (Loeve (1978), Section 32) to write

$$P\left(\sup_{j \ge n} |U_j - \hat{U}_j| > t\right) \le t^{-\nu} E|U_n - \hat{U}_n|^\nu.$$

Thus, by Theorem 5.3.2, the kth term in (2) is bounded by (check)

$$\varepsilon^{-\nu} \lambda_{2^{k+1}} E_F |U_{2^k} - \hat{U}_{2^k}|^\nu = O((k+1)^{-\delta\nu}).$$

Since $\delta\nu > 1$, the series in (2) is convergent. ∎

The foregoing result is given and utilized by Geertsema (1970).

5.3.4 The "Projection" of U_n for the General Case $\zeta_0 = \cdots = \zeta_{c-1} = 0 < \zeta_c$

(It is assumed that $E_F h^2 < \infty$.) Since $\zeta_d = 0$ for $d < c$, the variance formula for U-statistics (Lemma 5.2.1A) yields

$$\mathrm{Var}_F\{U_n\} = \frac{c!\binom{m}{c}^2 \zeta_c}{n^c} + O(n^{-c-1}), \qquad n \to \infty,$$

and thus

(1) $$\mathrm{Var}_F\{n^{(1/2)c}(U_n - \theta)\} \to c!\binom{m}{c}^2 \zeta_c, \qquad n \to \infty.$$

This suggests that in this case the random variable $n^{(1/2)c}(U_n - \theta)$ converges in distribution to a nondegenerate law.

Now, generalizing **5.3.1**, let us define the "projection" of U_n to be $\hat{U}_n$ given by

$$\hat{U}_n - \theta = \sum_{1 \le i_1 < \cdots < i_c \le n} E_F\{U_n | X_{i_1}, \ldots, X_{i_c}\} - \binom{n}{c}\theta.$$

Verify (Problem 5.P.11) that

$$\text{(2)} \quad \hat{U}_n - \theta = \frac{m(m-1)\cdots(m-c+1)}{n(n-1)\cdots(n-c+1)} \sum_{1 \le i_1 < \cdots < i_c \le n} \tilde{h}_c(X_{i_1}, \ldots, X_{i_c}).$$

Again (as in **5.3.2**), $U_n - \hat{U}_n$ is itself a U-statistic, based on the kernel

$$H(x_1, \ldots, x_m) = h(x_1, \ldots, x_m) - \sum_{1 \le i_1 < \cdots < i_c \le m} \tilde{h}_c(x_{i_1}, \ldots, x_{i_c}) - \theta,$$

with $E_F\{H\} = E_F\{H | X_1\} = \cdots = E_F\{H | X_1, \ldots, X_c\} = 0$, and thus $\zeta_c^{(H)} = 0$. Hence the variance formula for U-statistics yields

$$\text{(3)} \qquad E(U_n - \hat{U}_n)^2 = O(n^{-(c+1)}),$$

so that $E\{n^{(1/2)c}(U_n - \hat{U}_n)^2\} = O(n^{-1})$ and thus

$$n^{(1/2)c}(U_n - \hat{U}_n) \xrightarrow{p} 0.$$

Hence the limit law of $n^{(1/2)c}(U_n - \theta)$ may be found by obtaining that of $n^{(1/2)c}(\hat{U}_n - \theta)$. For the cases $c = 1$ and $c = 2$, this approach is carried out in Section 5.5.

Note that, more generally, for any even integer v, if $E_F H^v < \infty$ (implied by $E_F h^v < \infty$), then

$$\text{(4)} \qquad E|U_n - \hat{U}_n|^v = O(n^{-(1/2)v(c+1)}), \qquad n \to \infty,$$

The foregoing results may be extended easily to generalized U-statistics (Problem 5.P.12).

In the case under consideration, that is, $\zeta_{c-1} = 0 < \zeta_c$, the "projection" $\hat{U}_n - \theta$ corresponds to a term in the martingale representation of U_n given by Lemma 5.1.5A. Check (Problem 5.P.13) that $S_{0n} = \cdots = S_{c-1,n} = 0$ and

$$\hat{U}_n - \theta = \binom{m}{c}\binom{n}{c}^{-1} S_{cn}.$$

5.4 ALMOST SURE BEHAVIOR OF *U*-STATISTICS

The classical SLLN (Theorem 1.8B) generalizes to U-statistics:

Theorem A. *If* $E_F|h| < \infty$, *then* $U_n \xrightarrow{wp1} \theta$.

This result was first established by Hoeffding (1961), using the forward martingale structure of U-statistics given by Lemma 5.1.5A. A more direct proof, noted by Berk (1966), utilizes the *reverse* martingale representation of Lemma 5.1.5B. Since the classical SLLN has been generalized to reverse martingale sequences (see Doob (1953) or Loève (1978)), Theorem A is immediate.

For *generalized* k-sample U-statistics, Sen (1977) obtains strong convergence of U under the condition $E_F\{|h|(\log^+|h|^{k-1}\} < \infty$.

Under a slightly stronger moment assumption, namely $E_F h^2 < \infty$, Theorem A can be proved very simply. For, in this case, we have

$$E_F(U_n - \hat{U}_n)^2 = O(n^{-2})$$

as established in **5.3.2**. Thus $\sum_{n=1}^{\infty} E_F(U_n - \hat{U}_n)^2 < \infty$, so that by Theorem 1.3.5 $U_n - \hat{U}_n \xrightarrow{wp1} 0$. Now, as an application of the classical SLLN,

$$\hat{U}_n - \theta = \frac{m}{n} \sum_{i=1}^{n} \tilde{h}_1(X_i) \xrightarrow{wp1} mE_F\{\tilde{h}_1(X_1)\} = 0.$$

Thus $U_n \xrightarrow{wp1} \theta$. This argument extends to *generalized* U-statistics (Problem 5.P.14).

As an alternate proof, also restricted to the second order moment assumption, Theorem 5.3.3 may be applied for the part $U_n - \hat{U}_n \xrightarrow{wp1} 0$.

In connection with the strong convergence of U-statistics, the following rate of convergence is established by Grams and Serfling (1973). The argument uses Theorem 5.3.2 and the reverse martingale property to reduce to $\hat{U}_n$.

Theorem B. *Let* ν *be an even integer. If* $E_F h^\nu < \infty$, *then for any* $\varepsilon > 0$,

$$P\left(\sup_{k \geq n} |U_n - \theta| > \varepsilon\right) = O(n^{1-\nu}), \qquad n \to \infty.$$

The classical LIL may also be extended to U-statistics. As an exercise (Problem 5.P.15), prove

Theorem C. *Let* $h = h(x_1, \ldots, x_m)$ *be a kernel for* $\theta = \theta(F)$, *with* $E_F h^2 < \infty$ *and* $\zeta_1 > 0$. *Then*

$$\overline{\lim_{n\to\infty}} \frac{n^{1/2}(U_n - \theta)}{(2m^2\zeta_1 \log \log n)^{1/2}} = 1 \text{ wp1}.$$

5.5 ASYMPTOTIC DISTRIBUTION THEORY OF U-STATISTICS

Consider a kernel $h = h(x_1, \ldots, x_m)$ for unbiased estimation of $\theta = \theta(F) = E_F\{h\}$, with $E_F h^2 < \infty$. Let $0 = \zeta_0 \le \zeta_1 \le \cdots \le \zeta_m = \mathrm{Var}_F\{h\}$ be as defined in **5.2.1**. As discussed in **5.3.4**, in the case $\zeta_{c-1} = 0 < \zeta_c$, the random variable

$$n^{(1/2)c}(U_n - \theta)$$

has variance tending to a positive constant and its asymptotic distribution may be obtained by replacing U_n by its projection $\hat{U}_n$. In the present section we examine the limit distributions for the cases $c = 1$ and $c = 2$, which cover the great majority of applications. For $c = 1$, treated in **5.5.1**, the random variable $n^{1/2}(U_n - \theta)$ converges in distribution to a normal law. Corresponding rates of convergence are presented. For $c = 2$, treated in **5.5.2**, the random variable $n(U_n - \theta)$ converges in distribution to a weighted sum of (possibly infinitely many) independent χ_1^2 random variables.

5.5.1 The Case $\zeta_1 > 0$

The following result was established by Hoeffding (1948). The proof is left as an exercise (Problem 5.P.16).

Theorem A. *If* $E_F h^2 < \infty$ *and* $\zeta_1 > 0$, *then* $n^{1/2}(U_n - \theta) \xrightarrow{d} N(0, m^2\zeta_1)$, *that is,*

$$U_n \quad \text{is} \quad AN\left(\theta, \frac{m^2\zeta_1}{n}\right).$$

Example A. *The sample variance.* $\theta(F) = \sigma^2(F)$. As seen in **5.1.1** and **5.2.1**, $h(x_1, x_2) = \frac{1}{2}(x_1^2 + x_2^2 - 2x_1x_2)$, $\zeta_1 = (\mu_4 - \sigma^4)/4$, and

$$U_n = s^2 = \frac{1}{n-1}\sum_{i=1}^{n}(X_i - \bar{X})^2.$$

Assuming that F is such that $\sigma^4 < \mu_4 < \infty$, so that $E_F h^2 < \infty$ and $\zeta_1 > 0$, we obtain from Theorem A that

$$s^2 \quad \text{is} \quad AN\left(\sigma^2, \frac{\mu_4 - \sigma^4}{n}\right).$$

Compare Section 2.2, where the same conclusion was established for $m_2 = (n-1)s^2/n$.

In particular, suppose that F is binomial $(1, p)$. Then $\bar{X} = \hat{p}$, say, and (check) $s^2 = n\hat{p}(1 - \hat{p})/(n-1)$. Check that $\mu_4 - \sigma^4 > 0$ if and only if $p \neq \frac{1}{2}$. Thus Theorem A is applicable for $p \neq \frac{1}{2}$. (The case $p = \frac{1}{2}$ will be covered by Theorem 5.5.2.) ■

By routine arguments (Problem 5.P.18) it may be shown that a *vector* of several U-statistics based on the same sample is asymptotically multivariate normal. The appropriate limit covariance matrix may be found by the same method used in **5.2.1** for the computation of variances to terms of order $O(n^{-1})$.

It is also straightforward (Problem 5.P.19) to extend Theorem A to *generalized* U-statistics. In an obvious notation, for a k-sample U-statistic, we have

$$U \quad \text{is} \quad AN\left(\theta, \sum_{j=1}^{k} \frac{m_j^2 \zeta_{1j}}{n_j}\right),$$

provided that $n \sum m_j^2 \zeta_{1j}/n_j \geq B > 0$ as $n = \min\{n_1, \ldots, n_k\} \to \infty$.

Example B. *The Wilcoxon 2-sample statistic* (continuation of Example 5.1.3). Here $\theta = P(X \leq Y)$ and $h(x, y) = I(x \leq y)$. Check that $\zeta_{11} = P(X \leq Y_1, X \leq Y_2) - \theta^2$, $\zeta_{12} = P(X \leq Y, X_2 \leq Y) - \theta^2$. Under the null hypothesis that $\mathscr{L}(X) = \mathscr{L}(Y)$, we have $\theta = \frac{1}{2}$ and $\zeta_{11} = P(Y_3 \leq Y_1, Y_3 \leq Y_2) - \frac{1}{4} = \frac{1}{3} - \frac{1}{4} = \frac{1}{12} = \zeta_{12}$. In this case

$$U \quad \text{is} \quad AN\left(\frac{1}{2}, \frac{1}{12}\left(\frac{1}{n_1} + \frac{1}{n_2}\right)\right). \quad \blacksquare$$

The rate of convergence in the asymptotic normality of U-statistics has been investigated by Grams and Serfling (1973), Bickel (1974), Chan and Wierman (1977) and Callaert and Janssen (1978), the latter obtaining the sharpest result, as follows.

Theorem B. *If* $\nu = E|h|^3 < \infty$ *and* $\zeta_1 > 0$, *then*

$$\sup_{-\infty < t < \infty} \left| P\left(\frac{n^{1/2}(U_n - \theta)}{m\zeta_1^{1/2}} \leq t\right) - \Phi(t) \right| \leq C\nu(m^2\zeta_1)^{-3/2} n^{-1/2},$$

where C is an absolute constant.

5.5.2 The Case $\zeta_1 = 0 < \zeta_2$

For the function $\tilde{h}_2(x_1, x_2)$ associated with the kernel $h = h(x_1, \ldots, x_m)$ $(m \geq 2)$, we define an operator A on the function space $L_2(R, F)$ by

$$Ag(x) = \int_{-\infty}^{\infty} \tilde{h}_2(x, y) g(y) dF(y), \qquad x \in R, g \in L_2.$$

That is, A takes a function g into a new function Ag. In connection with any such operator A, we define the associated eigenvalues $\lambda_1, \lambda_2, \ldots$ to be the real numbers λ (not necessarily distinct) corresponding to the distinct solutions $g_1, g_2, \ldots$ of the equation

$$Ag - \lambda g = 0.$$

We shall establish

Theorem. *If* $E_F h^2 < \infty$ *and* $\zeta_1 = 0 < \zeta_2$, *then*

$$n(U_n - \theta) \xrightarrow{d} \frac{m(m-1)}{2} Y,$$

where Y *is a random variable of the form*

$$Y = \sum_{j=1}^{\infty} \lambda_j(\chi^2_{1j} - 1),$$

where $\chi^2_{11}, \chi^2_{12}, \ldots$ *are independent* χ^2_1 *variates, that is,* Y *has characteristic function*

$$E_F\{e^{itY}\} = \prod_{j=1}^{\infty} (1 - 2it\lambda_j)^{-1/2} e^{-it\lambda_j}.$$

Before developing the proof, let us illustrate the application of the theorem.

Example A. *The sample variance* (continuation of Examples 5.2.1A and 5.5.1A). We have $\tilde{h}_2(x, y) = \frac{1}{2}(x - y)^2 - \sigma^2$, $\zeta_1 = (\mu_4 - \sigma^4)/4$, and $\zeta_2 = \frac{1}{2}(\mu_4 + \sigma^4)$. Take now the case $\zeta_1 = 0$, that is, $\mu_4 = \sigma^4$. Then $\zeta_2 = \sigma^4 > 0$ and the preceding theorem may be applied. We seek values λ such that the equation

$$\int [\tfrac{1}{2}(x - y)^2 - \sigma^2] g(y) dF(y) = \lambda g(x)$$

has solutions g in $L_2(R, F)$. It is readily seen (justify) that any such g must be quadratic in form: $g(x) = ax^2 + bx + c$. Substituting this form of g in the equation and equating coefficients of x^0, x^1 and x^2, we obtain the system of equations

$$\frac{1}{2}\int y^2 g(y) dF(y) - \sigma^2 \int g(y) dF(y) = \lambda c, \qquad -\int y g(y) dF(y) = \lambda b,$$

$$\frac{1}{2}\int g(y) dF(y) = \lambda a.$$

Solutions (a, b, c, λ) depend upon F. In particular, suppose that F is binomial $(1, p)$, with $p = \frac{1}{2}$. Then (check) $\sigma^2 = \frac{1}{4}$, $\mu_4 = \sigma^4$, $\int y^k\, dF(y) = \frac{1}{2}$ for all k. Then (check) the system of equations becomes equivalently

$$a + b + 2c = 4a\lambda, \qquad a + b + c = -2b\lambda, \qquad a + b + c = (4c + 2a)\lambda.$$

It is then easily found (check) that $a = 0, b = -2c$, and $\lambda = -\frac{1}{4}$, in which case $g(x) = c(2x - 1)$, c arbitrary. The theorem thus yields, for this F,

$$n(s^2 - \tfrac{1}{4}) \xrightarrow{d} -\tfrac{1}{4}(\chi_1^2 - 1). \quad \blacksquare$$

Remark. Do s^2 and m_2 $(=(n-1)s^2/n)$ always have the same asymptotic distribution theory? Intuitively this would seem plausible, and indeed typically it is true. However, for F binomial $(1, \frac{1}{2})$, we have (Problem 5.P.22)

$$n(m_2 - \tfrac{1}{4}) \xrightarrow{d} -\tfrac{1}{4}\chi_1^2,$$

which differs from the above result for s^2. $\blacksquare$

Example B. $\theta(F) = \mu^2(F)$. We have $h(x_1, x_2) = x_1 x_2$ and

$$U_n = \frac{1}{\binom{n}{2}} \sum_{i<j} X_i X_j.$$

Check that $\zeta_1 = \mu^2\sigma^2$ and $\zeta_2 = \sigma^4 - 2\mu^2\sigma^2$. Assume that $0 < \sigma^2 < \infty$. Then we have the case $\zeta_1 > 0$ if $\mu \neq 0$ and the case $\zeta_1 = 0 < \zeta_2$ if $\mu = 0$. Thus

(i) If $\mu \neq 0$, Theorem 5.5.1A yields

$$U_n \text{ is } AN\left(\mu^2, \frac{4\mu^2\sigma^2}{n}\right);$$

(ii) If $\mu = 0$, the above theorem yields (check)

$$\frac{nU_n}{\sigma^2} \xrightarrow{d} \chi_1^2 - 1. \quad \blacksquare$$

Example C. (*continuation of Example* 5.1.1(ix)). Here find that $\zeta_1 > 0$ for any value of θ, $0 < \theta < 1$. Thus Theorem 5.5.1A covers all situations, and the present theorem has no role. $\blacksquare$

PROOF OF THE THEOREM. On the basis of the discussion in **5.3.4**, our objective is to show that the random variable

$$n(\hat{U}_n - \theta) = \frac{m(m-1)}{n-1} \sum_{1 \le i < j \le n} \tilde{h}_2(X_i, X_j)$$

converges in distribution to

$$\frac{m(m-1)}{2} Y,$$

where

$$Y = \sum_{j=1}^{\infty} \lambda_j (W_j^2 - 1),$$

with $W_1^2, W_2^2, \ldots$ being independent χ_1^2 random variables. Putting

$$T_n = \frac{1}{n} \sum_{i \neq j} \tilde{h}_2(X_i, X_j),$$

we have

$$n(\hat{U}_n - \theta) = \frac{m(m-1)}{2} \frac{n}{n-1} T_n.$$

Thus our objective is to show that

(*) $$T_n \xrightarrow{d} Y.$$

We shall carry this out by the method of characteristic functions, that is, by showing that

(**) $$E_F\{e^{ixT_n}\} \to E\{e^{ixY}\}, \qquad n \to \infty, \text{ each } x.$$

A special representation for $\tilde{h}_2(x, y)$ will be used. Let $\{\phi_j(\cdot)\}$ denote orthonormal eigenfunctions corresponding to the eigenvalues $\{\lambda_j\}$ defined in connection with $\tilde{h}_2$. (See Dunford and Schwartz (1963), pp. 905, 1009, 1083, 1087). Thus

$$E_F\{\phi_j(X)\phi_k(X)\} = \begin{cases} 1, & j = k \\ 0, & j \neq k, \end{cases}$$

and $\tilde{h}_2(x, y)$ may be expressed as the mean square limit of $\sum_{k=1}^K \lambda_k \phi_k(x)\phi_k(y)$ as $K \to \infty$. That is,

(1) $$\lim_{K \to \infty} E_F\left\{\left[\tilde{h}_2(X_1, X_2) - \sum_{k=1}^K \lambda_k \phi_k(X_1)\phi_k(X_2)\right]^2\right\} = 0,$$

and we write

(2) $$\tilde{h}_2(x, y) = \sum_{k=1}^{\infty} \lambda_k \phi_k(x)\phi_k(y).$$

Then (Problem 5.P.24(a)), in the same sense,

(3) $$\tilde{h}_1(x) = \sum_{k=1}^{\infty} \lambda_k \phi_k(x) E_F\{\phi_k(X)\}.$$

Therefore, since $\zeta_1 = 0$,

$$E_F\{\phi_k(X)\} = 0, \qquad \text{all } k.$$

Furthermore (Problem 5.P.24(b)),

(4) $$E_F\left\{\left[\tilde{h}_2(X_1, X_2) - \sum_{k=1}^K \lambda_k \phi_k(X_1)\phi_k(X_2)\right]^2\right\} = E_F\{\tilde{h}_2^2(X_1, X_2)\} - \sum_{k=1}^K \lambda_k^2,$$

whence (by (1))

$$\sum_{k=1}^{\infty} \lambda_k^2 = E_F\{\tilde{h}_2^2(X_1, X_2)\} < \infty.$$

In terms of the representation (2), T_n may be expressed as

$$T_n = \frac{1}{n} \sum_{i \neq j} \sum_{k=1}^{\infty} \lambda_k \phi_k(X_i)\phi_k(X_j).$$

Now put

$$T_{nK} = \frac{1}{n} \sum_{i \neq j} \sum_{k=1}^{K} \lambda_k \phi_k(X_i)\phi_k(X_j).$$

Using the inequality $|e^{iz} - 1| \leq |z|$, we have

$$\begin{aligned} |E\{e^{ixT_n}\} - E\{e^{ixT_{nK}}\}| &\leq E|e^{ixT_n} - e^{ixT_{nK}}| \\ &\leq |x|E|T_n - T_{nK}| \\ &\leq |x|[E(T_n - T_{nK})^2]^{1/2}. \end{aligned}$$

Next it is shown that

$$E(T_n - T_{nK})^2 \leq 2 \sum_{k=K+1}^{\infty} \lambda_k^2. \tag{5}$$

Observe that $T_n - T_{nK}$ is basically of the form of a U-statistic, that is,

$$T_n - T_{nK} = \frac{2}{n}\binom{n}{2} U_{nK},$$

where

$$U_{nK} = \binom{n}{2}^{-1} \sum_{i<j} g_K(X_i, X_j),$$

with

$$g_K(x, y) = \tilde{h}_2(x, y) - \sum_{k=1}^{K} \lambda_k \phi_k(x)\phi_k(y).$$

Justify (Problem 5.P.24(c)) that

$$E_F\{g_K(X_1, X_2)\} = 0 \tag{6a}$$

$$E_F\{g_K^2(X_1, X_2)\} = \sum_{k=K+1}^{\infty} \lambda_k^2, \tag{6b}$$

$$E_F\{g_K(x, X)\} \equiv 0. \tag{6c}$$

Hence $E\{U_{nk}\} = 0$ and, by Lemma 5.2.1A,

$$E\{U_{nk}^2\} = \binom{n}{2}^{-1} \sum_{k=K+1}^{\infty} \lambda_k^2.$$

Thus

$$E(T_n - T_{nK})^2 = (n-1)^2 \binom{n}{2}^{-1} \sum_{k=K+1}^{\infty} \lambda_k^2 \leq 2 \sum_{k=K+1}^{\infty} \lambda_k^2,$$

yielding (5).

Now fix x and let $\varepsilon > 0$ be given. Choose and fix K large enough that

$$|x| \left(2 \sum_{k=K+1}^{\infty} \lambda_k^2\right)^{1/2} < \varepsilon.$$

Then we have established that

$$(7) \qquad |E\{e^{ixT_n}\} - E\{e^{ixT_{nK}}\}| < \varepsilon, \qquad \text{all } n.$$

Next let us show that

$$(8) \qquad T_{nK} \xrightarrow{d} Y_K = \sum_{k=1}^{K} \lambda_k (W_k^2 - 1).$$

We may write

$$T_{nK} = \sum_{k=1}^{K} \lambda_k (W_{kn}^2 - Z_{kn}),$$

where

$$W_{kn} = n^{-1/2} \sum_{i=1}^{n} \phi_k(X_i)$$

and

$$Z_{kn} = n^{-1} \sum_{i=1}^{n} \phi_k^2(X_i).$$

From the foregoing considerations, it is seen that

$$E\{W_{kn}\} = 0, \qquad \text{all } k \text{ and } n,$$

and

$$\operatorname{Cov}\{W_{jn}, W_{kn}\} = \begin{cases} 1, & j = k, \\ 0, & j \neq k, \end{cases} \qquad \text{all } j, k \text{ and } n.$$

Therefore, by the Lindeberg–Lévy CLT,

$$(W_{1n}, \ldots, W_{Kn}) \xrightarrow{d} N(0, \mathbf{I}_{K \times K}).$$

Also, since $E_F\{\phi_k^2(X)\} = 1$, the classical SLLN gives

$$(Z_{1n}, \ldots, Z_{Kn}) \xrightarrow{wp1} (1, \ldots, 1).$$

Consequently (8) holds and thus

$$(9) \qquad |E\{e^{ixT_{nK}}\} - E\{e^{ixY_K}\}| < \varepsilon, \qquad \text{all } n \text{ sufficiently large.}$$

Finally, we show that

$$(10) \qquad |E\{e^{ixY_K}\} - E\{e^{ixY}\}| < \varepsilon[E(W_1^2 - 1)^2]^{1/2}, \qquad \text{all } n.$$

To accomplish this, let the random variables $W_1^2, W_2^2, \ldots$ be defined on a common probability space and represent Y as the limit in mean square of Y_K as $K \to \infty$. Then

$$\begin{aligned} |E\{e^{ixY_K}\} - E\{e^{ixY}\}| &\le |x|[E(Y - Y_K)^2]^{1/2} \\ &\le |x|[E(W_1^2 - 1)^2]^{1/2}\left[\sum_{k=K+1}^{\infty} \lambda_k^2\right]^{1/2}, \end{aligned}$$

yielding (10). Combining (7), (9) and (10), we have, for any x and any $\varepsilon > 0$, and for all n sufficiently large,

$$|E\{e^{ixT_n}\} - E\{e^{ixY}\}| \le \varepsilon\{2 + [E(W_1^2 - 1)^2]^{1/2}\},$$

proving (**). ■

This theorem has also been proved, independently, by Gregory (1977).

5.6 PROBABILITY INEQUALITIES AND DEVIATION PROBABILITIES FOR U-STATISTICS

Here we augment the convergence results of Sections 5.4 and 5.5 with exact exponential-rate bounds for $P(U_n - \theta \ge t)$ and with asymptotic estimates of moderate deviation probabilities

$$P\left(\frac{n^{1/2}(U_n - \theta)}{(m^2\zeta_1)^{1/2}} \ge c(\log n)^{1/2}\right).$$

5.6.1 Probability Inequalities for U-Statistics

For any random variable Y possessing a moment generating function $E\{e^{sY}\}$ for $0 < s < s_0$, one may obtain a probability inequality by writing

$$P(Y - E\{Y\} \ge t) = P(s[Y - E\{Y\} - t] \ge 0) \le e^{-st}E\{e^{s[Y - E\{Y\}]}\}$$

and minimizing with respect to $s \in (0, s_0]$. In applying this technique, we make use of the following lemmas. The first lemma will involve the function

$$f(x, y) = \frac{x}{x + y} e^{-y} + \frac{y}{x + y} e^{x}, \qquad x > 0, y > 0.$$

Lemma A. *Let* $E\{Y\} = \mu$ *and* $\text{Var}\{Y\} = \nu$.

(i) *If* $P(Y \le b) = 1$, *then*

$$E\{e^{s(Y-\mu)}\} \le f(s(b - \mu), s\nu/(b - \mu)), \ s > 0.$$

(ii) *If* $P(a \le Y \le b) = 1$, *then*

$$E\{e^{s(Y-\mu)}\} \le e^{(1/8)s^2(b-a)^2}, \ s > 0.$$

PROOF. (i) is proved in Bennett (1962), p. 42. Now, in the proof of Theorem 2 of Hoeffding (1963), it is shown that

$$qe^{-zp} + pe^{zq} \le e^{(1/8)z^2},$$

for $0 < p < 1$, $q = 1 - p$. Putting $p = y/(x + y)$ and $z = (x + y)$, we have

$$f(x, y) \le e^{(1/8)(x+y)^2},$$

so that (i) yields

$$E\{e^{s(Y-\mu)}\} \le e^{(1/8)s^2[(b-\mu)+\nu/(b-\mu)]^2}.$$

Now, as pointed out by Hoeffding (1963), $\nu = E(Y - \mu)^2 = E(Y - \mu)(Y - a) \le (b - \mu)E(Y - a) = (b - \mu)(\mu - a)$. Thus (ii) follows. ∎

The next lemma may be proved as an exercise (Problem 5.P.25).

Lemma B. *If* $E\{e^{sY}\} < \infty$ *for* $0 < s < s_0$, *and* $E\{Y\} = \mu$, *then*

$$E\{e^{s(Y-\mu)}\} = 1 + O(s^2), \qquad s \to 0.$$

In passing to U-statistics, we shall utilize the following relation between the moment generating function of a U-statistic and that of its kernel.

Lemma C. *Let* $h = h(x_1, \ldots, x_m)$ *satisfy*

$$\psi_h(s) = E_F\{e^{sh(X_1,\ldots,X_m)}\} < \infty, \qquad 0 < s \le s_0.$$

Then

$$E_F\{e^{sU_n}\} \le \psi_h^k\left(\frac{s}{k}\right), \qquad 0 < s \le s_0 k,$$

where $k = [n/m]$.

PROOF. By **5.1.6**, $U_n = (n!)^{-1} \sum_p W(X_{i_1}, \ldots, X_{i_n})$, where each $W(\cdot)$ is an average of $k = [n/m]$ I.I.D. random variables. Since the exponential function is convex, it follows by Jensen's inequality that

$$e^{sU_n} = e^{s(n!)^{-1} \Sigma_p W(\cdot)} \leq (n!)^{-1} \sum_p e^{sW(X_{i_1}, \ldots, X_{i_n})}.$$

Complete the proof as an exercise (Problem 5.P.26). ■

We now give three probability inequalities for U-statistics. The first two, due to Hoeffding (1963), require h to be bounded and give very useful explicit exponential-type bounds. The third, due to Berk (1970), requires less on h but asserts only an implicit exponential-type bound.

Theorem A. *Let* $h = h(x_1, \ldots, x_m)$ *be a kernel for* $\theta = \theta(F)$, *with* $a \leq h(x_1, \ldots, x_m) \leq b$. *Put* $\theta = E\{h(X_1, \ldots, X_m)\}$ *and* $\sigma^2 = \mathrm{Var}\{h(X_1, \ldots, X_m)\}$, *Then, for* $t > 0$ *and* $n \geq m$,

$$P(U_n - \theta \geq t) \leq e^{-2[n/m]t^2/(b-a)^2} \tag{1}$$

and

$$P(U_n - \theta \geq t) \leq e^{-[n/m]t^2/2[\sigma^2 + (1/3)(b-\theta)t]}. \tag{2}$$

PROOF. Write, by Lemmas A and C, with $k = [n/m]$ and $s > 0$,

$$\begin{aligned} P(U_n - \theta \geq t) &\leq E_F\{e^{s(U_n - \theta - t)}\} \leq e^{-st}\left[e^{-(s/k)\theta}\psi_h\left(\frac{s}{k}\right)\right]^k \\ &\leq e^{-st + (1/8)s^2(b-a)^2/k}. \end{aligned}$$

Now minimize with respect to s and obtain (1). A similar argument leads to

$$P(U_n - \theta \geq t) \leq \exp\left[\frac{-kt\left\{\left[1 + \frac{\sigma^2}{(b-\theta)t}\right]\log\left[1 + \frac{(b-\theta)t}{\sigma^2}\right] - 1\right\}}{(b-\theta)}\right]. \tag{2'}$$

It is shown in Bennett (1962) that the right-hand side of (2′) is less than or equal to that of (2). ■

(Compare Lemmas 2.3.2 and 2.5.4A.)

Theorem B. *Let* $h = h(x_1, \ldots, x_m)$ *be a kernel for* $\theta = \theta(F)$, *with*

$$E_F\{e^{sh(X_1, \ldots, X_m)}\} < \infty, \qquad 0 < s \leq s_0.$$

Then, for every $\varepsilon > 0$, *there exist* $C_\varepsilon > 0$ *and* $\rho_\varepsilon < 1$ *such that*

$$P(U_n - \theta \geq \varepsilon) \leq C_\varepsilon \rho_\varepsilon^n, \qquad \textit{all } n \geq m.$$

PROOF. For $0 < t \le s_0 k$, where $k = [n/m]$, we have by Lemma C that

$$P(U_n - \theta \ge \varepsilon) \le e^{-t\varepsilon}\left[e^{-(t/k)\theta}\psi_h\left(\frac{t}{k}\right)\right]^k$$
$$= [e^{-s\varepsilon}e^{-s\theta}\psi_h(s)]^k,$$

where $s = t/k$. By Lemma B, $e^{-s\theta}\psi_h(s) = 1 + O(s^2)$, $s \to 0$, so that

$$e^{-s\varepsilon}e^{-s\theta}\psi_h(s) = 1 - \varepsilon s + O(s^2), \qquad s \to 0,$$
$$< 1 \text{ for } s = s_\varepsilon \text{ sufficiently small.}$$

Complete the proof as an exercise. ■

Note that Theorems A and B are applicable for n small as well as for n large.

5.6.2 "Moderate Deviation" Probability Estimates for U-Statistics

A "moderate deviation" probability for a U-statistic is given by

$$q_n(c) = P\left(\frac{n^{1/2}(U_n - \theta)}{(m^2\zeta_1)^{1/2}} > c(\log n)^{1/2}\right),$$

where $c > 0$ and it is assumed that the relevant kernel h has finite second moment and $\zeta_1 > 0$. Such probabilities are of interest in connection with certain asymptotic relative efficiency computations, as will be seen in Chapter 10. Now the CLT for U-statistics tells us that $q_n(c) \to 0$. Indeed, Chebyshev's inequality yields a bound,

$$q_n(c) \le \frac{1}{c^2 \log n} = O((\log n)^{-1}).$$

However, this result and its analogues, $O((\log n)^{-(1/2)\nu})$, under ν-th order moment assumptions on h are quite weak. For, in fact, if h is bounded, then (Problem 5.P.29) Theorem 5.6.1A implies that for any $\delta > 0$

$$q_n(c) = O(n^{-[(1-\delta)2m\zeta_1/(b-a)^2]c^2}), \tag{1}$$

where $a \le h \le b$. Note also that if merely $E_F|h|^3 < \infty$ is assumed, then for c sufficiently small (namely, $c < 1$), the Berry-Esséen theorem for U-statistics (Theorem 5.5.1B) yields an *estimate*:

$$q_n(c) \sim 1 - \Phi(c(\log n)^{1/2}) \sim \frac{1}{(2\pi c^2 \log n)^{1/2}} n^{-(1/2)c^2}. \tag{*}$$

However, under the stronger assumption $E_F|h|^\nu < \infty$ for some $\nu > 3$, this approach does not yield greater latitude on the range of c. A more intricate analysis is needed. To this effect, the following result has been established by Funk (1970), generalizing a pioneering theorem of Rubin and Sethuraman (1965a) for the case U_n a sample mean.

Theorem. *If* $E_F|h|^\nu < \infty$, *where* $\nu > 2$, *then* (*) *holds for* $c^2 < \nu - 2$.

5.7 COMPLEMENTS

In **5.7.1** we discuss stochastic processes associated with a sequence of U-statistics and generalize the CLT for U-statistics. In **5.7.2** we examine the Wilcoxon one-sample statistic and prove assertions made in **2.6.5** for a particular confidence interval procedure. Extension of U-statistic results to the related V-statistics is treated in **5.7.3**. Finally, miscellaneous further complements and extensions are noted in **5.7.4**.

5.7.1 Stochastic Processes Associated with a Sequence of U-Statistics

Let $h = h(x_1, \ldots, x_m)$ be a kernel for $\theta = \theta(F)$, with $E_F(h^2) < \infty$ and $\zeta_1 > 0$. For the corresponding sequence of U-statistics, $\{U_n\}_{n \geq m}$, we consider two associated sequences of stochastic processes on the unit interval $[0, 1]$.

In one of these sequences of stochastic processes, the nth random function is based on $U_m, \ldots, U_n$ and summarizes the *past* history of $\{U_i\}_{i \leq n}$. In the other sequence of processes, the nth random function is based on $U_n, U_{n+1}, \ldots$ and summarizes the *future* history of $\{U_i\}_{i \geq n}$. Each sequence of processes converges in distribution to the *Wiener* process on $[0, 1]$, which we denote by $W(\cdot)$ (recall **1.11.4**).

The process pertaining to the *future* was introduced and studied by Loynes (1970). The nth random function, $\{Z_n(t), 0 \leq t \leq 1\}$, is defined by

$$Z_n(0) = 0;$$

$$Z_n(t_{nk}) = \frac{U_k - \theta}{(\mathrm{Var}\{U_n\})^{1/2}}, \ k \geq n, \qquad \text{where} \quad t_{nk} = \frac{\mathrm{Var}\{U_k\}}{\mathrm{Var}\{U_n\}};$$

$$Z_n(t) = Z_n(t_{nk}), \qquad t_{n,k+1} < t < t_{nk}.$$

For each n, the "times" $t_{nn}, t_{n,n+1}, \ldots$ form a sequence tending to 0 and $Z_n(\cdot)$ is a step function continuous from the left. We have

Theorem A (Loynes). *$Z_n(\cdot) \xrightarrow{d} W(\cdot)$ in $D[0, 1]$.*

This result generalizes Theorem 5.5.1A (asymptotic normality of U_n) and provides additional information such as

Corollary. *For $x > 0$,*

$$(1) \qquad \lim_{n\to\infty} P\left(\sup_{k\geq n}(U_k - \theta) \geq x(\mathrm{Var}\{U_n\})^{1/2}\right) = P\left(\sup_{0\leq t\leq 1} W(t) \geq x\right) = 2[1 - \Phi(x)]$$

and

(2) $$\lim_{n\to\infty} \mathrm{P}\left(\inf_{k\geq n}(\mathrm{U}_k - \theta) \leq -\mathrm{x}(\mathrm{Var}\{\mathrm{U}_n\})^{1/2}\right)$$
$$= \mathrm{P}\left(\inf_{0\leq t\leq 1} \mathrm{W}(t) \leq -\mathrm{x}\right) = 2[1 - \Phi(\mathrm{x})].$$

As an exercise, show that the *strong convergence* of U_n to θ follows from this corollary, under the assumption $E_F\{h^2\} < \infty$.

The process pertaining to the *past* has been dealt with by Miller and Sen (1972). Here the nth random function, $\{Y_n(t), 0 \leq t \leq 1\}$, is defined by

$$Y_n(0) = 0, \qquad 0 \leq t \leq \frac{m-1}{n};$$

$$Y_n\left(\frac{k}{n}\right) = \frac{k(U_k - \theta)}{(m^2\zeta_1)^{1/2}n^{1/2}}, \qquad k = m, m+1, \ldots, n;$$

$Y_n(t)$ defined elsewhere, $0 \leq t \leq 1$, by linear interpolation.

Theorem B (Miller and Sen). $\mathrm{Y}_n(\cdot) \xrightarrow{d} \mathrm{W}(\cdot)$ in $\mathrm{C}[0, 1]$.

This result likewise generalizes Theorem 5.5.1A and provides additional information such as

(3) $$\lim_{n\to\infty} P\left(\sup_{m\leq k\leq n} k(U_k - \theta) \geq x(m^2\zeta_1)^{1/2}n^{1/2}\right) = 2[1 - \Phi(x)], \qquad x > 0.$$

Comparison of (1) and (3) illustrates how Theorems A and B complement each other in the type of additional information provided beyond Theorem 5.5.1A.

See the Loynes paper for treatment of other random variables besides U-statistics. See the Miller and Sen paper for discussion of the use of Theorem B in the *sequential* analysis of U-statistics.

5.7.2 The Wilcoxon One-Sample Statistic as a *U*-Statistic

For testing the hypothesis that the median of a continuous symmetric distribution F is 0, that is, $\xi_{1/2} = 0$, the Wilcoxon one-sample test may be based on the statistic

$$\sum_{1\leq i<j\leq n} I(X_i + X_j > 0).$$

Equivalently, one may perform the test by estimating $G(0)$, where G is the distribution function $G(t) = P(\frac{1}{2}(X_1 + X_2) \leq t)$, with the null hypothesis to be rejected if the estimate differs sufficiently from the value $\frac{1}{2}$. In this way one may treat the related statistic

$$U_n = \frac{1}{\binom{n}{2}} \sum_{1 \le i < j \le n} I(X_i + X_j \le 0)$$

as an estimate of $G(0)$. This, of course, is a U-statistic (recall Example 5.1.1(ix)), so that we have the convenience of asymptotic normality (recall Example 5.5.2C-check as exercise).

In **2.6.5** we considered a related confidence interval procedure for $\xi_{1/2}$. In particular, we considered a procedure of Geertsema (1970), giving an interval

$$I_{Wn} = (W_{na_n}, W_{nb_n})$$

formed by a pair of the ordered values $W_{n1} \le \cdots \le W_{nN_n}$ of the $N_n = \binom{n}{2}$ averages $\frac{1}{2}(X_i + X_j)$, $1 \le i < j \le n$. We now show how the properties stated for I_{Wn} in **2.6.5** follow from a treatment of the U-statistic character of the random variable

$$G_n(x) = \frac{1}{\binom{n}{2}} \sum_{1 \le i < j \le n} I[\tfrac{1}{2}(X_i + X_j) \le x].$$

Note that G_n, considered as a function of x, represents a "sample distribution function" for the averages $\frac{1}{2}(X_i + X_j)$, $1 \le i < j \le n$. From our theory of U-statistics, we see that $G_n(x)$ is asymptotically normal. In particular, $G_n(\xi_{1/2})$ is asymptotically normal. The connection with the W_{nj}'s is as follows. Recall the Bahadur representation (**2.5.2**) relating order statistics X_{nk_n} to the sample distribution function F_n. Geertsema proves the analogue of this result for W_{nkn} and G_n. The argument is similar to that of Theorem 2.5.2, with the use of Theorem 5.6.1A in place of Lemma 2.5.4A.

Theorem. *Let* F *satisfy the conditions stated in* **2.6.5**. *Let*

$$\frac{k_n}{\binom{n}{2}} = \frac{1}{2} + o\left(\frac{\log n}{n^{1/2}}\right), \qquad n \to \infty.$$

Then

$$W_{nk_n} = \xi_{1/2} + \frac{\binom{n}{2}^{-1} k_n - G_n(\xi_{1/2})}{g(\xi_{1/2})} + R_n$$

where with probability 1

$$R_n = O(n^{-3/4} \log n), \qquad n \to \infty.$$

It is thus seen, via this theorem, that properties of the interval I_{Wn} may be derived from the theory of U-statistics.

5.7.3 Implications for V-Statistics

In connection with a kernel $h = h(x_1, \ldots, x_m)$, let us consider again the V-statistic introduced in **5.1.2**. Under appropriate moment conditions, the U-statistic and V-statistic associated with h are closely related in behavior, as the following result shows.

Lemma. *Let* r *be a positive integer. Suppose that*

$$E_F|h(X_{i_1}, \ldots, X_{i_m})|^r < \infty, \qquad \textit{all } i \le i_1, \ldots, i_m \le m.$$

Then

$$E|U_n - V_n|^r = O(n^{-r}).$$

PROOF. Check that

$$n^m(U_n - V_n) = (n^m - n_{(m)})(U_n - W_n),$$

where $n_{(m)} = n(n-1)\cdots(n-m+1)$ and W_n is the average of all terms $h(X_{i_1}, \ldots, X_{i_m})$ with at least one equality $i_a = i_b$, $a \ne b$. Next check that

$$n^m - n_{(m)} = O(n^{m-1}).$$

Finally, apply Minkowski's inequality. ■

Application of the lemma in the case $r = 2$ yields

$$n^{1/2}(U_n - V_n) \xrightarrow{p} 0,$$

in which case $n^{1/2}(U_n - \theta)$ and $n^{1/2}(V_n - \theta)$ have the same limit distribution, a useful relationship in the case $\zeta_1 > 0$. In fact, this latter result can actually be obtained under slightly weaker moment conditions on the kernel (see Bönner and Kirschner (1977).)

5.7.4 Further Complements and Extensions

(i) *Distribution-free estimation of the variance of a U-statistic* is considered by Sen (1960).

(ii) Consideration of U-statistics when the distribution of $X_1, X_2, \ldots$ are *not necessarily identical* may be found in Sen (1967).

(iii) *Sequential confidence intervals based on U-statistics* are treated by Sproule (1969a, b).

(iv) *Jackknifing* of estimates which are functions of U-statistics, in order to reduce bias and to achieve other properties, is treated by Arvesen (1969).

(v) Further results on *probabilities of deviations* (recall **5.6.2**) of U-statistics are obtained, via some further results on *stochastic processes* associated with U-statistics (recall **5.7.1**), by Sen (1974).

(vi) Consideration of U-statistics for *dependent* observations $X_1, X_2, \ldots$ arises in various contexts. For the case of *m-dependence*, see Sen (1963), (1965). For the case of *sampling without replacement* from a finite population, see Nandi and Sen (1963). For a treatment of the Wilcoxon 2-sample statistic in the case of samples from a *weakly dependent stationary process*, see Serfling (1968).

(vii) A somewhat different treatment of the case $\zeta_1 = 0 < \zeta_2$ has been given by Rosén (1969). He obtains asymptotic *normality* for U_n when the observations $X_1, \ldots, X_n$ are assumed to have a common distribution $F^{(n)}$ which behaves in a specified fashion as $n \to \infty$. In this treatment $F^{(n)}$ is constrained *not* to remain fixed as $n \to \infty$.

(viii) A general treatment of *symmetric* statistics exploiting an orthogonal expansion technique has been carried out by Rubin and Vitale (1980). For example, U-statistics and V-statistics are types of symmetric statistics. Rubin and Vitale provide a unified approach to the asymptotic distribution theory of such statistics, obtaining as limit random variable a weighted sum of infinite products of Hermite polynomials evaluated at $N(0, 1)$ variates.

5.P PROBLEMS

Section 5.1

1. Check the relations $E_F\{g_1(X_1)\} = 0$, $E_F\{g_2(x_1, X_2)\} = 0, \ldots$ in **5.1.5**.

2. Prove Lemma 5.1.5B.

Section 5.2

3. (i) Show that $\zeta_0 \le \zeta_1 \le \cdots \le \zeta_m$.

(ii) Show that $\zeta_1 \le \frac{1}{2}\zeta_2$. (Hint: Consider the function g_2 of **5.1.5**.)

4. Let $\{a_1, \ldots, a_m\}$ and $\{b_1, \ldots, b_m\}$ be two sets of m distinct integers from $\{1, \ldots, n\}$ with exactly c integers in common. Show that

$$E_F\{\tilde{h}(X_{a_1}, \ldots, X_{a_m})\tilde{h}(X_{b_1}, \ldots, X_{b_m})\} = \zeta_c.$$

5. In Lemma 5.2.1A, derive (iii) from (*).

6. Extend Lemma 5.2.1A(*) to the case of a generalized U-statistic.

7. Complete the details of proof for Lemma 5.2.2B.

8. Extend Lemma 5.2.2B to generalized U-statistics.

Section 5.3

9. The projection of a generalized U-statistic is defined as

$$\hat{U} = \sum_{j=1}^{k} \sum_{i=1}^{n_j} E_F\{U \mid X_i^{(j)}\} - (N-1)\theta,$$

where $N = n_1 + \cdots + n_k$. Define

$$\tilde{h}_{1j}(x) = E_F\{h(X_1^{(1)}, \ldots, X_{m_1}^{(1)}; \ldots; X_1^{(k)}, \ldots, X_{m_k}^{(k)}) \mid X_1^{(j)} = x\} - \theta.$$

Show that

$$\hat{U} - \theta = \sum_{j=1}^{k} \sum_{i=1}^{n_j} \frac{m_j}{n_j} \tilde{h}_{1j}(X_i^{(j)}).$$

10. (continuation) Show that $U_n - \hat{U}_n$ is a U-statistic based on a kernel H satisfying $E_F\{H\} = E_F\{H \mid X_1^{(j)}\} = 0$.

11. Verify relation (2) in **5.3.4**..

12. Extend (2), (3) and (4) of **5.3.4** to generalized U-statistics.

13. Let g_c and S_{cn} be as defined in **5.1.5**. Define a kernel G_c of order m by

$$G_c(x_1, \ldots, x_m) = \sum_{1 \le i_1 < \cdots < i_c \le m} g_c(x_{i_1}, \ldots, x_{i_c})$$

and let U_{cn} be the U-statistic corresponding to G_c. Show that

$$U_{cn} = \binom{m}{c}\binom{n}{c}^{-1} S_{cn}$$

and thus

$$U_n - \theta = \sum_{c=1}^{m} U_{cn}.$$

Now suppose that $\zeta_{c-1} = 0 < \zeta_c$. Show that $\hat{U}_n$ defined in **5.3.4** satisfies

$$\hat{U}_n - \theta = U_{cn}.$$

Section 5.4

14. For $E_F h^2 < \infty$, show strong convergence of generalized U-statistics.

15. Prove Theorem 5.4C, the LIL for U-statistics. (Hint: apply Theorem 5.3.3.)

Section 5.5

16. Prove Theorem 5.5.1A, the CLT for U-statistics.

17. Complete the details for Example 5.5.1A.

18. Extend Theorem 5.5.1A to a vector of several U-statistics defined on the same sample.

19. Extend Theorem 5.5.1A to generalized U-statistics (continuation of Problems 5.P.9, 10, 12).

20. Check the details of Example 5.5.1B.

21. Check the details of Example 5.5.2A.

22. (continuation) Show, for F binomial $(1, \frac{1}{2})$, that

$$n(m_2 - \tfrac{1}{4}) \xrightarrow{d} -\tfrac{1}{4}\chi_1^2.$$

(Hint: One approach is simply to apply the result obtained in Example 5.5.2A. Another approach is to write $m_2 = \hat{p} - \hat{p}^2$ and apply the methods of Chapter 3.)

23. Check the details of Example 5.5.2B.

24. Complete the details of proof of Theorem 5.5.2.

(a) Prove (3). (Hint: write $\tilde{h}_1(x) = E_F\{\tilde{h}_2(x, X_2)\}$ and use Jensen's inequality to show that

$$\lim_{K\to\infty} E_F\left\{\left[\tilde{h}_1(X_1) - \sum_{k=1}^{K} \lambda_k \phi_k(X_1)E_F\{\phi_k(X_2)\}\right]^2\right\}$$
$$\leq \lim_{K\to\infty} E_F\left\{\left[\tilde{h}_2(X_1, X_2) - \sum_{k=1}^{K} \lambda_k \phi_k(X_1)\phi_k(X_2)\right]^2\right\} = 0.$$

(b) Prove (4).

(c) Prove (6).

Section 5.6

25. Prove Lemma 5.6.1B. (Hint: Without loss assume $E\{Y\} = 0$. Show that $e^{sY} = 1 + sY + \frac{1}{2}s^2Z$, where $0 < Z < Y^2e^{s_0Y}$.)

26. Complete the proof of Lemma 5.6.1C.

27. Complete the proof of Theorem 5.6.1A.

28. Complete the proof of Theorem 5.6.1B.

29. In **5.6.2**, show that (1) follows from Theorem 5.6.1A and that (*) for $c \leq 1$ follows from Theorem 5.5.1B.

Section 5.7

30. Derive the strong convergence of U-statistics, under the assumption $E_F\{h^2\} < \infty$, from Corollary 5.7.1.

31. Check the claim of Example 5.5.2C.

32. Apply Theorem 5.7.2 to obtain properties of the confidence interval I_{Wn}.

33. Complete the details of proof of Lemma 5.7.3.

CHAPTER 6

Von Mises Differentiable Statistical Functions

Statistics which are representable as functionals $T(F_n)$ of the sample distribution F_n are called "statistical functions." For example, for the variance parameter σ^2 the relevant functional is $T(F) = \int [x - \int x\, dF(x)]^2\, dF(x)$ and $T(F_n)$ is the statistic m_2 considered in Section 2.2. The theoretical investigation of statistical functions as a class was initiated by von Mises (1947), who developed an approach for deriving the asymptotic distribution theory of such statistics. Further development is provided in von Mises (1964) and, using stochastic process concepts, by Filippova (1962).

Notions of *differentiability* of T play a key role in the von Mises approach, analogous to the treatment in Chapter 3 of transformations of asymptotically normal random vectors. We thus speak of "differentiable statistical functions." In typical cases, $T(F_n) - T(F)$ is asymptotically normal. Otherwise a higher "type" of distribution applies, in close parallel with the hierarchy of cases seen for U-statistics in Chapter 5.

This chapter develops the "differential approach" for deriving the *asymptotic distribution theory* of statistical functions. In the case of asymptotically *normal* $T(F_n)$, the related *Berry–Esséen rates* and *laws of iterated logarithm* are obtained also. Section 6.1 formulates the representation of statistics as functions of F_n and sketches the basic scheme for analysis of $T(F_n) - T(F)$ by reduction by the differential method to an appropriate approximating random variable V_n. Methodology for carrying out the *reduction* to V_n is provided in Section 6.2, and useful characterizations of the *structure* of V_n are provided in Section 6.3. These results are applied in Section 6.4 to obtain general results on the asymptotic distribution theory and almost sure behavior of statistical functions. A variety of examples are treated in Section 6.5. Certain complements are provided in Section 6.6, including discussion of some statistical interpretations of the derivative of a statistical function. Further applications of the development of this chapter will arise in Chapters 7, 8 and 9.

6.1 STATISTICS CONSIDERED AS FUNCTIONS OF THE SAMPLE DISTRIBUTION FUNCTION

We consider as usual the context of I.I.D. observations $X_1, X_2, \ldots$ on a distribution function F and denote by F_n the sample distribution function based on $X_1, \ldots, X_n$. Many important statistics may be represented as a function of F_n, say $T(F_n)$. Since F_n is a reasonable estimate of F, indeed converging to F in a variety of senses as seen in Section 2.1, we may expect $T(F_n)$ to relate to $T(F)$ in similar fashion, provided that the functional $T(\cdot)$ is sufficiently "well-behaved" in a neighborhood of F. This leads to consideration of F as a "point" in a collection $\mathcal{F}$ of distribution functions, and to consideration of notions of continuity, differentiability, and other regularity properties for functionals $T(\cdot)$ defined on $\mathcal{F}$. In this context von Mises (1947) introduced a *Taylor expansion* for $T(\cdot)$, whereby the difference $T(G) - T(F)$ may be represented in terms of the "derivatives" of $T(\cdot)$ and the "difference" $G - F$.

In **6.1.1** we look at examples of $T(F_n)$ and give an informal statement of von Mises' general proposition. In **6.1.2** the role of von Mises' Taylor expansion is examined.

6.1.1 First Examples and a General Proposition

Here we consider several examples of the broad variety of statistics which are amenable to analysis by the von Mises approach. Then we state a general proposition unifying the asymptotic distribution theory of the examples considered.

Examples. (i) For any function $h(x)$, the statistic

$$T_n = \int h(x)dF_n(x)\left(= n^{-1}\sum_{i=1}^{n} h(X_i)\right)$$

is a *linear* statistical function—that is, linear in the increments $dF_n(x)$. In particular, the *sample moments* $a_k = \int x^k \, dF_n(x)$ are linear statistical functions.

(ii) The *sample* kth *central moment*, $T_n = m_k = T(F_n)$, where

$$T(F) = \int\left[x - \int x\, dF(x)\right]^k dF(x).$$

(iii) *Maximum likelihood, minimum chi-square estimates* T_n are given by solving equations of the form $H(T, F_n) = 0$.

(iv) The *chi-squared statistic* is $T(F_n)$, where

$$T(F) = \sum_{i=1}^{k} p_i^{-1}\left(\int_{A_i} dF - p_i\right)^2,$$

where $\{A_i\}$ is a partition of R into k cells and $\{p_i\}$ is a set of specified (null-hypothesis) probabilities attached to the cells.

(v) The *generalized Cramér-von Mises test statistic*, considered in **2.1.2**, is given by $T(F_n)$, where $T(F) = \int w(F_0)(F - F_0)^2 \, dF_0$, for w and F_0 specified. ■

It turns out that examples (i), (ii) and (iii) are asymptotically normal (under appropriate conditions), example (iv) is asymptotically chi-squared, and example (v) is something still different (a weighted sum of chi-squared variates). Nevertheless, within von Mises' framework, these examples all may be viewed as special cases of a single unifying theorem, which is stated informally as follows.

Proposition (von Mises). *The type of asymptotic distribution of a differentiable statistical function* $\mathrm{T_n = T(F_n)}$ *depends upon which is the first nonvanishing term in the Taylor development of the functional* $\mathrm{T}(\cdot)$ *at the distribtion* F *of the observations. If it is the linear term, the limit distribution is normal (under the usual restrictions corresponding to the central limit theorem). In other cases, "higher" types of limit distribution result.*

More precisely, when the first nonvanishing term of the Taylor development of $T(\cdot)$ is the one of order m, the random variable $n^{m/2}[T(F_n) - T(F)]$ converges in distribution to a random variable with finite variance. For $m = 1$, the limit law is normal. (Actually, the normalization for the order m case can in some cases differ from $n^{m/2}$. See **6.6.5**.)

6.1.2 The Basic Scheme for Analysis of $T(F_n)$

In **6.2.1** a Taylor expansion of $T(F_n) - T(F)$ will be given:

$$T(F_n) - T(F) = d_1 T(F; F_n - F) + \frac{1}{2!} d_2 T(F; F_n - F) + \cdots.$$

Analysis of $T(F_n) - T(F)$ is to be carried out by reduction to

$$V_{mn} = \sum_{j=1}^{m} \frac{1}{j!} d_j T(F; F_n - F)$$

for an appropriate choice of m. The reduction step is performed by dealing with the remainder term $R_{mn} = T(F_n) - T(F) - V_{mn}$, and the properties of $T(F_n) - T(F)$ then are obtained from an *m-linear* structure typically possessed by V_{mn}.

In the case that $T(F_n)$ is *asymptotically normal*, we prove it by first showing that

(*) $$n^{1/2}R_{1n} \xrightarrow{p} 0.$$

Then it is checked that V_{1n} has the form of a sample mean of I.I.D. mean 0 random variables, so that $n^{1/2}V_{1n} \xrightarrow{d} N(0, \sigma^2(T, F))$ for a certain constant $\sigma^2(T, F)$, whereby

(1) $$n^{1/2}[T(F_n) - T(F)] \xrightarrow{d} N(0, \sigma^2(T, F)).$$

In this case the *law of the iterated logarithm* for $T(F_n) - T(F)$ follows by a similar argument replacing (*) by

(**) $$n^{1/2}R_{1n} = o((\log \log n)^{1/2}) wp1,$$

yielding

(2) $$\overline{\lim_{n\to\infty}} \frac{n^{1/2}[T(F_n) - T(F)]}{\sigma(T, F)(2 \log \log n)^{1/2}} = 1 \ wp1.$$

In addition, a *Berry–Esséen rate* for the convergence in (1) may be obtained through a closer study of R_{1n}. Invariably, standard methods applied to R_{1n} fail to lead to the best rate, $O(n^{-1/2})$. However, it turns out that if $T(F_n) - T(F)$ is approximated by V_{2n} instead of V_{1n}, the resulting ("smaller") remainder term R_{2n} behaves as needed for the standard devices to lead to the Berry–Esséen rate $O(n^{-1/2})$. Namely, one establishes

(***) $$P(|R_{2n}| > Bn^{-1}) = O(n^{-1/2})$$

for some constant $B > 0$, and obtains

(3) $$\sup_t \left| P\left(\frac{n^{1/2}[T(F_n) - T(F)]}{\sigma(T, F)} \le t\right) - \Phi(t)\right| = O(n^{-1/2}).$$

In the case that $P(V_{1n} = c) = 1$, that is, $V_{1n} = d_1T(F; F_n - F)$ is a degenerate random variable, the asymptotic distribution of $T(F_n)$ is found by finding the lowest m such that V_{mn} is *not* degenerate. Then a limit law for $n^{m/2}[T(F_n) - T(F)]$ is found by establishing $n^{m/2}R_{mn} \to 0$ and dealing with $n^{m/2}V_{mn}$. For $m > 1$, the case of widest practical importance is $m = 2$. Thus the random variable V_{2n} has *two* important roles—one for the case that $n[T(F_n) - T(F)]$ has a limit law, and one for the Berry–Esséen rate in the case that $T(F_n)$ is asymptotically normal.

Finally, we note that in general the *strong consistency* of $T(F_n)$ for estimation of $T(F)$ typically may be established by proving $R_{1n} \xrightarrow{wp1} 0$.

Methodology for handling the remainder terms R_{mn} is provided in **6.2.2**. The structure of the V_{mn} terms is studied in Section 6.3. These results are applied in Section 6.4 to obtain conclusions such as (1), (2), (3), etc.

6.2 REDUCTION TO A DIFFERENTIAL APPROXIMATION

The basic method of differentiating a functional $T(F)$ is described in **6.2.1** and applied to formulate a Taylor expansion of $T(F_n)$ about $T(F)$. In **6.2.2** various techniques of treating the remainder term in the Taylor expansion are considered.

6.2.1 Differentiation of Functionals $T(\cdot)$

Given two points F and G in the space $\mathscr{F}$ of all distribution functions, the "line segment" in $\mathscr{F}$ joining F and G consists of the set of distribution functions $\{(1-\lambda)F + \lambda G,\ 0 \le \lambda \le 1\}$, also written as $\{F + \lambda(G - F),\ 0 \le \lambda \le 1\}$. Consider a functional T defined on $F + \lambda(G - F)$ for all sufficiently small λ. If the limit

$$d_1 T(F; G - F) = \lim_{\lambda \to 0+} \frac{T(F + \lambda(G - F)) - T(F)}{\lambda}$$

exists, it is called the *Gâteaux differential of T at F in the direction of* G. Note that $d_1 T(F; G - F)$ is simply the ordinary right-hand derivative, at $\lambda = 0$, of the function $Q(\lambda) = T(F + \lambda(G - F))$ of the real variable λ. In general, we define the kth *order* Gâteaux differential of T at F in the direction of G to be

$$d_k T(F; G - F) = \frac{d^k}{d\lambda^k} T(F + \lambda(G - F))\bigg|_{\lambda = 0+},$$

provided the limit exists.

Example. Consider the functional

$$T(F) = \int \cdots \int h(x_1, \ldots, x_c) dF(x_1) \cdots dF(x_c),$$

where h is symmetric. Writing

$$T(F + \lambda(G - F))$$
$$= \sum_{j=0}^{c} \binom{c}{j} \lambda^{c-j} \int \cdots \int h(x_1, \ldots, x_c) \prod_{i=1}^{j} dF(x_i) \prod_{i=j+1}^{c} d[G(x_i) - F(x_i)]$$

and carrying out successive differentiations, we obtain

$$\frac{d^k}{d\lambda^k} T(F + \lambda(G - F)) = \sum_{j=0}^{c-k} \binom{c}{j} (c - j) \cdots (c - j - k + 1) \lambda^{c-j-k}$$
$$\times \int \cdots \int h(x_1, \ldots, x_c) \prod_{i=1}^{j} dF(x_i) \prod_{i=j+1}^{c} d[G(x_i) - F(x_i)]$$

and thus

$$d_k T(F; G - F) = c(c-1)\cdots(c-k+1)$$
$$\times \int\cdots\int h(x_1, \ldots, x_k, y_1, \ldots, y_{c-k}) \prod_{i=1}^{c-k} dF(y_i) \prod_{i=1}^{k} d[G(x_i) - F(x_i)]$$

for $k = 1, \ldots, c$, and $d_k T(F; G - F) = 0$, $k > c$.

In particular, for the *mean* functional $T(F) = \int x\, dF(x)$, we have

$$d_1 T(F; G - F) = \int h(x) d[G(x) - F(x)] = T(G) - T(F)$$

and $d_k T(F; G - F) = 0$ for $k > 1$.

For the *variance* functional, corresponding to

$$h(x_1, x_2) = \tfrac{1}{2}(x_1^2 + x_2^2 - 2x_1x_2),$$

we have (check)

$$d_1 T(F; G - F)$$
$$= \int x^2\, dG(x) - \int x^2\, dF(x) - 2\int x\, dF(x) \int x\, dG(x) + 2\left(\int x\, dF(x)\right)^2$$

and

$$d_2 T(F; G - F) = -2\left(\int x\, dG(x) - \int x\, dF(x)\right)^2. \quad ■$$

Suppose that the function $Q(\lambda)$ satisfies the usual assumptions for a Taylor expansion to be valid (the assumptions of Theorem 1.12.1A as extended in Remark 1.12.1(i)) with respect to the interval $0 \le \lambda \le 1$. (See Problem 6.P.2) Since $Q(0) = T(F)$, $Q(1) = T(G)$, $Q_+^{(1)}(0) = d_1 T(F; G - F)$, $Q_+^{(2)}(0) = d_2 T(F; G - F)$, etc., the Taylor expansion for $Q(\cdot)$ may be expressed as a *Taylor expansion for the functional* $T(\cdot)$:

$$(*) \qquad T(G) - T(F) = \sum_{k=1}^{m} \frac{1}{k!} d_k T(F; G - F)$$
$$+ \frac{1}{(m+1)!} \frac{d^{m+1}}{d\lambda^{m+1}} T(F + \lambda(G - F))\bigg|_{\lambda^*},$$

where $0 \le \lambda^* \le 1$. Note that even though we are dealing here with a functional on $\mathcal{F}$, sophisticated functional analysis is not needed at this stage, since the terms of the expansion may be obtained by routine calculus methods.

We are not bothering to state explicitly the conditions needed for (*) to hold formally, because in practice (*) is utilized rather *informally*, merely as a *guiding concept*. As discussed in **6.1.2**, our chief concern is

$$R_{mn} = T(F_n) - T(F) - \sum_{k=1}^{m} \frac{1}{k!} d_k T(F; F_n - F),$$

which may be investigated without requiring that R_{mn} have the form dictated by (*), and without requiring that (*) hold for G other than F_n.

6.2.2 Methods for Handling the Remainder Term R_{mn}

As discussed in **6.1.2**, the basic property that one would seek to establish for R_{mn} is

$$n^{m/2} R_{mn} \xrightarrow{p} 0, \tag{1}$$

In the case that the Taylor expansion of **6.2.1** is rigorous, it suffices for (1) to show that

$$n^{m/2} \sup_{0 \le \lambda \le 1} \left| \frac{d^{m+1}}{d\lambda^{m+1}} T(F + \lambda(F_n - F)) \right| \xrightarrow{p} 0. \tag{M}$$

This is the line of attack of von Mises (1947). Check (Problem 6.P.3), using Lemma 6.3.2B, that (M) holds for the functionals

$$T(F) = \int \cdots \int h(x_1, \ldots, x_c) dF(x_1) \cdots dF(x_c)$$

considered in Example 6.2.1.

An inconvenience of this approach is that (M) involves an order of differentiability higher than that of interest in (1). In order to avoid dealing with the unnecessarily complicated random variable appearing in (M), we may attempt a *direct* analysis of R_{mn}. Usually this works out to be very effective in practice.

Example A. (*Continuation of Example* 6.2.1). For the *variance* functional $T(F) = \iint h(x_1, x_2) dF(x_1) dF(x_2)$, where $h(x_1, x_2) = \frac{1}{2}(x_1^2 + x_2^2 - 2x_1x_2)$, we have (check)

$$T(G) - T(F) - d_1 T(F; G - F) = -(\mu_G - \mu_F)^2,$$

where μ_F and μ_G denote the means of F and G. Thus

$$R_{1n} = -(\bar{X} - \mu_F)^2.$$

It follows (check) by the Hartman and Wintner LIL (Theorem 1.10A) that

$$|R_{1n}| = O(n^{-1} \log \log n) \ wp1$$

and hence in particular

$$n^{1/2} R_{1n} \xrightarrow{p} 0$$

and

$$n^{1/2} R_{1n} = o((\log \log n)^{1/2}) wp1,$$

in conformity with (*) and (**) of **6.1.2.** ■

As a variant of the Taylor expansion idea, an alternative "guiding concept" consists of a differential for T at F in a sense stronger than the Gâteaux version. Let us formulate such a notion in close analogy with the differential of **1.12.2** for functions g defined on R^k. Let $\mathscr{D}$ be the linear space generated by differences $G - H$ of members of $\mathscr{F}$, the space of distribution functions. ($\mathscr{D}$ may be represented as $\{\Delta: \Delta = c(G - H), G, H \in \mathscr{F}, c \text{ real}\}$.) Let $\mathscr{D}$ be equipped with a norm $\|\cdot\|$. The functional T defined on $\mathscr{F}$ is said to have a *differential* at the point $F \in \mathscr{F}$ with respect to the norm $\|\cdot\|$ if there exists a functional $T(F; \Delta)$, defined on $\Delta \in \mathscr{D}$ and linear in the argument Δ, such that

$$\text{(D)} \qquad T(G) - T(F) - T(F; G - F) = o(\|G - F\|)$$

as $\|G - F\| \to 0$ ($T(F; \Delta)$ is called the "differential").

Remarks A. (i) To establish (D), it suffices (see Apostol (1957), p. 65) to verify it for all sequences $\{G_n\}$ satisfying $\|G_n - F\| \to 0$, $n \to \infty$.

(ii) By *linearity* of $T(F; \Delta)$ is meant that

$$T\left(F; \sum_{i=1}^{k} a_i \Delta_i\right) = \sum_{i=1}^{k} a_i T(F; \Delta_i)$$

for $\Delta_1, \ldots, \Delta_k \in \mathscr{D}$ and real $a_1, \ldots, a_k$.

(iii) In the general context of differentiation in Banach spaces, the differential $T(F; \Delta)$ would be called the *Fréchet derivative* of T (see Fréchet (1925), Dieudonné (1960), Luenberger (1969), and Nashed (1971)). In such treatments, the space on which T is defined is assumed to be a normed *linear* space. We intentionally avoid this assumption here, in order that T need only be defined at points F which are distribution functions. ■

It is evident from (D) that the differential approach approximates $T(F_n) - T(F)$ by the random variable $T(F; F_n - F)$, whereas the Taylor expansion approximates by $d_1 T(F; F_n - F)$. These approaches are in agreement, by the following result.

Lemma A. *If* T *has a differential at* F *with respect to* $\|\cdot\|$, *then, for any* G, $d_1T(F; G - F)$ *exists and*

$$d_1T(F; G - F) = T(F; G - F).$$

PROOF. Given G, put $F_\lambda = F + \lambda(G - F)$. Then $F_\lambda - F = \lambda(G - F)$ and thus $\|F_\lambda - F\| = \lambda\|G - F\| \to 0$ as $\lambda \to 0$ (G fixed). Therefore, by (D) and the linearity of $T(F; \Delta)$, we have

$$\begin{aligned} T(F_\lambda) - T(F) &= T(F; F_\lambda - F) + o(\|F_\lambda - F\|), \qquad \lambda \to 0, \\ &= \lambda T(F; G - F) + \lambda\, o(1), \qquad \lambda \to 0. \end{aligned}$$

Hence

$$\lim_{\lambda \to 0+} \frac{T(F_\lambda) - T(F)}{\lambda} = T(F; G - F). \quad \blacksquare$$

The role played by the differential in handling the remainder term R_{1n} is seen from the following result.

Lemma B. *Let* T *have a differential at* F *with respect to* $\|\cdot\|$. *Let* $\{X_i\}$ *be observations on* F *(not necessarily independent) such that* $n^{1/2}\|F_n - F\| = O_p(1)$, *Then* $n^{1/2}R_{1n} \xrightarrow{p} 0$.

PROOF. For any $\varepsilon > 0$, we have by (D) and Lemma A that there exists $\delta_\varepsilon > 0$ such that

$$|R_{1n}| < \varepsilon\|F_n - F\|$$

whenever $\|F_n - F\| < \delta_\varepsilon$. Let $\varepsilon_0 > 0$ be given. Then

$$P(n^{1/2}|R_{1n}| > \varepsilon_0) \le P\left(n^{1/2}\|F_n - F\| > \frac{\varepsilon_0}{\varepsilon}\right) + P(\|F_n - F\| > \delta_\varepsilon).$$

Complete the argument as an exercise (Problem 6.P.5). $\blacksquare$

Remarks B. (i) The use of (D) instead of (M) bypasses the higher-order remainder term but introduces the difficulty of handling a *norm.*

(ii) However, for the *sup-norm,* $\|g\|_\infty = \sup_x|g(x)|$, this enables us to take advantage of known stochastic properties of the *Kolmogorov–Smirnov distance* $\|F_n - F\|_\infty$. Under the usual assumption of I.I.D. observations $\{X_i\}$, the property $n^{1/2}\|F_n - F\|_\infty = O_p(1)$ required in Lemma B follows immediately from the Dvoretzky–Kiefer–Wolfowitz inequality (Theorem 2.1.3A).

(iii) The *choice of norm* in seeking to apply Lemma B must serve two somewhat conflicting purposes. The differentiability of T is more easily established if $\|\cdot\|$ is "large," whereas the property $n^{1/2}\|F_n - F\| = O_p(1)$ is more easily established if $\|\cdot\|$ is "small." Also, the two requirements differ in *type*, one being related to differential analysis, the other to stochastic analysis.

(iv) In view of Lemma A, the "*candidate*" differential $T(F; G - F)$ to be employed in establishing (D) is given by $d_1T(F; G - F)$, which is found by routine calculus methods as noted in **6.2.1**.

(v) Thus the choice of norm $\|\cdot\|$ in Lemma B plays no essential role in the actual application of the result, for the approximating random variable $d_1T(F; F_n - F)$ is defined and found *without* specification of any norm.

(vi) Nevertheless, the differential approach actually asserts *more*, for it characterizes $d_1T(F; F_n - F)$ as *linear* and hence as an *average* of random variables. That is, letting δ_x denote the distribution function degenerate at x, $-\infty < x < \infty$, and expressing F_n in the form

$$F_n = n^{-1}\sum_{i=1}^{n} \delta_{X_i},$$

we have

$$d_1T(F; F_n - F) = T(F; F_n - F) = T\left(F; n^{-1}\sum_{i=1}^{n}(\delta_{X_i} - F)\right)$$

$$= n^{-1}\sum_{i=1}^{n} T(F; \delta_{X_i} - F).$$

(vii) Prove (Problem 6.P.6) an analogue of Lemma B replacing the convergence $O_p(1)$ required for $n^{1/2}\|F_n - F\|$ by "$O((\log\log n)^{1/2})wp1$" and concluding "$n^{1/2}R_{1n} = o((\log\log n)^{1/2})wp1$." Justify that the requirement is met in the case of $\|\cdot\|_\infty$ and I.I.D. observations. ■

Remarks C. (i) In general the role of $d_1T(F; F_n - F)$ is to approximate $n^{1/2}[T(F_n) - T(F) - \mu(T, F)]$ by $n^{1/2}[d_1T(F; F_n - F) - \mu(T, F)]$, where $\mu(T, F) = E_F\{d_1T(F; F_n - F)\}$. Thus $\mu(T, F)$ may be interpreted as an asymptotic *bias* quantity. In typical applications, $\mu(T, F) = 0$. Note that when $d_1T(F; F_n - F)$ is *linear*, as in Remark B (vi) above, we have $\mu(T, F) = E_F\{T(F; \delta_{X_i} - F)\}$.

(ii) The formulation of the differential of T w.r.t. a norm $\|\cdot\|$ has been geared to the objective of handling R_{1n}. Analogous higher-order derivatives may be formulated in straightforward fashion for use in connection with R_{mn}, $m > 1$.

(iii) We have not concerned ourselves with the case that the functional T is defined only on a *subclass* of $\mathscr{F}$. The reason is that operationally we will utilize the differential only *conceptually* rather than strictly, as will be explained below. ■

Lemmas A and B and Remarks A, B, and C detail the use of a differential for T as a tool in establishing stochastic properties of R_{1n}. However, although appealing and useful as a *concept*, this form of differential is somewhat too narrow for the purposes of statistical applications. The following example illustrates the need for a less rigid formulation.

Example B (*continuation of Example A*). For the *variance* functional, in order to establish differentiability with respect to a norm $\|\cdot\|$, we must show that $L(G, F) \to 0$ as $\|G - F\| \to 0$, where

$$L(G, F) = -\frac{(\mu_G - \mu_F)^2}{\|G - F\|}.$$

Unfortunately, in the case of $\|\cdot\|_\infty$, it is found (check) by considering specific examples that $L(G, F)$ need not $\to 0$ as $\|G - F\|_\infty \to 0$. Thus (D) can fail to hold, so that T does not possess a differential at F with respect to $\|\cdot\|_\infty$. Hence Lemma B in its present form cannot be applied. However, we are able nevertheless to establish a *stochastic* version of (D). Write

$$L(F_n, F) = -[n^{1/2}(\bar{X} - \mu_F)] \cdot (\bar{X} - \mu_F) \cdot [n^{1/2}\|F_n - F\|_\infty]^{-1}.$$

By the CLT and the SLLN, the first factor is $O_p(1)$ and the second factor is $o_p(1)$. By Theorem 2.1.5A and subsequent discussion, $n^{1/2}\|F_n - F\|_\infty \xrightarrow{d} Z_F$, where Z_F is positive *wp*1 (we exclude the case that F is degenerate). Since the function $g(x) = 1/x$ is continuous *wp*1 with respect to the distribution of Z_F, the third factor in $L(F_n, F)$ is $O_p(1)$. It follows that $L(F_n, F) \xrightarrow{p} 0$. The proof of Lemma B carries through unchanged, yielding $n^{1/2}R_{1n} \xrightarrow{p} 0$ as desired. ■

It is thus useful to extend the concept of differential to stochastic versions. We call $T(F; \Delta)$ a *stochastic differential* for T with respect to $\|\cdot\|$ and $\{X_i\}$ if $\|F_n - F\| \xrightarrow{p} 0$ and relation (D) holds in the o_p sense for $G = F_n$. This suffices for proving $\xrightarrow{p}$ results about R_{1n}. For $\xrightarrow{wp1}$ results, we utilize a $\xrightarrow{wp1}$ version of the stochastic differential.

Although these stochastic versions broaden the scope of statistical application of the *concept* of differential, in practice it is more effective to analyze R_{1n} *directly*. A comparison of Examples A and B illustrates this point.

This is not to say, however, that manipulations with $\|F_n - F\|$ become entirely eliminated by a direct approach. Rather, by means of inequalities, useful upper bounds for $|R_{mn}|$ in terms of $\|F_n - F\|$ can lead to properties of $|R_{mn}|$ from those of $\|F_n - F\|$. Such an approach, which we term the *method of differential inequalities*, will be exploited in connection with M-estimates (Chapter 7) and L-estimates (Chapter 8).

We have discussed in detail how to prove $n^{1/2}R_{1n} \xrightarrow{p} 0$, for the purpose of approximating $n^{1/2}[T(F_n) - T(F)]$ in limit distribution by $n^{1/2}d_1T(F; F_n - F)$. Note that this purpose is equally well served by reduction to $T_F(F_n)d_1T(F; F_n - F)$, where $T_F(\cdot)$ is any auxiliary functional defined on $\mathscr{F}$ such that $T_F(F_n) \xrightarrow{p} 1$. That is, it suffices to prove

$$(*) \qquad n^{1/2}[T(F_n) - T(F) - T_F(F_n) \cdot d_1T(F; F_n - F)] \xrightarrow{p} 0$$

in place of $n^{1/2}R_{1n} \xrightarrow{p} 0$. We apply this scheme as follows. First compute $d_1T(F; F_n - F)$. Then select $T_F(\cdot)$ for convenience to make the left-hand side of (*) manageable and to satisfy $T_F(F_n) \xrightarrow{p} 1$. Then proceed to establish (*) by, for example, the method of differential inequalities noted above. We will apply this device profitably in connection with M-estimates (Chapter 7).

The foregoing considerations suggest an extension of the concept of differential. We call $T(F; \Delta)$ a *quasi-differential* with respect to $\|\cdot\|$ and $T_F(\cdot)$ if the definition of differential is satisfied with (D) replaced by

$$\text{(D1)} \qquad \lim_{\|G-F\| \to 0} T_F(G) = 1$$

and

$$\text{(D2)} \qquad T(G) - T(F) - T_F(G)T(F; G - F) = o(\|G - F\|).$$

6.3 METHODOLOGY FOR ANALYSIS OF THE DIFFERENTIAL APPROXIMATION

Here we examine the *structure* of the random variable V_{mn} to which consideration is reduced by the methods of Section 6.2. Under a multilinearity condition which typically is satisfied in applications, we may represent V_{mn} as a V-statistic and as a stochastic integral. In Section 6.4 we make use of these representations to characterize the asymptotic properties of $T(F_n) - T(F)$.

6.3.1 Multi-Linearity Property

In typical cases the kth order Gâteaux differential $d_kT(F; G - F)$ is *k-linear*: there exists a function $T_k[F; x_1, \ldots, x_k]$, $(x_1, \ldots, x_k) \in R^k$, such that

$$\text{(L)} \qquad d_kT(F; G - F) = \int \cdots \int T_k[F; x_1, \ldots, x_k] \prod_{i=1}^{k} d[G(x_i) - F(x_i)], \text{ all } G.$$

Remarks. (i) A review of **1.12.1** is helpful in interpreting the quantity $T_k[F; x_1, \ldots, x_k]$, which is the analogue of the kth order partial derivative,

$$\frac{\partial^k g(t_1, \ldots, t_l)}{\partial t_{i_1} \cdots \partial t_{i_k}},$$

as a function g defined on R^l. Thus $T_k[F; x_1, \ldots, x_k]$ may be interpreted as the kth order partial derivative of the functional $T(F)$, considered as a function of the arguments $\{dF(x), -\infty < x < \infty\}$, the partial being taken with all arguments except $dF(x_1), \ldots, dF(x_k)$ held fixed.

(ii) The function $T_1[F; x]$ may be found, within an additive constant, simply be evaluating $d_1 T(F; \delta_x - F)$. If (L) holds, then $d_1 T(F; \delta_x - F) = T_1[F; x] - \int T_1[F; x]dF(x)$.

(iii) If $T_k[F; x_1, \ldots, x_k]$ is *constant*, considered as a function of $x_1, \ldots, x_k$, then $d_k T(F; G - F) \equiv 0$ (all G). Note that a constant may be added to $T_k[F; x_1, \ldots, x_k]$ without altering its role in Condition (L).

(iv) If $d_1 T(F; G - F)$ is a *differential* for T at F with respect to a norm, then by definition $d_1 T(F; G - F)$ is linear and, as noted in Remark 6.2.2B (vi), we have

$$T_1[F; x] - \int T_1[F; x]dF(x) = T(F; \delta_x - F) = d_1 T(F; \delta_x - F). \quad \blacksquare$$

6.3.2 Representation as a V-Statistic

Under (L), the random variable $d_k T(F; F_n - F)$ may be expressed in the form of a V-statistic. This is seen from the following result.

Lemma A. *Let* F *be fixed and* $\mathrm{h}(\mathrm{x}_1, \ldots, \mathrm{x}_\mathrm{k})$ *be given. A functional of the form*

$$\mathrm{T}(\mathrm{G}) = \int \cdots \int \mathrm{h}(\mathrm{x}_1, \ldots, \mathrm{x}_\mathrm{k}) \prod_{\mathrm{i}=1}^{\mathrm{k}} \mathrm{d}[\mathrm{G}(\mathrm{x}_\mathrm{i}) - \mathrm{F}(\mathrm{x}_\mathrm{i})]$$

may be written as a functional of the form

$$\tilde{\mathrm{T}}(\mathrm{G}) = \int \cdots \int \tilde{\mathrm{h}}(\mathrm{x}_1, \ldots, \mathrm{x}_\mathrm{k}) \mathrm{dG}(\mathrm{x}_1) \cdots \mathrm{dG}(\mathrm{x}_\mathrm{k}),$$

where the definition of $\tilde{\mathrm{h}}$ *depends upon* F.

PROOF. For $k = 1$, take $\tilde{h}(x) = h(x) - \int h(x)dF(x)$. For $k = 2$, take

$$\tilde{h}(x_1, x_2) = h(x_1, x_2) - \int h(x_1, x_2)dF(x_1) - \int h(x_1, x_2)dF(x_2)$$
$$+ \iint h(x_1, x_2)dF(x_1)dF(x_2).$$

In general, take

$$\tilde{h}(x_1, \ldots, x_k) = h(x_1, \ldots, x_k) - \sum_{i=1}^{k} \int h(x_1, \ldots, x_k)dF(x_i)$$
$$+ \sum_{1 \le i < j \le k} \iint h(x_1, \ldots, x_k)dF(x_i)dF(x_j) - \cdots$$
$$+ (-1)^k \int \cdots \int h(x_1, \ldots, x_k)dF(x_1) \cdots dF(x_k). \quad \blacksquare$$

Remark. Check that $\int \tilde{h}(x_1, \ldots, x_k)dF(x_i) = 0$, $1 \le i \le k$. ■

It follows that under (L) there holds the representation

$$d_k T(F; F_n - F) = n^{-k} \sum_{i_1=1}^{n} \cdots \sum_{i_k=1}^{n} \tilde{T}_k[F; X_{i_1}, \ldots, X_{i_k}],$$

where $\tilde{T}_k[F; x_1, \ldots, x_k]$ is determined from $T_k[F; x_1, \ldots, x_k]$ as indicated in the above proof. Therefore, for the random variable

$$V_{mn} = \sum_{k=1}^{m} \frac{1}{k!} d_k T(F; F_n - F),$$

we have the representation (check)

$$V_{mn} = n^{-m} \sum_{i_1=1}^{n} \cdots \sum_{i_m=1}^{n} h(F; X_{i_1}, \ldots, X_{i_m}),$$

where $h(F; x_1, \ldots, x_m)$ is determined from $\tilde{T}_1, \tilde{T}_2, \ldots, \tilde{T}_m$.

Next we establish a further property of random variables having the structure given by Condition (L). The property is applied in Problem 6.P.3.

Lemma B. *Suppose that* $\mathrm{E_F}\{\mathrm{h}^2(\mathrm{X}_{\mathrm{i}_1}, \ldots, \mathrm{X}_{\mathrm{i}_m})\} < \infty$, *all* $1 \le \mathrm{i}_1, \ldots, \mathrm{i}_m \le \mathrm{m}$. *Then*

$$\text{(1)} \qquad \mathrm{E_F}\left\{\left(\int \cdots \int \mathrm{h}(\mathrm{x}_1, \ldots, \mathrm{x}_m) \prod_{i=1}^{m} \mathrm{d}[\mathrm{F}_n(\mathrm{x}_i) - \mathrm{F}(\mathrm{x}_i)]\right)^2\right\} = \mathrm{O}(\mathrm{n}^{-m}).$$

PROOF. By Lemma A,

$$\int \cdots \int h(x_1, \ldots, x_m) \prod_{i=1}^{m} d[F_n(x_i) - F(x_i)]$$
$$= \int \cdots \int \tilde{h}(x_1, \ldots, x_m)dF_n(x_1) \cdots dF_n(x_m)$$
$$= n^{-m} \sum_{i_1=1}^{n} \cdots \sum_{i_m=1}^{n} \tilde{h}(X_{i_1}, \ldots, X_{i_m}).$$

Thus the left-hand side of (1) is given by

$$(2) \quad n^{-2m} \sum_{i_1=1}^{n} \cdots \sum_{i_m=1}^{n} \sum_{j_1=1}^{n} \cdots \sum_{j_m=1}^{n} E\{\tilde{h}(X_{i_1}, \ldots, X_{i_m})\tilde{h}(X_{j_1}, \ldots, X_{j_m})\}.$$

By the remark following Lemma A, the typical term in (2) may be possibly nonzero only if the sequence of indices $i_1, \ldots, i_m, j_1, \ldots, j_m$ contains each member at least twice. The number of such cases is clearly $O(n^{-m})$. Thus (1) follows. ■

We have seen in **5.7.3** the close connection between U- and V-statistics. In particular, we showed that $E|U_n - V_n|^r = O(n^{-r})$ under rth moment assumptions on the kernel $h(x_1, \ldots, x_m)$. We now prove, for the case $m = 2$, an important further relation between U_n and V_n. The result will be of use in connection with V_{2n}.

Lemma C. *Suppose that* $h(x_1, x_2)$ *is symmetric in its arguments and satisfies* $E_F h^2(X_1, X_2) < \infty$ *and* $E_F|h(X_1, X_1)|^{3/2} < \infty$. *Then the corresponding* U- *and* V-*statistics* U_n *and* V_n *satisfy, for* $B > 2|E_F\{h(X_1, X_2) - h(X_1, X_1)\}|$,

$$P(|U_n - V_n| > Bn^{-1}) = o(n^{-1/2}).$$

PROOF. From the proof of Lemma 5.7.3, we have

$$U_n - V_n = n^{-1}(U_n - W_n),$$

where

$$W_n = n^{-1} \sum_{i=1}^{n} h(X_i, X_i).$$

Hence, using the fact that $B > 2|E_F\{h(X_1, X_2) - h(X_1, X_1)\}|$, we have

$$\begin{aligned} P(|U_n - V_n| > Bn^{-1}) &= P(|U_n - W_n| > B) \\ &\le P(|U_n - W_n - E\{U_n\} + E\{W_n\}| > \tfrac{1}{2}B) \\ &\le P\left(|U_n - E\{U_n\}| > \frac{B}{4}\right) + P\left(|W_n - E\{W_n\}| > \frac{B}{4}\right). \end{aligned}$$

The first term on the right is $O(n^{-1})$ by Chebyshev's inequality and Lemma 5.2.1A. For the second term, we use Theorem 4 of Baum and Katz (1965), which implies: *for* $\{Y_i\}$ I.I.D. *with* $E\{Y_1\} = 0$ *and* $E|Y_1|^r < \infty$, *where* $r \ge 1$, $P(|\bar{Y}| > \varepsilon) = o(n^{1-r})$ *for all* $\varepsilon > 0$. Applying the result with $r = \frac{3}{2}$, we have $o(n^{-1/2})$ for the second term on the right. ■

6.3.3 Representation as a Stochastic Integral

Under Condition (L), the random variable $d_m T(F; F_n - F)$ may be expressed as a stochastic integral, that is, in the form

$$\int \cdots \int h(x_1, \ldots, x_m) \prod_{i=1}^{m} d[F_n(x_i) - F(x_i)]$$

for a suitable kernel h. As in **2.1.3**, let us represent $\mathscr{L}\{X_1, \ldots, X_n\}$ as $\mathscr{L}\{F^{-1}(Y_1), \ldots, F^{-1}(Y_n)\}$, where $\{Y_i\}$ are independent uniform (0, 1) variates. Let $G_n(\cdot)$ denote the sample distribution function of $Y_1, \ldots, Y_n$, and consider the corresponding "empirical" stochastic process $Y_n(t) = n^{1/2}[G_n(t) - t]$, $0 \le t \le 1$. Thus

$$\mathscr{L}\{n^{m/2} d_m T(F; F_n - F)\}$$
$$= \mathscr{L}\left\{\int \cdots \int h(F^{-1}(t_1), \ldots, F^{-1}(t_m)) dY_n(t_1) \cdots dY_n(t_m)\right\},$$

so that the limit law of $n^{m/2} d_m T(F; F_n - F)$ may be found through an application of the convergence $Y_n(\cdot) \xrightarrow{d} W^0$ considered in **2.8.2**.

6.4 ASYMPTOTIC PROPERTIES OF DIFFERENTIABLE STATISTICAL FUNCTIONS

Application of the methodology of Sections 6.2 and 6.3 typically leads to approximation of $T(F)$ by a particular *V-statistic*,

$$V_{mn} = n^{-m} \sum_{i=1}^{n} \cdots \sum_{i_m=1}^{n} h(F; X_{i_1}, \ldots, X_{i_m}).$$

(In Section 6.5, as a preliminary to a treatment of several examples, we discuss how to "find" the kernel $h(F; x_1, \ldots, x_m)$ effectively in practice.) As discussed in **6.1.2**, under appropriate conditions on the remainder term $R_{mn} = T(F_n) - T(F) - V_{mn}$, the properties of $T(F_n) - T(F)$ are thus given by the corresponding properties of V_{mn}. In particular, **6.4.1** treats *asymptotic distribution theory*, **6.4.2** *almost sure behavior*, and **6.4.3** *the Berry–Esséen rate*.

6.4.1 Asymptotic Distribution Theory

Parallel to the asymptotic distribution theory of U-statistics (Section 5.5), we have a hierarchy of cases, corresponding to the following condition for the cases $m = 1, 2, \ldots$.

Condition A_m

(i) $\mathrm{Var}_F\{h(F; X_1, \ldots, X_k)\} = 0$ for $k < m$, >0 for $k = m$;

(ii) $n^{m/2} R_{mn} \xrightarrow{p} 0$.

For the case $m = 1$, the V-statistic V_{1n} is simply a sample mean, and by the CLT we have

Theorem A. *Consider a sequence of independent observations* $\{X_i\}$ *on the distribution* F. *Let* T *be a functional for which Condition* A_1 *holds. Put* $\mu(T, F) = E_F\{h(F; X_1)\}$ *and* $\sigma^2(T, F) = \text{Var}_F\{h(F; X_1)\}$. *Assume that* $0 < \sigma^2(T, F) < \infty$. *Then*

$$T(F_n) \quad \textit{is} \quad AN(T(F) + \mu(T, F), n^{-1}\sigma^2(T, F)).$$

Example (*Continuation of Examples* 6.2.1, 6.2.2A). For the *variance* functional we have

$$\begin{aligned} d_1T(F; F_n - F) &= \int x^2\, dF_n(x) - \int x^2\, dF(x) - 2\int x\, dF(x) \int x\, dF_n(x) + 2\left(\int x\, dF(x)\right)^2 \\ &= \int (x - \mu_F)^2\, dF_n(x) - \sigma_F^2 \\ &= \frac{1}{n}\sum_{i=1}^{n}[(X_i - \mu_F)^2 - \sigma_F^2], \end{aligned}$$

so that Condition (L) of **6.3.1** holds, and we approximate $T(F_n) - T(F)$ by V_{1n} based on $h(F; x) = (x - \mu_F)^2 - \sigma_F^2$. We have

$$\mu(T, F) = E_F h(F; X_1) = 0$$

and

$$\sigma^2(T, F) = \text{Var}_F\{h(F; X_1)\} = \mu_4(F) - \sigma_F^4.$$

Further, as seen in Example 6.2.2A, $n^{1/2}R_{1n} \xrightarrow{p} 0$. Thus the conditions of Theorem A are fulfilled, and we have

$$m_2 \quad \text{is} \quad AN\left(\sigma^2, \frac{\mu_4 - \sigma^4}{n}\right),$$

as seen previously in Section 2.2. ■

For the case $m = 2$, we have a result similar to Theorem 5.5.2 for U-statistics.

Theorem B. *Consider a sequence of independent observations* $\{X_i\}$ *on the distribution* F. *Let* T *be a functional for which Condition* A_2 *holds. Assume that* $h(F; x, y) = h(F; y, x)$ *and that* $E_F h^2(F; X_1, X_2) < \infty$, $E_F|h(F; X_1, X_1)|$

$< \infty$, *and* $E_F\{h(F; x, X_1)\} \equiv 0$ (*in* x). *Put* $\mu(T, F) = E_F h(F; X_1, X_2)$. *Denote by* $\{\lambda_i\}$ *the eigenvalues of the operator* A *defined on* $L_2(R, F)$ *by*

$$Ag(x) = \int_{-\infty}^{\infty} [h(F; x, y) - \mu(T, F)]g(y)dF(y), \qquad x \in R, \qquad g \in L_2(R, F).$$

Then

$$n[T(F_n) - T(F) - \mu(T, F)] \xrightarrow{d} \sum_{k=1}^{\infty} \lambda_k \chi_{1k}^2,$$

where χ_{1k}^2 (k = 1, 2, ...) *are independent* χ_1^2 *variates.*

Remark. Observe that the limit distribution has mean $\sum_1^\infty \lambda_k$, which is not necessarily 0. By Dunford and Schwartz (1963), p. 1087, $E_F\{h(F; X_1, X_1)\} = \sum_1^\infty \lambda_k$, which is thus finite by the assumption that $E_F|h(F; X_1, X_1)| < \infty$. This assumption is not made in the analogous result for U-statistics. ∎

PROOF. By Condition A_2, it suffices to show that

$$n\tilde{V}_n \xrightarrow{d} \sum_{k=1}^{\infty} \lambda_k \chi_{1k}^2,$$

where $\tilde{V}_n$ is the V-statistic based on the kernel $\tilde{h}(x, y) = h(F; x, y) - \mu(T, F)$. Consider also the associated U-statistic, $\tilde{U}_n = \binom{n}{2}^{-1} \sum_c \tilde{h}(X_i, X_j)$. As seen in the proof of Lemma 5.7.3, $\tilde{U}_n$ is related to $\tilde{V}_n$ through

$$n^2(\tilde{U}_n - \tilde{V}_n) = (n^2 - n_{(2)})(\tilde{U}_n - \tilde{W}_n),$$

where

$$\tilde{W}_n = n^{-1} \sum_{i=1}^{n} \tilde{h}(X_i, X_i).$$

Thus

$$n(\tilde{V}_n - \tilde{U}_n) = \tilde{W}_n - \tilde{U}_n.$$

Note that $E_F\{\tilde{h}(X_1, X_2)\} = 0$. Thus, by the strong convergence of U-statistics (Theorem 5.4A), $\tilde{U}_n \xrightarrow{wp1} 0$. Furthermore, by the SLLN and the above remark, $\tilde{W}_n \xrightarrow{wp1} \sum_1^\infty \lambda_k$. Therefore,

$$n(\tilde{V}_n - \tilde{U}_n) \xrightarrow{wp1} \sum_{k=1}^{\infty} \lambda_k.$$

Also, since $E_F h(F; x, X_1) \equiv 0$, $\tilde{U}_n$ satisfies the conditions of Theorem 5.5.2 and we have

$$n\tilde{U}_n \xrightarrow{d} \sum_{k=1}^{\infty} \lambda_k(\chi_{1k}^2 - 1),$$

completing the proof. ∎

For arbitrary m, a general characterization of the limit law of $n^{m/2}[T(F_n) - T(F)]$ has been given by Filippova (1962), based on the stochastic integral representation of **6.6.3**. Under Condition A_m, the limit law is that of a random variable of the form

$$B(g, W^0) = \int_0^1 \cdots \int_0^1 g(F; t_1, \ldots, t_m) dW^0(t_1) \cdots dW^0(t_m).$$

Alternatively, Rubin and Vitale (1980) characterize the limit law as that of a linear combination of products of Hermite polynomials of independent $N(0, 1)$ random variables. (Theorems A and B correspond to special cases of the characterizations, for $m = 1$ and $m = 2$, respectively.) These general characterizations also apply, in modified form, to the higher-order cases for U-statistics.

6.4.2 Almost Sure Behavior

Suppose simply that $R_{1n} \xrightarrow{wp1} 0$ and that $E_F|h(F; X_1)| < \infty$. Then $T(F_n) \xrightarrow{wp1} T(F) + \mu(T, F)$, where $\mu(T, F) = E_F h(F; X_1)$. Typically $\mu(T, F) = 0$, giving *strong consistency* of $T(F_n)$ for estimation of $T(F)$. Under higher-order moment assumptions, a *law of iterated logarithm* holds:

Theorem. *Suppose that* $R_{1n} = o(n^{-1/2}(\log\log n)^{1/2})$ wp1. *Put* $\mu(T, F) = E_F\{h(F; X_1)\}$ *and* $\sigma^2(T, F) = \mathrm{Var}_F\{h(F; X_1)\}$. *Assume that* $0 < \sigma^2(T, F) < \infty$. *Then*

$$\overline{\lim_{n\to\infty}} \frac{n^{1/2}[T(F_n) - T(F) - \mu(T, F)]}{\sigma(T, F)(2\log\log n)^{1/2}} = 1 \text{ wp1}.$$

Example (*Continuation of Examples* 6.2.1A *and* 6.4.1). For the *variance* functional the conditions of the above theorem have been established in previous examples. ■

6.4.3 Berry–Esseén Rate

We have seen (Theorem 6.4.1A) that asymptotic normality of $T(F_n) - T(F)$ may be derived by means of an approximation V_{1n} consisting (typically) of the first term of the Taylor expansion of **6.2.1** for $T(F_n) - T(F)$. A corresponding *Berry–Esséen rate* can be investigated through a closer analysis of the remainder term R_{1n}. For such purposes, a standard device is the following (Problem 6.P.9).

Lemma. *Let the sequence of random variables* $\{\xi_n\}$ *satisfy*

(*) $$\sup_t |P(\xi_n \le t) - \Phi(t)| = O(n^{-1/2}).$$

Then, for any sequences of random variables $\{\Delta_n\}$ *and positive constants* $\{a_n\}$,

$$(**)\quad \sup_t |P(\xi_n + \Delta_n \le t) - \Phi(t)| = O(n^{-1/2}) + O(a_n) + P(|\Delta_n| > a_n).$$

In applying the lemma, we obtain for $\xi_n + \Delta_n$ the *best* Berry–Esséen rate, $O(n^{-1/2})$, if we have $P(|\Delta_n| > Bn^{-1/2}) = O(n^{-1/2})$ for some constant $B > 0$. In seeking to establish such a rate for statistical functions, we could apply the lemma with $\xi_n = n^{1/2}V_{1n}$ and $\Delta_n = n^{1/2}R_{1n}$ and thus seek to establish that, for some constant $B > 0$, $P(|R_{1n}| > Bn^{-1}) = O(n^{-1/2})$. The following example illustrates the strength and limitations of this approach.

Example A (*Continuation of Examples* 6.4.1, 6.4.2). For the *variance* functional we have

$$\xi_n = n^{-1/2} \sum_{i=1}^{n} [(X_i - \mu)^2 - \sigma^2]$$

and

$$\Delta_n = -n^{1/2}(\bar{X} - \mu)^2.$$

Note that, by the classical Berry–Esséen theorem (**1.9.5**), (*) holds if $E|X_1|^6 < \infty$. However, the requirement $P(|\Delta_n| > Bn^{-1/2}) = O(n^{-1/2})$ takes the form

$$P(n(\bar{X} - \mu)^2 > B) = O(n^{-1/2}),$$

which *fails* to hold since $n(\bar{X} - \mu)^2$ has a nondegenerate limit distribution with support $(0, \infty)$. Thus we cannot obtain the best Berry–Esseen rate, $O(n^{-1/2})$, by dealing with R_{1n} in this fashion. However, we can do almost as well. By the classical Berry–Esséen theorem, we have (check)

$$P(n(\bar{X} - \mu)^2 > \sigma^2 \log n) = O(n^{-1/2}),$$

provided that $E|X_1|^3 < \infty$. Thus, with $a_n = \sigma^2(\log n)n^{-1/2}$, we have $P(|\Delta_n| > a_n) = O(n^{-1/2})$, so that (**) yields for $n^{1/2}(m_2 - \sigma^2)$ the Berry–Esséen rate $O(n^{-1/2}(\log n))$. Of course, for the closely related statistic s^2, we have already established the best rate $O(n^{-1/2})$ by U-statistic theory. Thus we anticipate that m_2 should also satisfy this rate. We will in fact establish this below, after first developing a more sophisticated method of applying the above lemma in connection with statistical functions. ■

The preceding example represents a case when the remainder term R_{1n} from approximation of $T(F_n) - T(F)$ by $V_{1n} = d_1T(F; F_n - F)$ is not quite small enough for the device of the above lemma to yield $O(n^{-1/2})$ as a Berry–Esséen rate. From consideration of other examples, it is found that this situation is quite typical. However, by taking as approximation the first *two* terms $V_{2n} = d_1T(F; F_n - F) + \frac{1}{2}d_2T(F; F_n - F)$ of the Taylor expansion for $T(F_n) - T(F)$, the remainder term becomes sufficiently reduced for the method of the lemma typically to yield $O(n^{-1/2})$ as the Berry–Esséen rate. In this regard, the approximating random variable is no longer a simple average. However, it typically is a V-statistic and hence approximately a U-statistic, enabling us to exploit the Berry–Esséen rate $O(n^{-1/2})$ established for U-statistics. We have

Theorem. *Let* $T(F_n) - T(F) = V_{2n} + R_{2n}$, *with*

$$V_{2n} = n^{-2}\sum_{i=1}^{n}\sum_{j=1}^{n} h(F; X_i, X_j),$$

where

(1) $$h(F; x, y) = h(F; y, x),\ E_F|h(F; X_1, X_2)|^3 < \infty,$$

and

$$E_F|h(F; X_1, X_1)|^{3/2} < \infty,$$

and, for some $A > 0$,

(2) $$P(|R_{2n}| > An^{-1}) = O(n^{-1/2}).$$

Put $\mu(T, F) = E_F\{h(F; X_1, X_2)\}$ *and* $\sigma^2(T, F) = 4\,\mathrm{Var}_F\{h_1(F; X_1)\}$, *where* $h_1(F; x) = E_F\{h(F; x, X_1)\}$. *Then*

(3) $$\sup_t \left| P\left(\frac{n^{1/2}[T(F_n) - T(F) - \mu(T, F)]}{\sigma(T, F)} \le t\right) - \Phi(t)\right| = O(n^{-1/2}).$$

PROOF. Let U_{2n} be the U-statistic corresponding to $h(F; x, y)$. By (1) and Lemma 6.3.2C, there exists $A > 0$ such that $P(|U_{2n} - V_{2n}| > An^{-1}) = o(n^{-1/2})$. Also, by (1) and Theorem 5.5.1B,

$$\sup_t \left| P\left(\frac{n^{1/2}(U_{2n} - \mu(T, F))}{\sigma(T, F)} \le t\right) - \Phi(t)\right| = O(n^{-1/2}).$$

Thus (check) the above lemma yields

$$\sup_t \left| P\left(\frac{n^{1/2}(V_{2n} - \mu(T, F))}{\sigma(T, F)} \le t\right) - \Phi(t)\right| = O(n^{-1/2}).$$

Then, by (2), a further application of the lemma yields (3) (check). ∎

Example B (*Continuation of Example A*). For the *variance* functional, check that

$$d_2T(F; G - F) = -2(\mu_G - \mu_F)^2,$$

so that

$$V_{2n} = \frac{1}{n}\sum_{i=1}^{n}[(X_i - \mu)^2 - \sigma^2] - (\bar{X} - \mu)^2$$

$$= \frac{1}{n^2}\sum_{i=1}^{n}\sum_{j=1}^{n}\{\tfrac{1}{2}(X_i - \mu)^2 + \tfrac{1}{2}(X_j - \mu)^2 - \sigma^2 - (X_i - \mu)(X_j - \mu)\}$$

$$= \frac{1}{n^2}\sum_{i=1}^{n}\sum_{j=1}^{n}[\tfrac{1}{2}(X_i - X_j)^2 - \sigma^2]$$

and (check)

$$R_{2n} = 0.$$

We thus apply the theorem with $h(F; x, y) = \frac{1}{2}(x - y)^2 - \sigma^2$. Check that the requirements on h are met if $E|X_1|^6 < \infty$ and that $\mu(T, F) = 0$ and $\sigma^2(T, F) = \mu_4 - \sigma^4$. Thus follows for m_2 the Berry–Esséen rate $O(n^{-1/2})$. ■

6.5 EXAMPLES

Illustration of the reduction methods of Section 6.2 and the application of Theorems 6.4.1A, B will be sketched for various examples: sample central moments, maximum likelihood estimates, minimum ω^2 estimates, sample quantiles, trimmed means, estimation of μ^2. Further use of the methods will be seen in Chapters 7, 8 and 9. See also Andrews *et al.* (1972) for some important examples of differentiation of statistical functions.

Remark (*On techniques of application*). In applying Theorem 6.4.1A, the key quantity involved in stating the conclusion of the theorem is $h(F; x)$. In the presence of relation (L) of **6.3.2**, we have (check)

$$h(F; x) = d_1 T(F; \delta_x - F)$$

and $\mu(T, F) = E_F\{h(F; X_1)\} = 0$. Thus, in order to state the "answer," namely that $T(F_n)$ is $AN(T(F), n^{-1}\sigma^2(T, F))$, with $\sigma^2(T, F) = E_F h^2(F; X_1)$, we need only evaluate

$$\left.\frac{dT(F + \lambda(\delta_x - F))}{d\lambda}\right|_{\lambda=0}, \qquad x \in R.$$

Of course, it remains to check that Condition A_1 of **6.4.1** holds.

In some cases we also wish to find $h(F; x, y)$, in order to apply Theorem 6.4.1B or Theorem 6.4.3. In this case it is usually most effective to evaluate $d_2 T(F; G - F)$, put $G = F_n$, and then by inspection find the function $\tilde{T}_2(F; x, y)$ considered in **6.3.2**. Then we have $h(F; x, y) = \frac{1}{2}[h(F; x) + h(F; y) + \tilde{T}_2(F; x, y)]$, as was illustrated in Example 6.4.3B. Alternatively, we can evaluate $d_1 T(F; F_n - F) + \frac{1}{2} d_2 T(F; F_n - F)$ and then by inspection recognize $h(F; x, y)$. ■

Example A *Sample central moments.* The kth central moment of a distribution F may be expressed as a functional as follows:

$$\mu_k = T(F) = \int_{-\infty}^{\infty} \left[x - \int_{-\infty}^{\infty} y\, dF(y)\right]^k dF(x).$$

The sample central moment is

$$m_k = T(F_n) = \int_{-\infty}^{\infty} (x - \bar{X})^k\, dF_n(x).$$

Put $\mu_F = \int x\, dF(x)$ and $F_\lambda = F + \lambda(G - F)$. Then $\mu_{F_\lambda} = \mu_F + \lambda(\mu_G - \mu_F)$. We have

$$T(F_\lambda) = \int (x - \mu_{F_\lambda})^k\, dF(x) + \lambda \int (x - \mu_{F_\lambda})^k\, d[G(x) - F(x)]$$

and (check)

$$\frac{dT(F_\lambda)}{d\lambda} = -k(\mu_G - \mu_F) \int (x - \mu_{F_\lambda})^{k-1}\, dF_\lambda(x) + \int (x - \mu_{F_\lambda})^k\, d[G(x) - F(x)]$$

and

$$\frac{d^2 T(F_\lambda)}{d\lambda^2} = k(k-1)(\mu_G - \mu_F)^2 \int (x - \mu_{F_\lambda})^{k-2}\, dF_\lambda(x) - 2k(\mu_G - \mu_F) \int (x - \mu_{F_\lambda})^{k-1}\, d[G(x) - F(x)].$$

Thus

$$d_1 T(F; G - F) = \int [(x - \mu)^k - k\mu_{k-1} x]\, d[G(x) - F(x)],$$

so that

$$h(F; x) = (x - \mu)^k - k\mu_{k-1} x - E_F\{(X - \mu)^k - k\mu_{k-1} X\}.$$

Thus the assertion of Theorem 6.4.1A is that

$$m_k \quad \text{is} \quad AN(\mu_k, n^{-1}\sigma^2(T, F)),$$

with

$$\sigma^2(T, F) = \mu_{2k} - \mu_k^2 - 2k\mu_{k-1}\mu_{k+1} + k^2\mu_{k-1}^2\mu_2.$$

This result was derived previously in Section 2.2. However, by the present technique we have cranked out the "answer" in a purely mechanical fashion. Of course, we must *validate* the "answer" by showing that $n^{1/2} R_{1n} \xrightarrow{p} 0$. Check that

$$R_{1n} = m_k - b_k + k\mu_{k-1}b_1,$$

where $b_j = n^{-1} \sum_{i=1}^n (X_i - \mu)^j$, $0 \le j \le k$, and thus (check)

$$R_{1n} = \sum_{j=0}^{k-1} \binom{k}{j}(-1)^j b_j b_1^{k-j} + k\mu_{k-1}b_1 = o_p(n^{-1/2}),$$

as required. To establish a related Berry–Esséen rate, check that

$$\begin{aligned} d_2 T(F; G - F) &= k(k-1)(\mu_G - \mu)^2\mu_{k-2} \\ &\quad - 2k(\mu_G - \mu)\int (x - \mu)^{k-1}\, d[G(x) - F(x)] \\ &= k \iint [(x - \mu)^{k-1}y + (y - \mu)^{k-1}x \\ &\quad - (k-1)\mu_{k-2}xy]d[G(x) - F(x)]d[G(y) - F(y)]. \end{aligned}$$

Complete the details. ■

Example B *Maximum likelihood estimation.* Under regularity conditions (**4.4.2**) on the family of distributions $\{F(x; \theta), \theta \in \Theta\}$ under consideration, the maximum likelihood estimate of θ is the solution of

$$\int g(\theta, x)dF_n(x) = 0,$$

where

$$g(\theta, x) = \frac{\dfrac{d}{d\theta} f(x; \theta)}{f(x; \theta)}$$

and

$$f(x; \theta) = \frac{d}{dx} F(x; \theta).$$

That is, the maximum likelihood estimate is $\theta(F_n)$, where $\theta(F)$ is the functional defined as the solution of

$$\text{(B1)} \qquad \int q(\theta, x)dF(x) = 0.$$

Under regularity condition on the family $\{F(\cdot, \theta), \theta \in \Theta\}$, we have

$$\text{(B2)} \qquad \int \frac{\frac{d^2}{d\theta^2} f(x;\theta)}{f(x;\theta)} dF(x;\theta) = \int \frac{d^2}{dx^2} f(x;\theta)dx = 0.$$

We find

$$\frac{d}{d\lambda}\theta(F + \lambda(\delta_{x_0} - F))\bigg|_{\lambda=0}$$

by implicit differentiation through the equation

$$H(\theta(F_\lambda), \lambda) = 0,$$

where $F_\lambda = F + \lambda(\delta_{x_0} - F)$ and $H(\theta, \lambda) = \int q(\theta, x)dF_\lambda(x)$. We have

$$\frac{\partial H}{\partial \theta}\bigg|_{\theta=\theta(F)} \cdot \frac{d\theta(F_\lambda)}{d\lambda}\bigg|_{\lambda=0} + \frac{\partial H}{\partial \lambda}\bigg|_{\lambda=0} = 0.$$

Thus

$$\text{(B3)} \qquad \frac{\partial \theta(F_\lambda)}{\partial \lambda}\bigg|_{\lambda=0} = -\frac{\partial H}{\partial \lambda}\bigg|_{\lambda=0} \bigg/ \frac{\partial H}{\partial \theta}\bigg|_{\theta=\theta(F)}.$$

Check (using (B1), (B2) and the fact that $\theta(F(\cdot\,;\theta_0)) = \theta_0$, each θ_0) that (B3) yields

$$\frac{d}{d\lambda}\theta(F + \lambda(\delta_{x_0} - F))\bigg|_{\lambda=0} = \frac{\frac{d}{d\theta} f(x_0;\theta)}{f(x_0;\theta)} \cdot \frac{1}{\int \frac{[(d/d\theta)f(x;\theta)]^2}{f(x;\theta)} dx}$$

and thus

$$h(F(\cdot\,;\theta); x) = \frac{\frac{d}{d\theta}\log f(x;\theta)}{\int \left[\frac{d}{d\theta}\log f(x;\theta)\right]^2 dF(x;\theta)}.$$

Therefore, the assertion of Theorem 6.4.1A is that (check)

$$\theta(F_n) \quad \text{is} \quad AN\left(\theta, \frac{1}{nE_{F(\cdot;\theta)}\left\{\left[\frac{d}{d\theta}\log f(X;\theta)\right]^2\right\}}\right),$$

as seen previously in **4.2.2.** ∎

Example C *Minimum ω^2 estimation.* The "minimum ω^2 estimate" of θ, in connection with a family of distributions $\{F(\cdot;\theta), \theta \in \Theta\}$, is the solution of

$$\int q(\theta, x, F_n)dx = 0,$$

where

$$q(\theta, x, G) = \frac{d}{d\theta}\{[G(x) - F(x;\theta)]^2 f(x;\theta)\}$$

and

$$f(x;\theta) = \frac{d}{dx}F(x;\theta).$$

That is, the ω^2-minimum estimate is $\theta(F_n)$, where $\theta(G)$ is the functional defined as the solution of

$$\int q(\theta, x, G)dx = 0.$$

By implicit differentiation as in Example B, check that

$$\frac{d}{d\lambda}\theta(F + \lambda(\delta_{x_0} - F))\bigg|_{\lambda=0} \frac{f(x_0;\theta)\frac{d}{d\theta}f(x_0;\theta)}{\int\left[\frac{d}{d\theta}f(x;\theta)\right]^2 dF(x;\theta)},$$

from which $\operatorname{Var}_F\{h(F(\cdot;\theta);X)\}$ may be found readily. ∎

Example D *Sample* p*th quantile.* Let $0 < p < 1$. The pth quantile of a distribution F is given by $\xi_p = T(F) = F^{-1}(p)$ and the corresponding sample pth quantile by $\hat{\xi}_{pn} = F_n^{-1}(p) = T(F_n)$. We have

$$\begin{aligned} T[F + \lambda(\delta_{x_0} - F)] &= \inf\{x: F(x) + \lambda(\delta_{x_0}(x) - F(x)) \geq p\} \\ &= \inf\{x: F(x) + \lambda[I(x \geq x_0) - F(x)] \geq p\} \\ &= \inf\left\{x: F(x) \geq \frac{p - \lambda I(x \geq x_0)}{1 - \lambda}\right\} \end{aligned}$$

(for λ sufficiently small). Confine attention now to the case that F has a positive density f in a neighborhood of $F^{-1}(p)$. Then, for any x_0 other that ξ_p. it is found (check) that

$$\text{(D1)} \qquad \left.\frac{dT[F + \lambda(\delta_{x_0} - F)]}{d\lambda}\right|_{\lambda=0} = \frac{p - \delta_{x_0}(\xi_p)}{f(\xi_p)} = \frac{p - I(x_0 \leq \xi_p)}{f(\xi_p)}.$$

The assertion of Theorem 6.4.1A is thus that

$$\hat{\xi}_{pn} \quad \text{is} \quad AN\left(\xi_p, \frac{p(1-p)}{nf^2(\xi_p)}\right),$$

as established previously in Section 2.3. In order to establish the validity of the assertion of the theorem, we find using (D1) that

$$\text{(D2)} \qquad R_{1n} = \hat{\xi}_{pn} - \xi_p - \frac{p - F_n(\xi_p)}{f(\xi_p)}$$

and we seek to establish that $n^{1/2}R_{1n} \xrightarrow{p} 0$. But R_{1n} is precisely the remainder term R_n in the Bahadur representation (Theorem 2.5.1) for $\hat{\xi}_{pn}$, and as noted in Remark 2.5.1(iv) Ghosh (1971) has shown that $n^{1/2}R^n \xrightarrow{p} 0$, provided $F'(\xi_p) > 0$. ■

Example E *α-trimmed mean.* Let F be symmetric and continuous. For estimation of the mean (=median), a competitor to the sample mean and the sample median is the "α-trimmed mean"

$$\bar{X}_{(\alpha)n} = \frac{1}{n - 2[\alpha n]} \sum_{k=[\alpha n]+1}^{n-[\alpha n]} X_{nk},$$

where $0 < \alpha < \frac{1}{2}$. This represents a compromise between the sample mean and the sample median, which represent the limiting cases as $\alpha \to 0$ and $\alpha \to \frac{1}{2}$, respectively. An asymptotically equivalent (in all typical senses) version of the α-trimmed mean is defined as

$$\bar{X}_{(\alpha)n} = T(F_n),$$

where

$$\text{(E1)} \qquad T(F) = \frac{1}{1-2\alpha}\int_{F^{-1}(\alpha)}^{F^{-1}(1-\alpha)} x\,dF(x) = \frac{1}{1-2\alpha}\int_{\alpha}^{1-\alpha} F^{-1}(p)dp.$$

We shall treat this version here. By application of (D1) in conjunction with (E1), we obtain

$$\left.\frac{dT[F + \lambda(\delta_{x_0} - F)]}{d\lambda}\right|_{\lambda=0} = \frac{1}{1-2\alpha}\int_{\alpha}^{1-\alpha}\frac{p - I(x_0 \le F^{-1}(p))}{f(F^{-1}(p))}\,dp$$

$$= \frac{1}{1-2\alpha}\int_{\alpha}^{1-\alpha}\frac{p - I(F(x_0) \le p)}{f(F^{-1}(p))}\,dp.$$

For the case $F(x_0) < \alpha$, this becomes

$$\frac{1}{1-2\alpha}\int_{\alpha}^{1-\alpha}\frac{p-1}{f(F^{-1}(p))}\,dp = \frac{1}{1-2\alpha}\int_{\alpha}^{1-\alpha}(p-1)dF^{-1}(p) = \frac{F^{-1}(\alpha) - c(\alpha)}{1-2\alpha},$$

where

$$c(\alpha) = \int_{\alpha}^{1-\alpha} F^{-1}(t)dt + \alpha F^{-1}(\alpha) + \alpha F^{-1}(1-\alpha).$$

The cases $F(x_0) > 1 - \alpha$ and $\alpha \le F(x_0) \le 1 - \alpha$ may be treated in similar fashion (check). Furthermore, the symmetry assumption yields

$$c(\alpha) = T(F) = \xi_{1/2}.$$

Thus we arrive at

$$\left.\frac{dT[F + \lambda(\delta_x - F)]}{d\lambda}\right|_{\lambda=0} = \begin{cases} \dfrac{1}{1-2\alpha}[F^{-1}(\alpha) - \xi_{1/2}], & x < F^{-1}(\alpha), \\ \dfrac{1}{1-2\alpha}(x - \xi_{1/2}), & F^{-1}(\alpha) \le x \le F^{-1}(1-\alpha), \\ \dfrac{1}{1-2\alpha}[F^{-1}(1-\alpha) - \xi_{1/2}], & x > F^{-1}(1-\alpha). \end{cases}$$

It follows that the assertion of Theorem 6.4.1A is

$$\overline{X}_{(\alpha)n} \text{ is } AN\left(\xi_{1/2}, \frac{1}{n}\frac{1}{(1-2\alpha)^2}\left[\int_{F^{-1}(\alpha)}^{F^{-1}(1-\alpha)}(x - \xi_{1/2})^2\,dF(x) + 2\alpha(F^{-1}(\alpha) - \xi_{1/2})^2\right]\right).$$ ■

Example F *Estimation of* μ^2. Consider

$$T(F) = E_F^2\{X\} = \left[\int_{-\infty}^{\infty} x\, dF(x)\right]^2 = \mu^2.$$

The corresponding statistical function is

$$T(F_n) = \bar{X}^2.$$

Derive $h(F; x)$ and $h(F; x_1, x_2)$, and apply Theorems 6.4.1A, B to obtain the asymptotic distribution theory for $\bar{X}^2$ in the cases $\mu \neq 0$, $\mu = 0$. (Compare Example 5.5.2B.) ■

6.6 COMPLEMENTS

Some useful *statistical interpretations* of the derivative of a statistical function are provided in **6.6.1**. Comments on the differential approach for analysis of statistical functions based on functionals of a *density* f are provided in **6.6.2**. Extension to the case of *dependent* X_i's is discussed in **6.6.3**. Normalizations other than $n^{m/2}$ are discussed in **6.6.4**.

6.6.1 Statistical Interpretations of the Derivative of a Statistical Function

In the case of a statistical function having nonvanishing first derivative (implying asymptotic normality, under mild restrictions), a variety of important features of the estimator may be characterized in terms of this derivative. Namely, the *asymptotic variance parameter*, and certain *stability properties* of the estimator under perturbation of the observations, may be characterized. These features are of special interest in studying *robustness* of estimators. We now make these remarks precise.

Consider observations $X_1, X_2, \ldots$ on a distribution F and a functional $T(\cdot)$. Suppose that T satisfies relation (L) at F, that is, $d_1T(F; G - F) = \int T_1[F; x]d[G(x) - F(x)]$, as considered in **6.3.1**, and put

$$h(F; x) = T_1[F; x] - \int T_1[F; x]dF(x).$$

The reduction methodology of Section 6.2 shows that the *error of estimation* in estimating $T(F)$ by $T(F_n)$ is given approximately by

$$\frac{1}{n}\sum_{i=1}^{n} h(F; X_i).$$

Thus $h(F; X_i)$ represents the approximate contribution, or "influence," of the observation X_i toward the estimation error $T(F_n) - T(F)$. This notion of interpreting $h(F; x)$ as a measure of "influence" toward error of estimation is due to Hampel (1968, 1974), who calls $h(F; x)$, $-\infty < x < \infty$, the *influence curve* of the estimator $T(F_n)$ for $T(F)$. Note that the curve may be defined directly by

$$h(F; x) = \left. \frac{dT[F + \lambda(\delta_x - F)]}{d\lambda} \right|_{\lambda=0}, \qquad -\infty < x < \infty.$$

(In the robust estimation literature, the notation $\Omega_F(x)$ or $IC(x; F, T)$ is sometimes used.)

In connection with the interpretation of $T[F; \cdot]$ as an "influence curve," Hampel (1974) identifies several key characteristics of the function. The "*gross-error-sensitivity*"

$$\gamma^* = \sup_x |h(F; x)|$$

measures the effect of contamination of the data by gross errors, whereby some of the observations X_i may have a distribution grossly different from F. Specifically, γ^* is interpreted as the worst possible influence which a fixed amount of contamination can have upon the estimator. The "*local-shift-sensitivity*"

$$\lambda^* = \sup_{x \neq y} \left| \frac{h(F; x) - h(F; y)}{x - y} \right|$$

measures the effect of "wiggling" the observations, that is, the local effects of rounding or grouping of the observations. The "*rejection point*" ρ^* is defined as the distance from the center of symmetry of a distribution to the point at which the influence curve becomes identically 0. Thus all observations farther away than ρ^* become completely rejected, that is, their "influence" is not only truncated but held to 0. This is of special interest in problems in which rejection of outliers is of importance.

Examples. The influence curve of the *sample mean* is

$$IC(x; T, F) = x - \mu_F, \qquad -\infty < x < \infty.$$

We note that in this case $\gamma^* = \infty$, indicating the extreme sensitivity of the sample mean to the influence of "wild" observations. The *α-trimmed mean*, for $0 < \alpha < \frac{1}{2}$, provides a correction for this deficiency. Its γ^* (see Example 6.5E) is $[F^{-1}(1 - \alpha) - T(F)]/(1 - 2\alpha)$. On the other hand, the sample mean has $\lambda^* = 1$ whereas the *sample median* has $\lambda^* = \infty$, due to irregularity of its influence curve

$$IC(x; T, F) = \frac{\text{sign}[F(x) - \frac{1}{2}]}{2f(F^{-1}(\frac{1}{2}))} \qquad (\text{sign } 0 = 0)$$

at the point $x = F^{-1/2}(\frac{1}{2})$. Also, contrary perhaps to intuition, the α-trimmed mean has $\rho^* = \infty$. However, Hampel (1978, 1974) and Andrews *et al.* (1972) discuss estimators which are favorable *simultaneously* with respect to γ^*, λ^* and ρ^* (see Chapter 7). ■

Further discussion of the influence curve and robust estimation is given by Huber (1972, 1977). Robustness principles dictate choosing $T(\cdot)$ to control $IC(x; T, F)$.

6.6.2 Functionals of Densities

An analogous theory of statistical functions can be developed with respect to parameters given as functionals of densities, say $T(f)$. For example, in **2.6.7** the efficacy parameter $\int f^2(x)dx$ arose in certain asymptotic relative efficiency considerations. A natural estimator of any such $T(f)$ is given by $T(f_n)$, where f_n is a density estimator of f such as considered in **2.1.8**. The differential approach toward analysis of $T(f_n) - T(f)$ is quite useful and can be formulated in analogy with the treatment of Sections 6.1–6.5. We merely mention here certain additional complications that must be dealt with. First, the structure of the sample *density* function f_n is typically not quite as simple as that of the sample distribution function F_n. Whereas $F_n(x)$ is the average at the nth stage of the random variables $I(X_1 \le x), I(X_2 \le x), \ldots,$ the estimator $f_n(x)$ is typically an average over a *double array* of random variables. This carries over to the approximating random variable $d_1T(f; f_n - f)$ here playing the role of $d_1T(F; F_n - F)$. Consequently, we need to use a double array CLT in deriving the asymptotic normality of $T(f_n)$, and we find that there does not exist a double array LIL at hand for us to exploit in deriving an LIL for $T(f_n) - T(f)$. Furthermore, unlike $F_n(x)$ as an estimator of $F(x)$, the estimator $f_n(x)$ is typically *biased* for estimation of $f(x)$. Thus $E_f T(f; f_n - f) \neq 0$ in typical cases, so that the analysis must deal with this type of term also.

See Beran (1977a, b) for minimum Hellinger distance estimation based on statistical functions of densities.

6.6.3 Dependent Observations $\{X_i\}$

Note that the asymptotic behavior of $T(F_n) - T(F)$ typically depends on the X_i's only through two elements,

$$V_{1n} = \frac{1}{n}\sum_{i=1}^{n} h(F; X_i)$$

and R_{1n}. Often R_{1n} can be handled via inequalities involving $\|F_n - F\|_\infty$ and the like. Thus, for example, the entire theory extends readily to any sequence $\{X_i\}$ of possibly dependent variables for which a CLT has been established and for which suitable stochastic properties of $\|F_n - F\|_\infty$ have been established.

6.6.4 Other Normalizations

For the random variable $T(F_n) - T(F)$ to have a nondegenerate limit law, the appropriate normalizing factor in the case that the first nonvanishing term in the Taylor expansion is the mth need not always be $n^{m/2}$. For example, in the case $m = 1$, we have a sum of I.I.D. random variables as $d_1 T(F; F_n - F)$, for which the correct normalization actually depends on the domain of attraction. For attraction to a stable law with exponent α, $0 < \alpha < 2$, the appropriate normalization is $n^{1/\alpha}$. See Gnedenko and Kolmogorov (1954) or Feller (1966).

6.6.5 Computation of Higher Gâteaux Derivatives

In the presence of Condition (L) of **6.3.1**, we have

$$\frac{dT(V_\lambda)}{d\lambda} = \frac{dT(V_{\lambda+z})}{dz}\bigg|_{z=0} = \frac{d}{dz} T\left(V_\lambda + \frac{z}{1-\lambda}(W - V_\lambda)\right)\bigg|_{z=0}$$

$$= \frac{1}{1-\lambda}\int_{-\infty}^{\infty} T_1[V_\lambda; x]d[W(x) - V_\lambda(x)]$$

$$= \int_{-\infty}^{\infty} T_1[V_\lambda; x]d[W(x) - V(x)].$$

Hence

$$\frac{d^2T(V_\lambda)}{d\lambda^2}\bigg|_{\lambda=0} = \int_{-\infty}^{\infty} \frac{d}{d\lambda} T_1[V_\lambda; x]\bigg|_{\lambda=0} d[W(x) - V(x)],$$

etc.

6.P PROBLEMS

Section 6.2

1. Check the details for Example 6.2.1.
2. Formulate and prove the extended form of Theorem 1.12.1A germane to the Taylor expansion discussed in **6.2.1**.
3. (Continuation of Example 6.2.1) Show, applying Lemma 6.3.2B, that

$$\sup_{0\le\lambda\le 1}\left|\frac{d^k}{d\lambda^k} T(F + \lambda(F_n - F))\right| = O_p(n^{-(1/2)k}).$$

4. Complete the details of Example 6.2.2A.
5. Complete the argument for Lemma 6.2.2B.
6. Verify Remarks 6.2.2B (ii), (vii).
7. Check the claim of Example 6.2.2B.

Section 6.3

8. Complete the details for the representation of V_{mn} as a V-statistic **(6.3.2)**.

Section 6.4

9. Prove Lemma 6.4.3.
10. Check details of Examples 6.4.3A, B.
11. Complete details of proof of Theorem 6.4.3.

Section 6.5

12. Supply the missing details for Example 6.5A (sample central moments).
13. Supply details for Example 6.5B (maximum likelihood estimate).
14. Supply details for Example 6.5C (minimum ω^2 estimate).
15. Supply details for Example 6.5D (sample pth quantile).
16. Supply details for Example 6.5E (α-trimmed mean).
17. Supply details for Example 6.5F (estimation of μ^2).
18. Apply Theorem 6.4.3 to obtain the Berry–Esséen theorem for the sample pth quantile (continuation of Problem 15 above).

Section 6.6

19. Provide details for **6.6.5**.

CHAPTER 7

M-Estimates

In this chapter we briefly consider the asymptotic properties of statistics which are obtained as *solutions of equations*. Often the equations correspond to some sort of *minimization* problem, such as in the cases of *maximum likelihood* estimation, *least squares* estimation, and the like. We call such statistics "*M-estimates*." (Recall previous discussion in **4.3.2**.)

A treatment of the class of *M*-estimates could be carried out along the lines of the classical treatment of maximum likelihood estimates, as in **4.2.2**. However, for an important subclass of *M*-estimates, we shall apply certain specialized methods introduced by Huber (1964). Also, as a general approach, we shall formulate *M*-estimates as statistical functions and apply the methods of Chapter 6. Section 7.1 provides a general formulation and various examples. The asymptotic properties of *M*-estimates, namely consistency and asymptotic normality with related rates of convergence, are derived in Section 7.2. Various complements and extensions are discussed in Section 7.3.

Two closely related competing classes of statistics, L-estimates and *R*-estimates, are treated in Chapters 8 and 9. In particular, see Section 9.3.

7.1 BASIC FORMULATION AND EXAMPLES

A general formulation of *M*-estimation is presented in **7.1.1**. The special case of *M*-estimation of a *location* parameter, with particular attention to *robust* estimators, is studied in **7.1.2**.

7.1.1 General Formulation of *M*-Estimation

Corresponding to any function $\psi(x, t)$, we may associate a functional T defined on distribution functions F, $T(F)$ being defined as a solution t_0 of the equation

(*) $$\int \psi(x, t_0)dF(x) = 0.$$

We call such a $T(\cdot)$ the *M-functional corresponding to* ψ. For a sample $X_1, \ldots, X_n$ from F, the *M-estimate corresponding to* ψ is the "statistical function" $T(F_n)$, that is, a solution T_n of the equation

$$(**) \qquad \sum_{i=1}^{n} \psi(X_i, T_n) = 0.$$

In our theorems for such parameters and estimates, we have to allow for the possibility that (*) or (**) has multiple solutions.

When the ψ function defining an M-functional has the form $\psi(x, t) = \tilde{\psi}(x - t)$ for some function $\tilde{\psi}$, $T(F)$ is called a *location parameter*. This case will be of special interest.

In typical cases, the equation (*) corresponds to *minimization* of some quantity

$$\int \rho(x, t_0) dF(x),$$

the function ψ being given by

$$\psi(x, t) = c \frac{\partial}{\partial t} \rho(x, t)$$

for some constant c, in the case of $\rho(x, \cdot)$ sufficiently smooth.

In a particular estimation problem, the parameter of interest θ may be represented as $T(F)$ for various choices of ψ. The corresponding choices of $T(F_n)$ thus represent competing estimators. Quite a variety of ψ functions can thus arise for consideration. It is important that our theorems cover a very broad class of such functions.

Example *Parametric Estimation.* Let $\mathscr{F}_0 = \{F(\cdot\,; \theta), \theta \in \Theta\}$ represent a "parametric" family of distributions. Let $\psi = \psi(x, t)$ be a function such that

$$\int \psi(x, \theta) dF(x; \theta) = 0, \qquad \theta \in \Theta,$$

that is, for $F = F(\cdot\,; \theta)$ the solution of (*) coincides with θ. In this case the corresponding M-functional T satisfies $T(F(\cdot\,; \theta)) = \theta$, $\theta \in \Theta$, so that a natural estimator of θ is given by $\hat{\theta} = T(F_n)$. Different choices of ψ lead to different estimators. For example, if the distributions $F(\cdot\,; \theta)$ have densities or mass functions $f(\cdot\,; \theta)$, then the *maximum likelihood* estimator corresponds to

$$\rho(x, \theta) = -\log f(x; \theta),$$

$$\psi(x, \theta) = -\frac{\partial}{\partial \theta} \log f(x; \theta).$$

We have studied maximum likelihood estimation in this fashion in Example 6.5B. Likewise, in Example 6.5C, we examined *minimum* ω^2 estimation, corresponding to a different ψ.

A *location parameter problem* is specified by supposing that the members of $\mathscr{F}_0$ are of the form $F(x;\theta) = F_0(x-\theta)$, where F_0 is a fixed distribution thus generating the family $\mathscr{F}_0$. It then becomes appropriate from invariance considerations to restrict attention to ψ of the form $\psi(x, t) = \tilde{\psi}(x - t)$.

In classical parametric location estimation, the distribution F_0 is assumed known. In *robust* estimation, it is merely assumed that F_0 belongs to a neighborhood of some specified distribution such as Φ. (See Example 7.1.2E.) ■

In considering several possible ψ for a given estimation problem, the corresponding *influence curves* are of interest (recall **6.6.1**). Check (Problem 7.P.1) that the Gâteaux differential of an M-functional is

$$d_1 T(F; G - F) = -\frac{\int \psi(x, T(F))dG(x)}{\lambda_F'(T(F))},$$

provided that $\lambda_F'(T(F)) \neq 0$, where we define

$$\lambda_F(t) = \int \psi(x, t)dF(x), \qquad -\infty < t < \infty.$$

Thus the influence curve of (the M-functional corresponding to) ψ is

$$IC(x; F, T) = -\frac{\psi(x, T(F))}{\lambda_F'(T(F))}, \qquad -\infty < x < \infty. \tag{1}$$

Note that IC is *proportional* to ψ. Thus the principle of M-estimation possesses the nice feature that desired properties for an influence curve may be achieved simply by choosing a ψ with the given properties. This will be illustrated in some of the examples of **7.1.2**.

Further information immediate from (1) is that, under appropriate regularity conditions,

$$T(F_n) \quad \text{is} \quad AN(T(F), n^{-1}\sigma^2(T, F)),$$

where typically

$$\sigma^2(T, F) = \frac{\int \psi^2(x, T(F))dF(x)}{[\lambda_F'(T(F))]^2}.$$

This is seen from Theorem 6.4.1A (see Remark 6.5) and the fact that $\int IC(x, F, T)dF(x) = 0$. A detailed treatment is carried out in **7.2.2**. (In some cases $\sigma^2(T, F)$ comes out differently from the above.)

7.1.2 Examples Apropos to Location Parameter Estimation

The following examples illustrate the wide variety of ψ functions arising for consideration in the contexts of efficient and robust estimation. We consider the M-functional $T(F)$ to be a solution of

$$\int \psi(x - t_0)dF(x) = 0$$

and the corresponding M-estimate to be $T(F_n)$.

Example A *The least squares estimate.* Corresponding to minimization of $\sum_1^n (X_i - \theta)^2$, the relevant ψ function is

$$\psi(x) = x, \qquad -\infty < x < \infty.$$

For this ψ, the M-functional T is the *mean* functional and the M-estimate is the *sample mean.* ■

Example B *The least absolute values estimate.* Corresponding to minimization of $\sum_1^n |X_i - \theta|$, the relevant ψ function is

$$\psi(x) = \begin{cases} -1, & x < 0, \\ 0, & x = 0, \\ 1, & x > 0. \end{cases}$$

Here the corresponding M-functional is the *median* functional and the corresponding M-estimate the *sample median.* ■

Example C *The maximum likelihood estimate.* For the parametric location model considered in Example 7.1.1, let F_0 have density f_0 and take

$$\psi(x) = -\frac{f_0'(x)}{f_0(x)}, \qquad -\infty < x < \infty.$$

The corresponding M-estimate is the maximum likelihood estimate. Note that this choice of ψ depends on the particular F_0 generating the model. ■

Example D *A form of trimmed mean.* Huber (1964) considers minimization of $\sum_1^n \rho(X_i - \theta)$, where

$$\rho(x) = \begin{cases} x^2, & |x| \le k, \\ k^2, & |x| > k. \end{cases}$$

The relevant ψ is

$$\psi(x) = \begin{cases} x, & |x| \le k, \\ 0, & |x| > k. \end{cases}$$

The corresponding M-estimator T_n is a type of trimmed mean. In the case that no X_i satisfies $|X_i - T_n| = k$, it turns out to be the sample mean of the X_i's satisfying $|X_i - T_n| < k$. (Problem 7.P.2) Note that this estimator eliminates the "influence" of outliers. ■

Example E *A form of trimmed mean.* Huber (1964) considers minimization mization of $\sum_1^n \rho(X_i - \theta)$, where

$$\rho(x) = \begin{cases} \frac{1}{2}x^2, & |x| \le k, \\ k|x| - \frac{1}{2}k^2, & |x| > k. \end{cases}$$

The relevant ψ is

$$\psi(x) = \begin{cases} -k, & x < -k, \\ x, & |x| \le k \\ k, & x > k. \end{cases}$$

The corresponding M-estimator T_n is a type of Winsorized mean. It turns out to be the sample mean of the modified X_i's, where X_i becomes replaced by $T_n \pm k$, whichever is nearer, if $|X_i - T_n| > k$ (Problem 7.P.3). This estimator limits, but does not entirely eliminate, the influence of outliers. However, it has a smoother IC then the ψ of Example D. The $\rho(\cdot)$ of the present example represents a compromise between least squares and least absolute values estimation. It also represents the *optimal* choice of ρ, in the *minimax* sense, for *robust* estimation of θ in the *normal* location model. Specifically, let C denote the set of all symmetric contaminated mornal distributions $F = (1 - \varepsilon)\Phi + \varepsilon H$, where $0 \le \varepsilon < 1$ is fixed and H varies over all symmetric distributions. Huber (1964) defines a *robust* M-estimator ψ to be the ψ_0 which minimaxes the asymptotic variance parameter $\sigma^2(T, F)$, that is,

$$\sup_F \sigma^2(T_{\psi_0}, F) = \inf_\psi \sup_F \sigma^2(T_\psi, F).$$

Here F ranges through C, ψ ranges over a class of "nice" ψ functions, and $\sigma^2(T_\psi, F)$ is as given in **7.1.1**. For the given C, the optimal ψ_0 corresponds to the above form, for k defined by $\int_{-k}^k \phi(t)dt + 2\phi(k)/k = 1/(1 - \varepsilon)$. The ψ functions of this form are now known as "Hubers." Note that the IC function is *continuous*, *nondecreasing*, and *bounded*. ■

Example F. Hampel (1968, 1974) suggested a modification of the "Hubers" in order to satisfy qualitative criteria such as low *gross-error-sensitivity*, small *local-shift-sensitivity*, etc., as discussed in **6.6.1**. He required $\psi(x)$ to return to 0 for $|x|$ sufficiently large:

$$\psi(x) = \begin{cases} x, & 0 \le x \le a, \\ a, & a \le x \le b, \\ a\left(\dfrac{c - x}{c - b}\right), & b \le x \le c, \\ 0, & x > c, \end{cases}$$

and $\psi(x) = -\psi(-x)$, $x < 0$. This M-estimator has the property of completely rejecting outliers while giving up very little efficiency (compared to the Hubers) at the normal. M-estimators of this type are now known as "Hempels." ■

Example G *A smoothed "Hampel"*. One of many varieties of smoothed "Hampels" is given by

$$\psi(x) = \begin{cases} \sin ax, & 0 \le x < \dfrac{\pi}{a}, \\ 0, & x > \dfrac{\pi}{a}, \end{cases}$$

and $\psi(x) = -\psi(-x)$, $x < 0$. See Andrews *et al* (1972). ■

Remarks. (i) *Further examples*, and small sample size comparisons, are provided by Andrews *et al.* (1972).

(ii) *Construction of robust M-estimates.* For *high efficiency* at the model distribution F_0, one requires that the influence function be *roughly proportional* to $-f'_0(x)/f(x)$. For protection against *outliers*, one requires that the influence function be *bounded*. For protection against the effects of *round-off and grouping*, one requires the influence function to be *reasonably continuous in x*. In order to *stabilize the asymptotic variance* of the estimate under small changes in F_0, one requires the influence function to be *reasonably continuous as a function of F*. These requirements are apropos for any kind of estimator. However, in the case of M-estimators, they translate directly into similar requirements on the ψ function. One can thus find a suitable M-estimator simply by *defining* the ψ function appropriately. ■

7.2 ASYMPTOTIC PROPERTIES OF *M*-ESTIMATES

We treat *consistency* in **7.2.1**, *asymptotic normality* and the *law of the iterated logarithm* in **7.2.2**, and *Berry–Esséen rates* in **7.2.3**.

7.2.1 Consistency

As in **7.1.1**, we consider a function $\psi(x, t)$ and put $\lambda_F(t) = \int \psi(x, t)dF(x)$. Given that the "parametric" equation $\lambda_F(t) = 0$ has a root t_0 and the "empirical" equation $\lambda_{F_n}(t) = 0$ has a root T_n, under what conditions do we have $T_n \xrightarrow{wp1} t_0$? (Here, as usual, we consider a sample $X_1, \ldots, X_n$ from F, with sample distribution function F_n.) As may be seen from the examples of **7.1.2**, many ψ functions of special interest are of the form $\psi(x, t) = \tilde{\psi}(x - t)$, where either $\tilde{\psi}$ is monotone or $\tilde{\psi}$ is continuous and bounded. These cases, among others, are covered by the following two lemmas, based on Huber (1964) and Boos (1977), respectively.

Lemma A. *Let* t_0 *be an isolated root of* $\lambda_F(t) = 0$. *Let* $\psi(x, t)$ *be monotone in* t. *Then* t_0 *is unique and any solution sequence* $\{T_n\}$ *of the empirical equation* $\lambda_{F_n}(t) = 0$ *converges to* t_0 wp1. *If, further,* $\psi(x, t)$ *is continuous in* t *in a neighborhood of* t_0, *then there exists such a solution sequence.*

PROOF. Assume that $\psi(x, t)$ is nonincreasing in t. Then $\lambda_F(t)$ and $\lambda_{F_n}(t)$, each n, are nonincreasing functions of t. Since $\lambda_F(t)$ is monotone, $\lambda_F(t_0) = 0$, and t_0 is an isolated root, t_0 is the unique root. Let $\varepsilon > 0$ be given. Then $\lambda_F(t_0 + \varepsilon) < 0 < \lambda_F(t_0 - \varepsilon)$. Now, by the SLLN, $\lambda_{F_n}(t) \xrightarrow{wp1} \lambda_F(t)$, each t. Therefore,

$$\lim_{n\to\infty} P(\lambda_{F_m}(t_0 + \varepsilon) < 0 < \lambda_{F_m}(t_0 - \varepsilon), \text{ all } m \geq n) = 1.$$

Complete the argument as an exercise. ■

Remark A. Note that t_0 need not actually be a root of $\lambda_F(t) = 0$. It suffices that $\lambda_F(t)$ change sign uniquely in a neighborhood of t_0. Then we still have, for any $\varepsilon > 0$, $\lambda_F(t_0 + \varepsilon) < 0 < \lambda_F(t_0 - \varepsilon)$, and the assertions on $\{T_n\}$ follow as above. ■

For example, by the above lemma the sample mean, the sample median, and the Hubers (Examples 7.1.2A, B, E) are, under suitable restrictions, consistent estimators of the corresponding location parameters. However, for the Hampels (Example 7.1.2F), we need a result such as the following.

Lemma B. *Let* t_0 *be an isolated root of* $\lambda_F(t) = 0$. *Let* $\psi(x, t)$ *be continuous in* t *and bounded. Then the empirical equation* $\lambda_{F_n}(t) = 0$ *has a solution sequence* $\{T_n\}$ *which converges to* t_0 wp1.

PROOF. Justify that $\lambda_F(t)$ and $\lambda_{F_n}(t)$, each n, are continuous functions of t. Then complete the proof as with Lemma A. ■

Corollary. *For an M-functional* T *based on* ψ, *let* F *be such that* $T(F)$ *is an isolated root of* $\lambda_F(t) = 0$. *Suppose that* $\psi(x, t)$ *is continuous in* t *and either* *monotone in* t *or bounded. Then the empirical equation* $\lambda_{F_n}(t) = 0$ *admits a strongly consistent estimation sequence* T_n *for* T(F).

Remark B. In application of Lemma B in cases when the empirical equation $\lambda_{F_n}(t) = 0$ may have multiple solutions, there is the difficulty of *identifying* a consistent solution sequence $\{T_n\}$. Thus in practice one needs to go further than Lemma B and establish consistency for a particular solution sequence obtained by a specified algorithm.

For example, Collins (1976) considers a robust model for location in which the underlying F is governed by the standard normal density on an interval $t_0 \pm d$ and may be arbitrary elsewhere. He requires that ψ be continuous with continuous derivative, be skew-symmetric, and vanish outside an interval $[-c, c], c < d$. He establishes consistency for T_n the Newton method solution of $\lambda_{F_n}(t) = 0$ starting with the sample median.

Portnoy (1977) assumes that F has a symmetric density f satisfying certain regularity properties, and requires ψ to be bounded and have a bounded and a.s. (Lebesgue) uniformly continuous derivative. He establishes consistency for T_n the solution of $\lambda_{F_n}(t) = 0$ nearest to any given consistent estimator $\tilde{T}_n$. ∎

7.2.2 Asymptotic Normality and the LIL

Let $\psi(x, t)$ be given, put $\lambda_F(t) = \int \psi(x, t)dF(x)$, and let $t_0 = T(F)$ be a solution of $\lambda_F(t) = 0$. Based on $\{X_i\}$ I.I.D. from F, let $T_n = T(F_n)$ be a consistent (for t_0) solution sequence of $\lambda_{F_n}(t) = 0$. Conditions for consistency were given in **7.2.1**. Here we investigate the nature of further conditions under which

$$\text{(AN)} \qquad n^{1/2}(T_n - t_0) \xrightarrow{d} N(0, \sigma^2(T, F)),$$

with $\sigma^2(T, F)$ given by either $\int \psi^2(x, t_0)dF(x)/[\lambda_F'(t_0)]^2$ (as in **7.1.1**) or $\int \psi^2(x, t_0)dF(x)/[\int (\partial\psi(x, t)/\partial t)|_{t=t_0})dF(x)]^2$, depending upon the assumptions on $\psi(x, t)$.

In some cases we are able also to conclude

$$\text{(LIL)} \qquad \overline{\lim_{n\to\infty}} \frac{n^{1/2}(T_n - t_0)}{\sigma(T, F)(2 \log \log n)^{1/2}} = 1 \ wp1.$$

Three theorems establishing (AN) will be given. Theorem A, parallel to Lemma 7.2.1A, is based on Huber (1964) and deals with $\psi(x, t)$ monotone in t. In the absence of this monotonicity, we can obtain (AN) under differentiability restrictions on $\psi(x, \cdot)$, by an extension of the classical treatment of maximum likelihood estimation (recall **4.2.2**). For example, conditions such as

$$\left|\frac{\partial^2\psi(x, \theta)}{\partial\theta^2}\right| < M(x), \qquad \text{with } \sup_{\theta\in\Theta} E_\theta M(X) < \infty,$$

play a role. A development of this type is indicated by Rao (1973), p. 378. As a variant of this approach, based on Huber (1964), Theorem B requires a condition somewhat weaker than the above. Finally, Theorem C, based on Boos (1977), obtains (AN) by a rather different approach employing methods of Chapter 6 in conjunction with stochastic properties of $\|F_n - F\|_\infty$. Instead of differentiability restrictions on $\psi(x, \cdot)$, a condition is imposed on the variation of the function $\psi(\cdot, t) - \psi(\cdot. t_0)$, as $t \to t_0$. The approaches of Theorems B and C also lead to (LIL) in straightforward fashion.

We now give Theorem A. Note that its assumptions include those of Lemma 7.1.2A.

Theorem A. *Let* t_0 *be an isolated root of* $\lambda_F(t) = 0$. *Let* $\psi(x, t)$ *be monotone in* t. *Suppose that* $\lambda_F(t)$ *is differentiable at* $t = t_0$, *with* $\lambda'_F(t_0) \neq 0$. *Suppose that* $\int \psi^2(x, t)dF(x)$ *is finite for* t *in a neighborhood of* t_0 *and is continuous at* $t = t_0$. *Then any solution sequence* T_n *of the empirical equation* $\lambda_{F_n}(t) = 0$ *satisfies (AN), with* $\sigma^2(T, F)$ *given by* $\int \psi^2(x, t_0)dF(x)/[\lambda'_F(t_0)]^2$. *(That* $T_n \xrightarrow{wp1} t_0$ *is guaranteed by Lemma* 7.2.1A.)

PROOF. Assume that $\psi(x, t)$ is nonincreasing in t, so that $\lambda_{F_n}(t)$ is nonincreasing. Thus (justify)

$$P(\lambda_{F_n}(t) < 0) \leq P(T_n \leq t) \leq P(\lambda_{F_n}(t) \leq 0).$$

Therefore, to obtain (AN), it suffices (check) to show that

$$\lim_{n\to\infty} P(\lambda_{F_n}(t_{z,n}) < 0) = \lim_{n\to\infty} P(\lambda_{F_n}(t_{z,n}) \leq 0) = \Phi(z), \qquad \text{each } z,$$

where $t_{z,n} = t_0 + z\sigma n^{-1/2}$, with $\sigma = \sigma(T, F)$. Equivalently (check), we wish to show that

$$\lim_{n\to\infty} P\left(n^{-1/2}\sum_{i=1}^{n} Y_{ni} \leq \frac{-n^{1/2}\lambda_F(t_{z,n})}{s_{z,n}}\right) = \Phi(z), \qquad \text{each } z,$$

where $s_{z,n}^2 = \mathrm{Var}_F\{\psi(X_1, t_{z,n})\}$ and

$$Y_{ni} = \frac{[\psi(X_i, t_{z,n}) - \lambda_F(t_{z\,n})]}{s_{z,n}}, \qquad 1 \leq i \leq n.$$

Justify, using the assumptions of the theorem, that $n^{1/2}\lambda_F(t_{z,n}) \to \lambda'_F(t_0)z\sigma$ and that $s_{z,n} \to -\lambda'_F(t_0)\sigma$, as $n \to \infty$. Thus $-n^{1/2}\lambda_F(t_{z,n})/s_{z,n} \to z$, $n \to \infty$, and it thus suffices (why?) to show that

$$\lim_{n\to\infty} P\left(n^{-1/2}\sum_{i=1}^{n} Y_{ni} \leq z\right) = \Phi(z), \qquad \text{each } z.$$

Since Y_{ni}, $1 \le i \le n$, are I.I.D. with mean 0 and variance 1, each n, we may apply the double array CLT (Theorem 1.9.3). The "uniform asymptotic negligibility" condition is immediate in the present case, so it remains to verify the Lindeberg condition

$$\lim_{n\to\infty} \int_{|y|>n^{1/2}\varepsilon} y^2\, dF_{Y_{n1}}(y) = 0, \qquad \text{every} \quad \varepsilon > 0,$$

or equivalently (check)

$$(1) \qquad \lim_{n\to\infty} \int_{|\psi(x, t_{z,n})|>n^{1/2}\varepsilon} \psi^2(x, t_{z,n}) dF(x) = 0, \qquad \text{every} \quad \varepsilon > 0.$$

For any $\eta > 0$, we have for n sufficiently large that

$$\psi(x, t_0 + \eta) \le \psi(x, t_{z,n}) \le \psi(x, t_0 - \eta), \qquad \text{all } x,$$

and thus, putting $u(x) = \max\{|\psi(x, t_0 - \eta)|, |\psi(x, t_0 + \eta)|\}$, that

$$\int_{|\psi(x, t_{z,n})|>n^{1/2}\varepsilon} \psi^2(x, t_{z,n}) dF(x) \le \int_{u(x)>n^{1/2}\varepsilon} u^2(x) dF(x).$$

Hence (1) follows (why?). ∎

Example A *The Sample* pth *Quantile.* Let $0 < p < 1$. Suppose that F is differentiable at ξ_p and $F'(\xi_p) > 0$. Take $\psi(x, t) = \psi(x - t)$, where

$$\psi(x) = \begin{cases} -1, & x < 0, \\ 0, & x = 0, \\ p/(1-p), & x > 0. \end{cases}$$

Check that for t in a neighborhood of ξ_p, we have

$$\lambda_F(t) = \frac{[p - F(t)]}{(1-p)}$$

and thus

$$\lambda_F(\xi_p) = 0$$

and

$$\lambda'_F(\xi_p) = \frac{-F'(\xi_p)}{(1-p)} < 0.$$

Check that the remaining conditions of the theorem hold and that $\sigma^2(T, F) = p(1-p)/[F'(\xi_p)]^2$. Thus (AN) holds for any solution sequence T_n of $\lambda_{F_n}(t) = 0$. In particular, for $T_n = \hat{\xi}_{pn}$ as considered in Section 2.3, we again obtain Corollary 2.3.3A. ∎

Example B *The Hubers* (continuation of Example 7.1.2E). Take $\psi(x, t) = \psi(x - t)$, where

$$\psi(x) = \begin{cases} -k, & x < -k, \\ x, & |x| \le k, \\ k, & x > k. \end{cases}$$

Verify, using Theorem A, that any solution sequence T_n of $\lambda_{F_n}(t) = 0$ satisfies (AN) with

$$\sigma^2(T, F) = \frac{[\int_{t_0-k}^{t_0+k} x^2\, dF(x) + k^2 \int_{-\infty}^{t_0-k} dF(x) + k^2 \int_{t_0+k}^{\infty} dF(x)]}{(\int_{t_0-k}^{t_0+k} dF(x))^2}. \quad \blacksquare$$

The next theorem trades monotonicity of $\psi(x, \cdot)$ for smoothness restrictions, and also assumes (implicitly) conditions on $\psi(x, t)$ sufficient for existence of a consistent estimator T_n of t_0. Note that the variance parameter $\sigma^2(T, F)$ is given by a different formula than in Theorem A. The proof of Theorem B will use the following easily proved (Problem 7.P.9) lemma giving simple extensions of the classical WLLN and SLLN.

Lemma A. *Let* $g(x, t)$ *be continuous at* t_0 *uniformly in* x. *Let* F *be a distribution function for which* $\int |g(x, t_0)| dF(x) < \infty$. *Let* $\{X_i\}$ *be I.I.D.* F *and suppose that*

$$T_n \xrightarrow{p} t_0. \tag{1}$$

Then

$$\frac{1}{n}\sum_{i=1}^{n} g(X_i, T_n) \xrightarrow{p} E_F g(X, t_0). \tag{2}$$

Further, if the convergence in (1) *is* wp1, *then so is that in* (2).

Theorem B. *Let* t_0 *be an isolated root of* $\lambda_F(t) = 0$. *Let* $\partial\psi(x, t)/\partial t$ *be continuous at* $t = t_0$ *uniformly in* x. *Suppose that* $\int (\partial\psi(x, t)/\partial t)|_{t_0} dF(x)$ *is finite and nonzero, and that* $\int \psi^2(x, t_0) dF(x) < \infty$. *Let* T_n *be a solution sequence of* $\lambda_{F_n}(t) = 0$ *satisfying* $T_n \to t_0$. *Then* T_n *satisfies* (AN) *with* $\sigma^2(T, F) = \int \psi^2(x, t_0) dF(x)/[\int \partial\psi(x, t)/\partial t)|_{t_0} dF(x)]^2$.

PROOF. Since $\psi(x, t)$ is differentiable in t, so is the function $\sum_1^n \psi(X_i, t)$, and we have

$$\sum_{i=1}^{n} \psi(X_i, T_n) - \sum_{i=1}^{n} \psi(X_i, t_0) = (T_n - t_0) \sum_{i=1}^{n} \left.\frac{\partial\psi(X_i, t)}{\partial t}\right|_{t=\tilde{T}_n},$$

where $|\tilde{T}_n - t_0| \le |T_n - t_0|$. Since $\lambda_{F_n}(T_n) = 0$, we thus have

$$n^{1/2}(T_n - t_0) = \frac{-A_n}{B_n},$$

where

$$A_n = n^{-1/2} \sum_{i=1}^{n} \psi(X_i, t_0)$$

and

$$B_n = n^{-1} \sum_{i=1}^{n} \left. \frac{\partial \psi(X_i, t)}{\partial t} \right|_{t=\tilde{T}_n}.$$

Complete the proof using the CLT and Lemma A. ■

Remark A. A variant of Theorem B, due to Boos (1977), relaxes the *uniform* continuity of $g(x, t) = 2\psi(x, t)/\partial t$ at $t = t_0$ to just continuity, but imposes the additional conditions that the function $g(\cdot, t) - g(\cdot, t_0)$ have variation $O(1)$ as $t \to t_0$ and that the function $\int g(x, t)dF(x)$ be continuous at $t = t_0$. This follows by virtue of a corresponding variant of Lemma A (see Problem 7.P.14). ■

Example C *The maximum likelihood estimate of a location parameter* (continuation of Example 7.1.2C). Here $\psi(x) = -f'_0(x)/f_0(x)$ is not necessarily monotone. However, under further regularity conditions on f_0, Theorem A is applicable and yields (check) asymptotic normality with $\sigma^2(T, F) = 1/I(F_0)$, where $I(F_0) = \int (f'_0/f_0)^2 \, dF_0$. ■

The next theorem bypasses differentiability restrictions on $\psi(x, t)$, except what is implied by differentiability of $\lambda_F(t)$. The following lemma will be used. Denote by $\|\cdot\|_V$ the *variation* norm,

$$\|h\|_V = \lim_{\substack{a \to -\infty \\ b \to \infty}} V_{a,b}(h),$$

where

$$V_{a,b}(h) = \sup \sum_{i=1}^{k} |h(x_i) - h(x_{i-1})|,$$

the supremum being taken over all partitions $a = x_0 < \cdots < x_k = b$ of the interval $[a, b]$.

Lemma B. *Let the function* H *be continuous with* $\|\mathrm{H}\|_\mathrm{V} < \infty$ *and the function* K *be right-continuous with* $\|\mathrm{K}\|_\infty < \infty$ *and* $\mathrm{K}(\pm\infty) = 0$. *Then*

$$\left| \int \mathrm{H} \, d\mathrm{K} \right| \leq \|\mathrm{H}\|_\mathrm{V} \cdot \|\mathrm{K}\|_\infty.$$

PROOF. Apply integration by parts to write $\int H \, dK = \int K \, dH$, using the fact that $|H(\pm\infty)| < \infty$. Then check that $|\int K \, dH| \leq \|K\|_\infty \cdot \|H\|_V$ (or see Natanson (1961), p. 232). ■

Theorem C. *Let* t_0 *be an isolated root of* $\lambda_F(t) = 0$. *Let* $\psi(x, t)$ *be continuous in* x *and satisfy*

(V) $$\lim_{t \to t_0} \|\psi(\cdot, t) - \psi(\cdot, t_0)\|_V = 0.$$

Suppose that $\lambda_F(t)$ *is differentiable at* $t = t_0$, *with* $\lambda_F'(t_0) \neq 0$. *Suppose that* $\int \psi^2(x, t_0)dF(x) < \infty$. *Let* T_n *be a solution sequence of* $\lambda_{F_n}(t) = 0$ *satisfying* $T_n \xrightarrow{p} t_0$. *Then* T_n *satisfies* (AN) *with* $\sigma^2(T, F) = \int \psi^2(x, t_0)dF(x)/[\lambda_F'(t_0)]^2$.

PROOF. The differential methodology of Chapter 6 will be applied and, in particular, the *quasi-differential* notion of **6.2.2** will be exploited. As noted in **7.1.1**,

$$d_1 T(F; G - F) = \frac{-\lambda_G(t_0)}{\lambda_F'(t_0)}.$$

In order to deal with $T(G) - T(F) - d_1 T(F; G - F)$, it is useful to define the function

$$h(t) = \frac{\lambda_F(t) - \lambda_F(t_0)}{t - t_0}, \qquad t \neq t_0,$$
$$= \lambda_F'(t_0), \qquad t = t_0.$$

Thus

$$T(G) - T(F) - d_1 T(F; G - F) = \frac{\lambda_F(T(G)) - \lambda_F(t_0)}{h(T(G))} + \frac{\lambda_G(t_0)}{\lambda_F'(t_0)}.$$

Unfortunately, this expression is not especially manageable. However, the quasi-differential device is found to be productive using the auxiliary functional $T_F(G) = \lambda_F'(t_0)/h(T(G))$. We thus have

$$T(G) - T(F) - T_F(G)d_1 T(F; G - F) = \frac{\lambda_F(T(G)) - \lambda_F(t_0) + \lambda_G(t_0)}{h(T(G))},$$

with $T_F(G) \to 1$ as $T(G) \to T(F)$. Assuming $\lambda_G(T(G)) = 0$, we may write

$$\lambda_F(T(G)) - \lambda_F(t_0) - \lambda_G(t_0) = -\int [\psi(x, T(G)) - \psi(x, t_0)]d[G(x) - F(x)].$$

Specializing to $G = F_n$ and $T(G) = T_n$, we thus have

(1) $$T_n - t_0 - T_F(F_n)d_1 T(F; F_n - F) = -\frac{\int [\psi(x, T_n) - \psi(x, t_0)]d[F_n(x) - F(x)]}{h(T_n)}.$$

Check that Lemma B is applicable and, with the convergence $T_n \xrightarrow{p} t_0$, yields that the right-hand side of (1) is $o_p(\|F_n - F\|_\infty)$. It follows (why?) that

$$n^{1/2}[T_n - t_0 - T_F(F_n)d_1T(F; F_n - F)] \xrightarrow{p} 0.$$

Finally, check that $n^{1/2}T(F_n)d_1T(F; F_n - F) \xrightarrow{d} N(0, \sigma^2(T, F))$. ■

Examples D. Consider M-estimation of a *location* parameter, in which case $\psi(x, t)$ may be replaced by $\psi(x - t)$. The regularity conditions on ψ imposed by Lemmas 7.2.1A, B (for existence of a *consistent* M-estimation sequence) and by Theorem C above (for *asymptotic normality*) are that ψ be continuous, either bounded or monotone, and satisfy

$$(*) \qquad \lim_{b \to 0} \|\psi(\cdot - b) - \psi(\cdot)\|_V = 0.$$

These requirements are met by typical ψ considered in robust estimation: "least pth power" estimates corresponding to $\psi(x) = |x|^{p-1}\,\mathrm{sgn}(x)$, provided that $1 < p \le 2$; the Hubers (Example 7.1.2E); the Hampels (Example 7.1.2F); the smoothed Hampel (Example 7.1.2G). In checking (*), a helpful relation is $\|H\|_V = \int |H'(x)|dx$, for H an absolutely continuous function. ■

Remark B. LIL for M-Estimates. Under the conditions of either Theorem B or Theorem C, with the convergence of T_n to t_0 strengthened to $wp1$, T_n satisfies (LIL). This is readily seen by minor modification in the proofs of these results (Problem 7.P.16). ■

7.2.3 Berry–Esséen Rates

The approach of **6.4.3** may be applied. For simplicity let us confine attention to the case $\psi(x, t) = \psi(x - t)$. As an exercise (Problem 7.P.17), augment the development in the proof of Theorem 7.2.2C by evaluating $d_2T(F; F_n - F)$ and showing that the remainder $R_{2n} = T_n - t_0 - d_1T(F; F_n - F) - \frac{1}{2}d_2T(F; F_n - F)$ may be expressed in the form $R_{2n} = A_n + B_n + C_n + D_n$, where

$$A_n = -\frac{\int [\psi(x - T_n) - \psi(x - t_0)]d[F_n(x) - F(x)]}{h(T_n)},$$

$$B_n = \left[\frac{1}{\lambda_F'(t_0)} - \frac{1}{h(T_n)}\right] \int \psi(x - t_0)d[F_n(x) - F(x)],$$

$$C_n = \frac{\int \psi(x - t_0)d[F_n(x) - F(x)] \int \psi'(x - t_0)d[F_n(x) - F(x)]}{\lambda_F'(t_0)},$$

and

$$D_n = \frac{\lambda_F''(t_0)\{\int \psi(x - t_0)d[F_n(x) - F(x)]\}^2}{[\lambda_F'(t_0)]^2}.$$

A brute force treatment of these quantities separately leads to

$$|R_{2n}| \le C\|F_n - F\|_\infty^3$$

under moderate restrictions on ψ, ψ', ψ'' and on $\lambda_F'(t), \lambda_F''(t)$ and $\lambda_F'''(t)$ for t in a neighborhood of t_0. By application (check) of the Dvoretzky–Kiefer–Wolfowitz inequality (Theorem 2.1.3A), $P(|R_{2n}| > Cn^{-1}) = O(n^{-1/2})$, so that Theorem 6.4.3 yields the Berry–Esséen rate $O(n^{-1/2})$ for the asymptotic normality of T_n.

For other discussion of the Berry–Esséen rate for M-estimates, see Bickel (1974).

7.3 COMPLEMENTS

7.3.1 Information Inequality; Most Efficient M-Estimation

Assume regularity conditions permitting the following interchange of order of integration and differentiation:

$$(1) \qquad \frac{d}{dt}\int \psi(x, t)dF(x)\bigg|_{t=t_0} = \int \frac{\partial\psi(x, t)}{\partial t}\bigg|_{t=t_0} dF(x).$$

Then the two forms of $\sigma^2(T, F)$ in **7.2.2** agree. Assume also that F has a density f with derivative f'. Further, consider now the case that $\psi(x, t) = \psi(x - t)$. Then, by integration by parts, (1) yields

$$\lambda_F'(t_0) = -\int \psi'(x - t_0)f(x)dx = -\int \psi(x - t_0)f'(x)dx.$$

Hence

$$\sigma^2(T_\psi, F) = \frac{\int \psi^2(x - t_0)f(x)dx}{[\int \psi(x - t_0)f'(x)dx]^2}$$

and thus, by the Schwarz inequality, (check)

$$(*) \qquad \sigma^2(T_\psi, F) \ge \frac{1}{\int (f'/f)^2\, dF} = \frac{1}{I(F)},$$

which is again the "information inequality" discussed in **4.1.3**. This lower bound is achieved if and only if $\psi(x - t_0)$ is of the form $af'(x)/f(x)$ for some constant a. To make this more transparent, suppose that $F(x) = F_0(x - t_0)$, making $t_0 = T(F)$ a location parameter in the location model generated by a distribution F_0 (recall Example 7.1.1). Then equality in (*) is achieved if and only if $\psi = \psi_0$, where

$$\psi_0 = \frac{af'_0}{f_0}$$

for some constant a, that is, if ψ is the maximum likelihood estimator. That is, the *most efficient* M-estimator is the maximum likelihood estimator. Now compare the *most robust* estimator (**7.3.2**).

7.3.2 "Most Robust" *M*-Estimation

Let F_0 in the location model $F(x;\theta) = F_0(x - \theta)$ be unknown but assumed to belong to a class C of distributions (as in Example 7.1.2E). In some cases (see Huber (1964)) there exists a unique "least favorable" $\tilde{F} \in C$, in the sense that

$$\sigma^2(T_{\tilde{\psi}}, \tilde{F}) \geq \sigma^2(T_{\tilde{\psi}}, F), \qquad \text{all } F \in C,$$

where $\tilde{\psi} = -\tilde{f}'/\tilde{f}$ (the $\tilde{\psi}$ yielding efficient M-estimation of θ when $\tilde{F}$ is the underlying distribution F_0). But, by (*),

$$\sigma^2(T_{\tilde{\psi}}, \tilde{F}) \leq \sigma^2(T_{\psi}, \tilde{F}), \qquad \text{all } \psi.$$

Hence

$$\sup_F \sigma^2(T_{\tilde{\psi}}, F) = \inf_{\psi} \sup_F \sigma^2(T_{\psi}, F).$$

Thus the M-estimator corresponding to $\tilde{\psi}$ is *most robust* in the sense of minimaxing the asymptotic variance. We see that the "most robust" M-estimator has both a maximum likelihood and a minimax interpretation. For the contaminated normal class C of Example 7.1.2E, the least favorable $\tilde{F}$ has density $\tilde{f}(x) = (1 - \varepsilon)(2\pi)^{-1/2} \exp \rho(x)$.

7.3.3 The Differential of an *M*-Functional

The proof of Theorem 7.2.2C showed that, under the conditions of the theorem, $T(F;\Delta) = -\int \psi(x, t_0) d\Delta(x)/\lambda'_F(t_0)$ is (recall **6.2.2**) a *quasi-differential* with respect to $\|\cdot\|_\infty$ and $T_F(\cdot)$. If in addition we require that $\|\psi(\cdot, t_0)\|_V < \infty$, then $T(F;\Delta)$ is a *strict* differential w.r.t. $\|\cdot\|_\infty$ (Problem 7.P.19).

7.3.4 One-Step *M*-Estimators

Consider solving the empirical equation $\lambda_{F_n}(t) = 0$ by Newton's method starting with some consistent estimator $\tilde{T}_n$ (for the solution t_0 of $\lambda_F(t) = 0$). The first iteration has the form

$$T_n^{(1)} = \tilde{T}_n - \frac{\lambda_{F_n}(\tilde{T}_n)}{\lambda'_{F_n}(\tilde{T}_n)},$$

with

$$\lambda'_{F_n}(\tilde{T}_n) = n^{-1} \sum_{i=1}^{n} \frac{\partial \psi(X_i, t)}{\partial t}\bigg|_{t=\tilde{T}_n}.$$

Now check that

$$T_n^{(1)} - t_0 = \tilde{T}_n - t_0 - \frac{\lambda_{F_n}(\tilde{T}_n)}{\lambda'_{F_n}(\tilde{T}_n)} = A_n + B_n - C_n,$$

where

$$A_n = \frac{-\lambda_{F_n}(t_0)}{\lambda'_{F_n}(\tilde{T}_n)},$$

$$B_n = \frac{-[\lambda_{F_n}(\tilde{T}_n) - \lambda_{F_n}(t_0) - \lambda'_{F_n}(t_0)(\tilde{T}_n - t_0)]}{\lambda'_{F_n}(\tilde{T}_n)},$$

and

$$C_n = \frac{[\lambda'_{F_n}(\tilde{T}_n) - \lambda'_{F_n}(t_0)](\tilde{T}_n - t_0)}{\lambda'_{F_n}(\tilde{T}_n)}.$$

Assume the conditions of Theorem 7.2.2B and also

(1) $$n^{1/2}(\tilde{T}_n - t_0) = O_p(1), \qquad n \to \infty.$$

Then immediately (justify)

$$n^{1/2} A_n \xrightarrow{d} N(0, \sigma^2(T, F))$$

and

$$n^{1/2} C_n \xrightarrow{p} 0.$$

Find additional conditions on ψ such that

$$n^{1/2} B_n \xrightarrow{p} 0,$$

and thus conclude that $n^{1/2}(T_n^{(1)} - t_0) \xrightarrow{d} N(0, \sigma^2(T, F))$, in which case the performance of the "one-step" is the same as the "full iterate."

7.3.5 Scaling

As discussed in Huber (1977), in order to make a location M-estimate scale-invariant, one must introduce a location-invariant scale estimate s_n, and then take T_n to be the solution of

$$\sum_{i=1}^{n} \psi\left(\frac{X_i - T_n}{s_n}\right) = 0.$$

In this case, if S_n estimates $\sigma(F)$, then T_n estimates $T(F)$ defined as the solution of

$$\int \psi\left(\frac{x - T(F)}{\sigma(F)}\right) dF(x) = 0.$$

A recommended choice of s_n is the mean absolute deviation (MAD),

$$s_n = \text{median of } \{|X_1 - m|, \ldots, |X_n - m|\},$$

where $m =$ median of $\{X_1, \ldots, X_n\}$. Another old favorite is the sample interquartile range (discussed in **2.3.6**). The results of this chapter extend to this formulation of M-estimation.

7.3.6 Bahadur Representation for *M*-Estimates

Let T_n be as defined in **7.3.5**. Under various regularity conditions on ψ and F, Carroll (1978) represents T_n as a linear combination of the scale estimate s_n and the average of n bounded random variables, except for a remainder term $O(n^{-1}(\log \log n))wp1$.

7.3.7 *M*-Estimates for Regression

See Huber (1973).

7.3.8 Multiparameter *M*-Estimates

See Huber (1977).

7.3.9 Connections Between *M*-Estimates and *L*- and *R*-Estimates

See Chapter 9.

7.P PROBLEMS

Section 7.1

1. Derive the IC for an M-estimate, as given in **7.1.1**. (Hint: use the method of Example 6.5B.)

2. Verify the characterization of the M-estimator of Example 7.1.2D as a form of trimmed mean. Exemplify.

3. Verify the characterization of the M-estimator of Example 7.1.2E as a form of Winsorized mean. Exemplify.

Section 7.2

4. Complete the proofs of Lemmas 7.2.1A, B.
5. Does Remark 7.2.1A apply to Lemma B also?
6. Complete the details of proof of Theorem 7.2.2A.

7. Supply details for Example 7.2.2A (the sample pth quantile).
8. Supply details for Example 7.2.2B (the Hubers).
9. Prove Lemma 7.2.2A. Hint: write

$$\left|\frac{1}{n}\sum_{i=1}^{n} g(X_i, T_n) - E_F g(X, t_0)\right| \leq \left|\int [g(x, T_n) - g(x, t_0)]dF_n(x)\right| + \left|\int g(x, t_0)d[F_n(x) - F(x)]\right|.$$

10. Complete the proof of Theorem 7.2.2B.
11. Check details of Example 7.2.2C (m.l.e. of location parameter).
12. Check details of proof of Lemma 7.2.2B.
13. Supply details for the proof of Theorem 7.2.2C.
14. Prove the variant of Lemma 7.2.2A noted in Remark 7.2.2A. (Hint: apply Lemma 7.2.2B.)
15. Check the claims of Examples 7.2.2D.
16. Verify Remark 7.2.2B (LIL for M-estimates).
17. Provide details in **7.2.3** (Berry–Esséen rates for M-estimates).

Section 7.3

18. Details for **7.3.1–2**.
19. Details for **7.3.3**.
20. Details for **7.3.4**.

CHAPTER 8

L-Estimates

This chapter deals briefly with the asymptotic properties of statistics which may be represented as *linear combinations of order statistics*, termed "*L-estimates*" here. This class of statistics is computationally more appealing than the M-estimates, yet competes well from the standpoints of robustness and efficiency. It also competes well against R-estimates (Chapter 9).

Section 8.1 provides the basic formulation and a variety of examples illustrating the scope of the class. Asymptotic properties, focusing on the case of asymptotically *normal* L-estimates, are treated in Section 8.2. Four different methodological approaches are examined.

8.1 BASIC FORMULATION AND EXAMPLES

A general formulation of L-estimation is presented in **8.1.1**. The special case of *efficient parametric* L-estimation of location and scale parameters is treated in **8.1.2**. *Robust* L-estimation is discussed in **8.1.3**. From these considerations it will be seen that the theoretical treatment of L-estimates must serve a very wide scope of practical possibilities.

8.1.1 General Formulation and First Examples

Consider independent observations $X_1, \ldots, X_n$ on a distribution function F and, as usual, denote the ordered values by $X_{n1} \leq \cdots \leq X_{nn}$. As discussed in **2.4.2**, many important statistics may be expressed as *linear* functions of the ordered values, that is, in the form

(1) $$T_n = \sum_{i=1}^{n} c_{ni} X_{ni}$$

for some choice of constants $c_{n1}, \ldots, c_{nn}$. We term such statistics "*L-estimates*." Simple examples are the sample mean $\bar{X}$, the extremes X_{n1} and X_{nn}, and the sample range $X_{nn} - X_{n1}$. From the discussion of **2.4.3** and **2.4.4**, it is clear that the asymptotic distribution theory of L-statistics takes quite different forms, depending on the character of the coefficients $\{c_{ni}\}$. The present development will attend only to cases in which T_n is asymptotically *normal.*

Examples A. (i) The *sample* pth *quantile*, $\hat{\xi}_{pn}$, may be expressed in the form (1) with $c_{ni} = 1$ if $i = np$ or if $np \neq [np]$ and $i = [np] + 1$, and $c_{ni} = 0$ otherwise.

(ii) *Gini's mean difference*,

$$\frac{2}{n(n-1)} \sum_{1 \le i < j \le n} |X_i - X_j|,$$

considered previously in **5.1.1** as a U-statistic for unbiased estimation of the dispersion parameter $\theta = E_F|X_1 - X_2|$, may be represented as an L-estimate as follows (supply missing steps):

$$\begin{aligned}\frac{2}{n(n-1)} \sum_{1 \le i < j \le n} |X_i - X_j| &= \frac{2}{n(n-1)} \sum_{1 \le i < j \le n} |X_{ni} - X_{nj}| \\ &= \frac{2}{n(n-1)} \sum_{i=1}^{n-1} \sum_{j=i+1}^{n} (X_{nj} - X_{ni}) \\ &= \frac{2}{n(n-1)} \sum_{i=1}^{n} (2i - n - 1) X_{ni},\end{aligned}$$

which is of form (1) with $c_{ni} = 2(2i - n - 1)/n(n-1)$. ■

A convenient subclass of (1) broad enough for all typical applications is given by

$$(1') \qquad T'_n = \frac{1}{n} \sum_{i=1}^{n} J\left(\frac{i}{n+1}\right) X_{ni} + \sum_{j=1}^{m} a_j X_{n,[np_j]}.$$

Here $J(u)$, $0 \le u \le 1$, represents a *weights-generating* function. It is assumed that $0 < p_1 < \cdots < p_m < 1$ and that $a_1, \ldots, a_m$ are nonzero constants. Thus T'_n is of form (1) with c_{ni} given by $n^{-1}J(i/(n+1))$ plus an additional contribution a_j if $i = [np_j]$ for some $j \in \{1, \ldots, m\}$. Typically, J is a fairly smooth function. Thus L-estimates of form (1') are sums of two special types of L-estimate, one type weighting all the observations according to a reasonably smooth weight function, the other type consisting of a weighted sum of a fixed number of quantiles. In many cases, of course, the statistic of interest is just a single one of these types. Also, in many cases, the initial statistic of interest is modified slightly to bring it into the convenient form (1'). For example, the sample pth quantile $T_n = \hat{\xi}_{pn}$ is replaced by $T'_n = X_{n,[np]}$, given by (1') with the first term absent and the second term corresponding to $m = 1$, $p_1 = p$,

$a_1 = 1$. Similarly, Gini's mean difference T_n may be replaced by $T'_n = [(n+1)/(n-1)]T_n$, which is of form (1′) with $J(u) = 4u - 2, 0 \le u \le 1$, and with the second term absent. In such cases, in order that conclusions obtained for T'_n may be applied to T_n, a separate analysis showing that $T_n - T'_n$ is negligible in an appropriate sense must be carried out.

Examples B. (i) *The α-trimmed mean* (previously considered in Example 6.5E). Let $0 < \alpha < 1$. Then

$$T_n = \frac{1}{n - 2[n\alpha]} \sum_{i=[n\alpha]+1}^{n-[n\alpha]} X_{ni}$$

is of form (1). Asymptotically equivalent to T_n is T'_n of form (1′) with $J(u) = 1/(1-2\alpha)$ for $\alpha < u < 1 - \alpha$ and $= 0$ elsewhere, and with $m = 0$.

(ii) *The α-Winsorized mean.* Let $0 < \alpha < \frac{1}{2}$. Then

$$T_n = \frac{1}{n}\left([n\alpha]X_{n,[n\alpha]+1} + \sum_{i=[n\alpha]+1}^{n-[n\alpha]} X_{ni} + [n\alpha]X_{n,n-[n\alpha]}\right)$$

is asymptotically equivalent to T'_n of form (1′) with $J(u) = 1$ for $\alpha < u < 1 - \alpha$ and $=0$ elsewhere, and with $m = 2$, $p_1 = \alpha$, $p_2 = 1 - \alpha$, $a_1 = a_2 = \alpha$.

(iii) *The interquartile range* (recall **2.3.6**) is essentially of form (1′) with $J(u) \equiv 0$ and $m = 2$, $p_1 = \frac{1}{4}$, $p_2 = \frac{3}{4}$, $a_1 = \frac{1}{2}$, $a_2 = \frac{1}{2}$. ■

As these examples illustrate, a given statistic such as the interquartile range may have two asymptotically equivalent formulations as an L-estimate. Further, even the form (1′) has its variations. In place of $J(i/(n+1))$, some authors use $J(i/n)$, which is a little neater but makes the definition of $J(u)$ at $u = 0$ a more troublesome issue. Some authors use

$$n \int_{(i-1)/n}^{i/n} J(u)du$$

in place of $J(i/(n+1))$. In this case, we may express the first term of (1′) in the form

$$\sum_{i=1}^{n} \left[\int_{(i-1)/n}^{i/n} J(u)du\right] F_n^{-1}\left(\frac{i}{n}\right) = \int_0^1 F_n^{-1}(t)J(t)dt.$$

This requires that J be integrable, but lends itself to formulation of L-estimates as *statistical functions.* Thus, using this version of weights in the first term in (1′) and modifying the second term by putting $F_n^{-1}(p_j)$ in place of $X_{n,[np_j]}$, $1 \le j \le m$, we obtain the closely associated class of L-estimates

$$T''_n = T(F_n), \tag{1″}$$

where $T(\cdot)$ denotes the functional

$$(*) \qquad T(F) = \int_0^1 F^{-1}(t)J(t)dt + \sum_{j=1}^{m} a_j F^{-1}(p_j).$$

More generally, a wide class of L-estimates may be represented as statistical functions $T(F_n)$, in terms of functionals of the form

$$T(F) = \int_0^1 F^{-1}(t)dK(t),$$

where $K(\cdot)$ denotes a linear combination of distribution functions on the interval [0, 1].

Not only does the functional representation help us see what an L-estimate is actually estimating, but also it brings into action the useful heuristic tool of influence curve analysis. From Example 6.5D and **6.6.1**, the influence curve of the tth quantile $F^{-1}(t)$ is

$$IC(x; F^{-1}(t), F) = \frac{t - I(x \le F^{-1}(t))}{f(F^{-1}(t))}, \qquad -\infty < x < \infty.$$

(See also Problem 8.P.2.) Thus the functional $T_2(\cdot)$ given by the *second* term of (*) has influence curve

$$IC(x; T_2, F) = \sum_{j=1}^{m} a_j \frac{p_j - I(x \le F^{-1}(p_j))}{f(F^{-1}(p_j))}.$$

Let us now deal with the functional T_1 given by the *first* term of (*). Putting $K_1(t) = \int_0^t J(u)du$, we have (Problem 8.P.3)

Lemma A. *If* $\int_0^1 \mathrm{F}^{-1}(\mathrm{t})\mathrm{J}(\mathrm{t})\mathrm{dt}$ *is finite, then*

$$\int_0^1 \mathrm{F}^{-1}(\mathrm{t})\mathrm{J}(\mathrm{t})\mathrm{dt} = \int_{-\infty}^{\infty} \mathrm{x}\, \mathrm{dK}_1(\mathrm{F}(\mathrm{x})).$$

We thus obtain (Problem 8.P.4), for $K_1(\cdot)$ a linear combination of distribution functions, and in particular for $K_1(t) = \int_0^t J(u)du$,

Lemma B. $\mathrm{T}_1(\mathrm{G}) - \mathrm{T}_1(\mathrm{F}) = -\int_{-\infty}^{\infty} [\mathrm{K}_1(\mathrm{G}(\mathrm{x})) - \mathrm{K}_1(\mathrm{F}(\mathrm{x}))]\mathrm{dx}.$

Applying Lemma B, we may obtain the Gâteaux differential of T_1 at F (see Problem 8.P.5 for details) and in particular the influence curve

$$IC(x; T_1, F) = -\int_{-\infty}^{\infty} [I(x \le y) - F(y)]J(F(y))dy.$$

The influence curve of the functional T given by (*) is thus

$$IC(x; T, F) = IC(x; T_1, F) + IC(x; T_2, F).$$

Note that the second term, when present, gives the curve jumps of sizes $a_j/f(F^{-1}(p_j))$ at the points $x = F^{-1}(p_j)$, $1 \le j \le m$.

The relevant asymptotic normality assertion for $T(F_n)$ may now be formulated. Following the discussion of Remark 6.5, we note that $E_F\{IC(X; T, F)\} = 0$ and we define $\sigma^2(T, F) = \text{Var}_F\{IC(X; T, F)\}$. We thus anticipate that $T(F_n)$ is $AN(T(F), n^{-1}\sigma^2(T, F))$. The detailed treatment is provided in Section 8.2.

Clearly, the L-estimates tend to be more attractive computationally than the M-estimates. In particular, L-estimation is thus more appealing computationally than maximum likelihood estimation. Does this mean that efficiency must be sacrificed to gain this ease of computation? No, it turns out in classical parametric estimation problems that the constants c_{ni} may be selected so that T_n has the *same* asymptotic variance as the maximum likelihood estimate. In **8.1.2** we consider a number of specific examples of such problems. Furthermore, Bickel and Lehmann (1975) compare M-, L- and R-estimates for location estimation in the case of asymmetric F and conclude that L-estimates offer the best compromise between the competing demands of efficiency at the parametric model and robustness in a nonparametric neighborhood of the parametric model. In particular, the trimmed means are recommended. In **8.1.3** we consider robust L-estimation.

Fixed sample size analysis of L-estimation seems to have begun with Lloyd (1952), who developed estimators which are unbiased and of minimum variance (for each n) in the class of statistics consisting of linear transformations of statistics T_n of form (1). See David (1970), Chapter 6, for details and further references. See Sarhan and Greenberg (1962) for tabulated values.

An asymptotic analysis was developed by Bennett (1952), who derived asymptotically optimal c_{ni}'s (J functions) by an approach not involving considerations of asymptotic normality. Some of his results were obtained independently by Jung (1955).

The asymptotic analysis has become linked with the question of asymptotic normality by several investigators, notable results earliest being given by Chernoff, Gastwirth and Johns (1967). Among other things, they demonstrate that Bennett's estimators are asymptotically efficient. Alternate methods of proving asymptotic normality have been introduced by Stigler (1969), Shorack (1969), and Boos (1979). We discuss these various approaches in **8.2.1**–**8.2.4**, and corresponding strong consistency and LIL results will be noted. The related Berry–Esséen rates will be discussed in **8.2.5**, with special attention to results of Bjerve (1977), Helmers (1977), and Boos and Serfling (1979).

8.1.2 Examples in Parametric Location and Scale Estimation

Let the distribution of $X_1, \ldots, X_n$ be a member $F(x; \theta_1, \theta_2)$ of a specified location and scale parameter family $\mathscr{F} = \{F(x; \theta_1, \theta_2), (\theta_1, \theta_2) \in \Theta\}$, where

$$F(x; \theta_1, \theta_2) = F\left(\frac{x - \theta_1}{\theta_2}\right)$$

with density

$$f(x; \theta_1, \theta_2) = \frac{1}{\theta_2} f\left(\frac{x - \theta_1}{\theta_2}\right),$$

and F is a specified distribution with density f. For example, if $F = \Phi$, then $\mathscr{F}$ is a family of normal distributions. One or both of θ_1 and θ_2 may be unknown. The problem under consideration is that of estimation of each unknown parameter by an L-estimate, that is, by a statistic of the convenient form

$$T_n = \frac{1}{n} \sum_{i=1}^{n} J\left(\frac{i}{n+1}\right) X_{ni},$$

with the J function selected optimally. Furthermore, solutions are desired in both the cases of censored and uncensored data (censored data arises in connection with life-testing experiments, or in connection with outlier-rejection procedures). We will consider several examples from Chernoff, Gastwirth and Johns (1967).

Assume that $\mathscr{F}$ satisfies regularity conditions (recall Section 4.2) sufficient for the asymptotic covariance matrix of the normalized maximum likelihood estimates of θ_1 and θ_2 to coincide with the inverse of the information matrix

$$\mathbf{I}_{\boldsymbol{\theta}} = \left[E\left\{\frac{\partial \log f(X; \theta_1, \theta_2)}{\partial \theta_i} \cdot \frac{\partial \log f(X; \theta_1, \theta_2)}{\partial \theta_j}\right\}\right]_{2 \times 2}.$$

Defining

$$L_1(x) = -\frac{f'(x)}{f(x)}, \qquad L_2(x) = -\left[1 + \frac{xf'(x)}{f(x)}\right],$$

and assuming that f'' exists and $x^2 f'(x) \to 0$ as $|x| \to \infty$, the matrix $\mathbf{I}_{\boldsymbol{\theta}}$ may be written as $\theta_2^{-2}\mathbf{I}_F$, where

$$\mathbf{I}_F = \begin{bmatrix} \int_{-\infty}^{\infty} L_1'(x) f(x) dx & \int_{-\infty}^{\infty} L_2'(x) f(x) dx \\ \int_{-\infty}^{\infty} xL_1'(x) f(x) dx & \int_{-\infty}^{\infty} xL_2'(x) f(x) dx \end{bmatrix} = \begin{bmatrix} I_{11} & I_{12} \\ I_{21} & I_{22} \end{bmatrix}.$$

The problem is to find J functions such that estimates of the form

$$T_n = \frac{1}{n}\sum_{i=1}^{n} J\left(\frac{i}{n+1}\right)X_{ni}\left(+\sum_{j=1}^{m} a_j X_{n,[np_j]} \text{ perhaps}\right)$$

have asymptotic covariance matrix $n^{-1}\theta_2^2 I_F^{-1}$.

Example A *Uncensored case, scale known.* For estimation of the location parameter θ_1 when the scale parameter θ_2 is known, the efficient J function is found to be

$$J(u) = I_{11}^{-1}L_1'(F^{-1}(u)).$$

It is established that the corresponding L-estimate T_n is $AN(\mu_1, n^{-1}\sigma_1^2)$, where $\mu_1 = \theta_1 + I_{11}^{-1}I_{12}\theta_2$ and $\sigma_1^2 = \theta_2^2 I_{11}^{-1}$. It follows that $T_n - I_{11}^{-1}I_{12}\theta_2$ is an asymptotically efficient estimator of the location parameter θ_1, when the scale parameter is known. In particular:

(i) For the *normal* family $\mathscr{F}$ based on $F = \Phi$, the appropriate weight function is, of course, simply $J(u) \equiv 1$;

(ii) For the *logistic* family $\mathscr{F}$ based on

$$F(x) = (1 + e^{-x})^{-1}, \qquad -\infty < x < \infty,$$

the appropriate weight function is

$$J(u) = 6u(1-u), \qquad 0 \le u \le 1;$$

(iii) For the *Cauchy* family $\mathscr{F}$ based on

$$F(x) = \frac{1}{\pi}[\tan^{-1}(x) + \tfrac{1}{2}\pi], \qquad -\infty < x < \infty,$$

the appropriate weight function is

$$J(u) = \frac{\sin 4\pi(u - \frac{1}{2})}{\tan \pi(u - \frac{1}{2})}. \quad \blacksquare$$

Example B *Uncensored case, location known.* For estimation of θ_2 when θ_1 is known, the efficient J function is found to be

$$J(u) = I_{12}^{-1}L_2'(F^{-1}(u)).$$

It is established that the corresponding L-estimate T_n is $AN(\mu_2, n^{-1}\sigma_2^2)$, where $\mu_2 = \theta_2 + I_{22}^{-1}I_{12}\theta_1$ and $\sigma_2^2 = \theta_2^2 I_{22}^{-1}$. It follows that $T_n - I_{22}^{-1}I_{12}\theta_1$ is an asymptotically efficient estimator of θ_2 when θ_1 is known. In particular, for the normal, logistic, and Cauchy families considered in Example A, the corresponding appropriate weight functions are

$$J(u) = \Phi^{-1}(u),$$

$$J(u) = \frac{9}{\pi^2 + 3}\left[2u - 1 + 2u(1 - u)\log\left(\frac{u}{1 - u}\right)\right],$$

and

$$J(u) = 8\,\frac{\tan \pi(u - \frac{1}{2})}{\sec^2 \pi(u - \frac{1}{2})},$$

respectively. ■

Example C *Uncensored case, location and scale both unknown.* In this case the vector $(T_n^{(1)}, T_n^{(2)})$ corresponding to

$$[J_1(u), J_2(u)] = [L_1'(F^{-1}(u)), L_2'(F^{-1}(u))]\mathbf{I}_F^{-1}$$

is $AN\ ((\theta_1, \theta_2), n^{-1}\theta_2^2 \mathbf{I}_F^{-1})$. ■

Example D *Censored case, location and scale both unknown.* In the case of symmetric two-sided censoring of the upper 100p % and lower 100p % observations, it is found that the asymptotically efficient estimate of the *location* parameter is formed by using weights specified by

$$J(u) = I_{11}^{-1} L_1'(F^{-1}(u))$$

for the uncensored observations and additional weight

$$w = I_{11}^{-1}[p^{-1}f^2(F^{-1}(p)) - f'(F^{-1}(p))]$$

for the largest and smallest uncensored observations. For the *normal* family we have, putting $\xi_p = \Phi^{-1}(p)$,

$$I_{11} = \int_{-\xi_p}^{\xi_p} x^2\phi(x)dx + 2p^{-1}\phi^2(\xi_p) = 1 - 2p + 2\xi_p\phi(\xi_p) + 2p^{-1}\phi^2(\xi_p),$$

$$\begin{aligned} J(u) &= I_{11}^{-1}, \qquad p < u < 1 - p, \\ &= 0, \qquad \text{otherwise}, \end{aligned}$$

and

$$w = I_{11}^{-1}[p\phi^2(\xi_p) + \xi_p\phi(\xi_p)].$$

As a numerical example, for $p = 0.05$ we have $I_{11} = 0.986$ and $w = 0.0437$, yielding the efficient L-estimate

$$T_n = \frac{1}{0.986n}\sum_{i=[0.05n]+1}^{n-[0.05n]} X_{ni} + 0.0437(X_{n,[0.05n]+1} + X_{n,n-[0.05n]}).$$

Note the similarity to the p-Winsorized mean (Example 8.1.1B (ii)). ■

8.1.3 Examples in Robust Estimation

For *robust* L-estimation, the influence curve $IC(x; T, F)$ derived in **8.1.1** should be *bounded* and reasonably *smooth*. This curve is the sum of the two curves

$$IC(x; T_1, F) = -\int_{-\infty}^{\infty} [I(y \geq x) - F(y)]J(F(y))dy$$

and

$$IC(x; T_2, F) = -\sum_{j=1}^{m} \frac{a_j[I(F^{-1}(p_j) \geq x) - p_j]}{f(F^{-1}(p_j))}.$$

The first curve is smooth, having derivative $J(F(x))$ (Problem 8.P.6), but can be unbounded. To avoid this, robust L-estimation requires that $J(u)$ vanish outside some interval (a, b), $0 < a < b < 1$. The second curve is bounded, but has m discontinuities.

Example A *The "Gastwirth".* In the Monte Carlo study by Andrews *et al.* (1972), favorable properties were found for the L-estimate

$$0.3F_n^{-1}(\tfrac{1}{3}) + 0.4F_n^{-1}(\tfrac{1}{2}) + 0.3F_n^{-1}(\tfrac{2}{3}),$$

proposed by Gastwirth (1966). ■

Example B *The α-trimmed mean* (see Example 8.1.1B(i)). This is the L-estimate $T(F_n)$, where

$$T(F) = \frac{1}{1 - 2\alpha} \int_{\alpha}^{1-\alpha} F^{-1}(t)dt.$$

For F symmetric about $F^{-1}(\frac{1}{2})$, the influence curve is (recall Example 6.5E)

$$IC(x; T, F) = \begin{cases} \dfrac{1}{1-2\alpha}[F^{-1}(\alpha) - F^{-1}(\tfrac{1}{2})], & x < F^{-1}(\alpha), \\ \dfrac{1}{1-2\alpha}[x - F^{-1}(\tfrac{1}{2})], & F^{-1}(\alpha) \leq x \leq F^{-1}(1-\alpha), \\ \dfrac{1}{1-2\alpha}[F^{-1}(1-\alpha) - F^{-1}(\tfrac{1}{2})], & x > F^{-1}(1-\alpha). \end{cases}$$

Thus this L-estimate behaves the same as a certain M-estimate, the "Huber" (Example 7.2.2B) with $k = F^{-1}(1 - \alpha)$. ■

Example C *The α-Winsorized mean* (see Example 8.1.1B(ii)). This is $T(F_n)$ based on

$$T(F) = \int_{\alpha}^{1-\alpha} F^{-1}(t)dt + \alpha F^{-1}(\alpha) + \alpha F^{-1}(1-\alpha).$$

Find its influence curve (Problem 8.P.8). ■

Example D *The smoothly trimmed mean.* Stigler (1973) provides an example showing that the trimmed mean has *nonnormal* asymptotic distribution if the trimming is at non-unique quantiles of F. As one remedy, he introduces the "smoothly trimmed mean," corresponding to a J function of the form

$$\begin{aligned} J(u) &= 0, \qquad u < \tfrac{1}{2}\alpha, \\ &= (u - \tfrac{1}{2}\alpha)\left(\frac{2c}{\alpha}\right), \qquad \tfrac{1}{2}\alpha \le u \le \alpha, \\ &= c, \qquad \alpha < u < 1 - \alpha, \\ &= (1 - \tfrac{1}{2}\alpha - u)\left(\frac{2c}{\alpha}\right), \qquad 1 - \alpha \le u \le 1 - \tfrac{1}{2}\alpha, \\ &= 0, \qquad u > \tfrac{1}{2}\alpha. \end{aligned}$$ ■

8.2 ASYMPTOTIC PROPERTIES OF L-ESTIMATES

In this section we exhibit *asymptotic normality* of L-estimates under various restrictions on J and F. Four different methodological approaches will be considered, in **8.2.1–8.2.4**, respectively. *Consistency* and *LIL* results will also be noted, along the way. In **8.2.5** we consider *Berry–Esséen* rates.

8.2.1 The Approach of Chernoff, Gastwirth and Johns (1967)

Chernoff, Gastwirth and Johns (1967) deal with L-estimates in the general form

$$(1) \qquad T_n = n^{-1} \sum_{i=1}^{n} c_{ni} h(X_{ni})$$

where h is some measurable function. (This includes as a special case the formulation of Section 8.1, given by $h(x) = x$ and replacing c_{ni} in (1) by nc_{ni}.) For the purpose of deriving *distribution theory* for T_n, we may assume that $X_1, X_2, \ldots$ are given by $F^{-1}(U_1), F^{-1}(U_2), \ldots,$ where $U_1, U_2, \ldots$ are independent *uniform* (0, 1) variates. Thus $X_{ni} = F^{-1}(U_{ni}), 1 \le i \le n$. Also, put

$$V_{ni} = -\log(1 - U_{ni}), \qquad 1 \le i \le n.$$

It is readily seen that the V_{ni} are the order statistics of a sample from the *negative exponential* distribution, $G(x) = 1 - \exp(-x)$, $x > 0$. Thus, putting $\tilde{H} = h \circ F^{-1} \circ G$, the composition of h, F^{-1}, and G, we have the representation

$$T_n = n^{-1} \sum_{i=1}^{n} c_{ni} \tilde{H}(V_{ni}).$$

We now apply differentiability of $\tilde{H}$ in conjunction with the following special representation of the V_{ni} (see, e.g., David (1970)).

Lemma A. *The* V_{ni} *may be represented in distribution as*

$$V_{ni} = \frac{Z_1}{n} + \cdots + \frac{Z_i}{n - i + 1}, \qquad 1 \le i \le n,$$

where $Z_1, \ldots, Z_n$ *are independent random variables with distribution* G.

Assumption A. $\tilde{H}(v)$ *is continuously differentiable for* $0 < v < \infty$. *Define* $\tilde{v}_{ni} = E\{V_{ni}\}$ *and note that*

$$\tilde{v}_{ni} = \frac{1}{n} + \cdots + \frac{1}{n - i + 1}, \qquad 1 \le i \le n.$$

Now apply Lemma A and Assumption A to write

$$\tilde{H}(V_{ni}) - \tilde{H}(\tilde{v}_{ni}) = \tilde{H}'(\tilde{v}_{ni}) \sum_{j=1}^{i} \frac{Z_j}{n - j + 1} + G_{ni}(V_{ni}),$$

where

$$G_{ni}(v) = \frac{\tilde{H}(v) - \tilde{H}(\tilde{v}_{ni})}{v - \tilde{v}_{ni}} - \tilde{H}'(\tilde{v}_{ni}), \qquad v \ne \tilde{v}_{ni},$$

$$= 0, \; v = \tilde{v}_{ni}.$$

We thus have the representation (check)

$$T_n = \mu_n + Q_n + R_n, \tag{2}$$

where

$$\mu_n = n^{-1} \sum_{i=1}^{n} c_{ni} \tilde{H}(\tilde{v}_{ni}),$$

$$Q_n = n^{-1} \sum_{i=1}^{n} \alpha_{ni}(Z_i - 1),$$

with

$$\alpha_{ni} = \frac{1}{n - i + 1} \sum_{j=1}^{n} c_{nj} \tilde{H}'(\tilde{v}_{nj}),$$

and

$$R_n = n^{-1} \sum_{i=1}^{n} c_{ni}(V_{ni} - \tilde{v}_{ni}) G_{ni}(V_{ni}).$$

Here μ_n *is nonrandom,* Q_n *can be shown asymptotically normal by standard central limit theory, and* R_n *is a remainder which is found to be asymptotically negligible. Note that* Q_n *has variance* σ_n^2/n*, where*

$$\sigma_n^2 = n^{-1} \sum_{i=1}^{n} \alpha_{ni}^2.$$

The following further assumptions are needed. First we state an easily proved (Problem 8.P.9) preliminary.

Lemma B. *The random variables* U_{ni} *and* V_{ni} *can be simultaneously bounded in probability: given* $\varepsilon > 0$*, there exists*

$$u_{ni}(\varepsilon), \qquad u^{ni}(\varepsilon), \qquad v_{ni}(\varepsilon), \qquad v^{ni}(\varepsilon)$$

such that

$$P(u_{ni}(\varepsilon) < U_{ni} < u^{ni}(\varepsilon), \qquad 1 \le i \le n) \ge 1 - \varepsilon$$

and

$$P(v_{ni}(\varepsilon) < V_{ni} < v^{ni}(\varepsilon), \qquad 1 \le i \le n) \ge 1 - \varepsilon,$$

with

$$v_{ni}(\varepsilon) = -\log[1 - u_{ni}(\varepsilon)], \qquad v^{ni}(\varepsilon) = -\log[1 - u^{ni}(\varepsilon)]$$

and

$$u_{ni}(\varepsilon) < \frac{i}{n+1} < u^{ni}(\varepsilon).$$

Assumption B. *For each* $\varepsilon > 0$,

$$\sum_{i=1}^{n} |c_{ni}| g_{ni}(\varepsilon) \left[\frac{i}{n - i + 1} \right]^{1/2} = o(n\sigma_n),$$

where

$$g_{ni}(\varepsilon) = \sup_{v_{ni}(\varepsilon) < v < v^{ni}(\varepsilon)} |G_{ni}(v)|.$$

Assumption C. $\max_{1 \le i \le n} |\alpha_{ni}| = o(n^{1/2}\sigma_n)$.

Theorem (Chernoff, Gastwirth and Johns). *Under Assumptions A, B, and C,*

$$T_n \textit{ is } AN(\mu_n, n^{-1}\sigma_n^2).$$

PROOF (Sketch). It can be shown by a characteristic function argument (Problem 8.P.10(a)) that $\mathscr{L}(n^{1/2}Q_n/\sigma_n) \to N(0, 1)$ if and only if Assumption C holds, no matter what the values of the constants $\{\alpha_{ni}\}$. Further, it can be shown under Assumption B that $R_n = o_p(n^{-1/2}\sigma_n)$. (See CGJ (1967) for details). ■

For the special case

$$T_n = n^{-1}\sum_{i=1}^{n} J\left(\frac{i}{n+1}\right)h(X_{ni}) + \sum_{j=1}^{m} a_j h(X_{n,[np_j]}),$$

CGJ (1967) give special conditions under which T_n is $AN(\mu, n^{-1}\sigma^2)$, where

$$\mu = \int_0^1 J(u)H(u)du + \sum_{j=1}^{m} a_j H(p_j)$$

and

$$\sigma^2 = \int_0^1 \alpha^2(u)du,$$

where

$$\alpha(u) = \frac{1}{1-u}\left\{\int_u^1 J(w)H'(w)(1-w)dw + \sum_{p_j \ge u} a_j(1-p_j)H'(p_j)\right\}$$

and $H = h \circ F^{-1}$. See also **8.2.5**.

8.2.2 The Approach of Stigler (1969, 1974)

We have seen the method of *projection* used in Chapter 5 and we will see it again in Chapter 9. Stigler deals with L-estimates in the form

$$S_n = \sum_{i=1}^{n} c_{ni} X_{ni},$$

by approximating the L-estimate by its projection

$$\hat{S}_n = \sum_{i=1}^{n} c_{ni} \hat{X}_{ni},$$

where $\hat{X}_{ni}$ is the projection of the order statistic X_{ni}. To express this projection, we introduce the notation

$$\psi(t) = \frac{d}{dt} F^{-1}(t) = \frac{1}{f(F^{-1}(t)}$$

and

$$g_{ni}(u) = n\binom{n-1}{i-1}u^{i-1}(1-u)^{n-i}, \qquad 0 < u < 1.$$

The latter is the density of U_{ni}, the ith order statistic of a sample from uniform (0, 1). Stigler (1969) proves

Lemma. *There is some* $n_0 = n_0(F)$ *such that for* $i \geq n_0$ *and* $n - i + 1 \geq n_0$,

$$\hat{X}_{ni} = n^{-1} \sum_{k=1}^{n} \int_0^{F(X_k)} \psi(u) g_{ni}(u) du + nE\{X_{n-1,i-1}\} - (n-1)E\{X_{ni}\}.$$

(In particular, since $\mathscr{L}\{X_{ni}\} = \mathscr{L}\{F^{-1}(U_{ni})\}$, $E\{X_{ni}\} = \int_0^1 F^{-1}(u) g_{ni}(u) du$.)

Stigler develops conditions under which

$$\frac{E(S_n - \hat{S}_n)^2}{\sigma^2(S_n)} \to 0,$$

so that for S_n $AN(E\{S_n\}, \sigma^2(S_n))$ it suffices to deal with $\hat{S}_n$ by standard central limit theory. Noting that

$$\hat{S}_n = n^{-} \sum_{k=1}^{n} Z_{nk} + \Delta_n,$$

where Δ_n is nonrandom and $Z_{nk} = \sum_{i=1}^n c_{ni} \int_0^{F(X_k)} \psi(u) g_{ni}(u) du$, it suffices to verify the Lindeberg condition for $\sum_{k=1}^n (Z_{nk} - EZ_{nk})$. (See details in Stigler (1969).) As noted by Stigler (1974), his assumptions leading to S_n $AN(E\{S_n\}, \sigma^2(S_n))$ may be characterized informally as follows:

(i) the extremal order statistics do not contribute too much to S_n;
(ii) the tails of the population distribution are smooth and the population density is continuous and positive over its support:
(iii) the variance of S_n is of the same order as that of $\sum |c_{ni}| X_{ni}$.

Stigler (1974) confines attention to the case

$$S_n = n^{-1} \sum_{i=1}^{n} J\left(\frac{i}{n+1}\right) X_{ni}$$

and strengthens the condition (ii) (through assumptions on J) in order essentially to be able to dispense with conditions (i) and (ii). He establishes several results. (See also Stigler (1979).)

Theorem A. *Suppose that* $E\{X^2\} < \infty$, *and that* J *is bounded and continuous a.e.* F^{-1}. *Suppose that*

$$\sigma^2(J, F) = \int_{-\infty}^{\infty}\int_{-\infty}^{\infty} J(F(x))J(F(y))[F(\min(x, y)) - F(x)F(y)]dx\,dy$$

is positive. Then

$$S_n \quad is \quad AN(E\{S_n\}, \sigma^2(S_n)).$$

Also,

$$\lim_{n\to\infty} n\sigma^2(S_n) = \sigma^2(J, F).$$

Theorem B. *Suppose that* $\int [F(x)(1 - F(x))]^{1/2}\,dx < \infty$ *and that* $J(u) = 0$ *for* $0 < u < \alpha$ *and* $1 - \alpha < u < 1$, *is bounded, and satisfies a Lipschitz condition of order* $>\frac{1}{2}$ *(except possibly at a finite number of points of* F^{-1} *measure* 0). *Then*

$$\lim_{n\to\infty} n^{1/2}[E\{S_n\} - \mu(J, F)] = 0,$$

where

$$\mu(J, F) = \int_0^1 F^{-1}(u)J(u)du.$$

(As noted by Stigler, if F has regularly varying tails (see Feller (1966), p. 268) with a finite exponent, then the conditions $E\{X^2\} < \infty$ and

$$\int [F(x)(1 - F(x))]^{1/2}\,dx < \infty$$

are equivalent.)

Under the combined conditions of Theorems A and B, we have that S_n is $AN(\mu(J, F), n^{-1}\sigma^2(J, F))$. Further, if J puts no weight on the extremes, the tail restrictions on F can be dropped (see Stigler's Theorem 5 and Remark 3):

Theorem C. *Suppose that* $J(u)$ *is bounded and continuous a.e.* F^{-1}, $=0$ *for* $0 < u < \alpha$ *and* $1 - \alpha < u < 1$, *and is Lipschitz of order* $>\frac{1}{2}$ *except at a finite set of points of* F^{-1} *measure* 0. *Then*

$$S_n \quad is \quad AN(\mu(J, F), n^{-1}\sigma^2(J, F)).$$

Example. The α-trimmed mean satisfies the preceding, provided that the αth and $(1-\alpha)$th quantiles of F are unique. ■

For *robust* L-estimation, it is quite appropriate to place the burden of restrictions on J rather than F.

8.2.3 The Approach of Shorack (1969, 1972)

Shorack (1969, 1972) considers L-estimates in the form

$$(1) \qquad T_n = n^{-1} \sum_{i=1}^{n} c_{ni} g(X_{ni})$$

and, without loss of generality, assumes that $X_1, X_2, \ldots$ are *uniform* (0, 1) variates. In effect Shorack introduces a signed measure ν on (0, 1) such that T_n estimates $\mu = \int_0^1 g \, d\nu$. He introduces a sequence of signed measures ν_n which approach ν in a certain sense and such that ν_n puts mass $n^{-1}c_{ni}$ at $i/n, 1 \le i \le n$, and 0 elsewhere. Thus

$$T_n = \int_0^1 g \circ F_n^{-1} \, d\nu_n.$$

He then introduces the stochastic process $L_n(t) = n^{1/2}[g \circ F_n^{-1}(t) - g(t)]$, $0 \le t \le 1$, and considers

$$n^{1/2}(T_n - \mu) = \int_0^1 L_n \, d\nu + n^{1/2} \int_0^1 g \circ F_n^{-1} \, d(\nu_n - \nu).$$

By establishing negligibility of the second term, treating the convergence of the stochastic process $L_n(\cdot)$, and treating the convergence of the functional $\int L_n \, d\nu$ over $L_n(\cdot)$, the asymptotic distribution of T_n is derived. His results yield the following examples.

Example A. Let $\{X_i\}$ be I.I.D. F (F arbitrary), with $E|X|^r < \infty$ for some $r > 0$. Let

$$T_n = n^{-1} \sum_{i=1}^{n} J(t_{ni}) X_{ni},$$

where $\max_{1 \le i \le n} |t_{ni} - i/n| \to 0$ as $n \to \infty$ and where for some $a > 0$

$$a\left[\min\left(\frac{i}{n}, 1 - \frac{i}{n}\right)\right] \le t_{ni} \le 1 - a\left[\min\left(\frac{i}{n}, 1 - \frac{i}{n}\right)\right], \qquad 1 \le i \le n.$$

Suppose that J is continuous except at a finite number of points at which F^{-1} is continuous, and suppose that

$$|J(t)| \le M[t(1-t)]^{-(1/2)+1/r+\delta}, \qquad 0 < t < 1,$$

for some $\delta > 0$. Let J_n be a function on $[0, 1]$ equal to $J(t_{ni})$ for $(i-1)/n < t \le i/n$ and $1 \le i \le n$ with $J_n(0) = J(t_{n1})$. Then

$$n^{1/2}(T_n - \mu(J_n, F)) \xrightarrow{d} N(0, \sigma^2(J, F)),$$

where $\mu(J_n, F) = \int_0^1 F^{-1}(t)J_n(t)dt$ and

$$\sigma^2(J, F) = \int_0^1 \int_0^1 [\min(s, t) - st]J(s)J(t)dF^{-1}(s)dF^{-1}(t).$$

It is desirable to replace $\mu(J_n, F)$ by $\mu(J, F) = \int_0^1 F^{-1}(t)J(t)dt$. This may be done if J' exists and is continuous on $(0, 1)$ with

$$|J(t)| \le M[t(1-t)]^{-(1/2)+1/r+\delta}, \qquad 0 < t < 1,$$

for some $\delta > 0$, and the "max-condition" is strengthened to

$$n \max_{1 \le i \le n} \left| t_{ni} - \frac{i}{n} \right| = O(1). \quad \blacksquare$$

Example A1. Let $X_1, \ldots, X_n$ be a sample from $\Phi = N(0, 1)$. For integral $r > 0$, an estimator of $E\{X^{r+1}\}$ is given by

$$T_n = n^{-1} \sum_{i=1}^{n} \left[\Phi^{-1}\left(\frac{i}{n+1}\right)\right]^r X_{ni}.$$

By Example A,

$$n^{1/2}(T_n - E\{X^{r+1}\}) \xrightarrow{d} N(0, E\{X^{2r+2}\} - E^2\{X^{r+1}\}). \quad \blacksquare$$

Example B *The α-trimmed mean.* Let $X_1, \ldots, X_n$ be a sample from $F_\theta = F(\cdot - \theta)$, where F is any distribution symmetric about 0. Let $0 < \alpha < \frac{1}{2}$. For n even define

$$T_n = \sum_{i=[n\alpha]+1}^{(1/2)n} \frac{[2(i - [n\alpha] - 1](X_{ni} + X_{n,n-i+1})}{2(\frac{1}{2}n - [n\alpha])^2}.$$

(Omit n odd.) Then

$$n^{1/2}(T_n - \theta) \xrightarrow{d} \left[\frac{4}{(1-2\alpha)^2}\right] 2^{1/2} \int_\alpha^{1/2} (t - \alpha)W(t)dF^{-1}(t),$$

where $W(\cdot)$ is the Wiener process **(1.11.4)**. $\blacksquare$

Note that Shorack requires J to be smooth but not necessarily bounded, and requires little on F. He also deals with more general J functions under additional restrictions on F.

Wellner (1977a, b) follows the Shorack set-up and establishes *almost sure* results for T_n given by (1). Define J_n on $[0, 1]$ by $J_n(0) = c_{n1}$ and $J_n(t) = c_{ni}$ for $(i-1)/n < t \le i/n$. Set $\mu_n = \int_0^1 J_n(t)g(t)dt$ and $\mu = \int_0^1 J(t)g(t)dt$.

***Assumption* 1.** *The function* g *is left continuous on* (0, 1) *and is of bounded variation on* $(\theta, 1-\theta)$ *for all* $\theta > 0$. *For fixed* b_1, b_2 *and* M,

$$|J(t)| \le Mt^{-b_1}(1-t)^{-b_2}, \qquad 0 < t < 1,$$

and the same bound holds for $J_n(\cdot)$, *each* n. *Further,*

$$|g(t)| \le Mt^{-1+b_1+\delta}(1-t)^{-1+b_2+\delta}, \qquad 0 < t < 1,$$

for some $\delta > 0$, *and* $\int_0^1 t^{1-b_1-(1/2)\delta}(1-t)^{1-b_2-(1/2)\delta}\, d|g| < \infty$. ■

***Assumption* 2.** $\lim_{n\to\infty} J_n(t) = J(t)$, $t \in (0, 1)$.

***Theorem*.** *Under Assumption* 1, $T_n - \mu_n \xrightarrow{wp1} 0$. *If also Assumption* 2 *holds, then* $T_n \xrightarrow{wp1} \mu$.

Example A* (parallel to Example A above). Let $\{X_i\}$, F, T_n, and $\{t_{ni}\}$ be as in Example A. Suppose that

$$|J(t)| \le M[t(1-t)]^{-1+1/r+\delta}, \qquad 0 < t < 1,$$

for some $\delta > 0$, and that J is continuous except at finitely many points. Then

$$T_n \xrightarrow{wp1} \mu(J, F) = \int_0^1 F^{-1}(t)J(t)dt. \quad ■$$

Note that the requirements on J in Example A* are milder than in Example A. Wellner also develops the *LIL* for T_n given by (1). For this, however, the requirements on J follow exactly those of Example A.

8.2.4 The Differentiable Statistical Function Approach

Consider the functional $T(F) = T_1(F) + T_2(F)$, where

$$T_1(F) = \int_0^1 F^{-1}(t)J(t)dt,$$

with J such that $K(t) = \int_0^t J(u)du$ is a linear combination of distribution functions on (0, 1), and

$$T_2(F) = \sum_{j=1}^m a_j F^{-1}(p_j).$$

We consider here the L-estimate given by the statistical function $T(F_n)$. Applying the methods of Chapter 6, we obtain *asymptotic normality* and the *LIL* in relatively straightforward fashion.

From **8.1.1** it is readily seen that

$$d_1 T(F; F_n - F) = n^{-1} \sum_{i=1}^{n} h(F; X_i),$$

where

$$h(F; x) = -\int_{-\infty}^{\infty} [I(y \geq x) - F(y)]J(F(y))dy + \sum_{j=1}^{m} a_j \frac{p_j - I(F^{-1}(p_j) \geq x)}{f(F^{-1}(p_j))}.$$

Note that $E_F\{h(F; X)\} = 0$. Put $\sigma^2(T, F) = \text{Var}_F\{h(F; X)\}$. If $0 < \sigma^2(T, F) < \infty$, we obtain that $T(F_n)$ is $AN(T(F), n^{-1}\sigma^2(T, F))$ if we can establish $n^{1/2}R_{1n} \xrightarrow{p} 0$, where $R_{1n} = \Delta_{1n} + \Delta_{2n}$, with

$$\Delta_{in} = T_i(F_n) - T_i(F) - d_1 T_i(F; F_n - F), \qquad i = 1, 2.$$

Now, in Example 6.5D, we have already established

$$n^{1/2}\Delta_{2n} \xrightarrow{p} 0,$$

provided that $F'(F^{-1}(p_j)) > 0, j = 1, \ldots, m$. It remains to deal with Δ_{1n}. By Lemma 8.1.1B, we have (check)

$$\text{(1)} \quad \Delta_{1n} = -\int_{-\infty}^{\infty} \{K(F_n(x)) - K(F(x)) - J(F(x))[F_n(x) - F(x)]\}dx$$

$$= -\int_{-\infty}^{\infty} W_{F_n, F}(x) \cdot [F_n(x) - F(x)]dx,$$

where we define

$$W_{G,F}(x) = \frac{K(G(x)) - K(F(x))}{G(x) - F(x)} - J(F(x)), \qquad G(x) \neq F(x),$$
$$= 0, \qquad G(x) = F(x).$$

Via (1), Δ_{1n} may be handled by any of several natural approaches, each involving different trade-offs between restrictions on J and restrictions on F. For example, (1) immediately implies

$$\text{(2A)} \qquad |\Delta_{1n}| \leq \|W_{F_n, F}\|_{L_1} \cdot \|F_n - F\|_\infty,$$

where $\|h\|_\infty = \sup_x |h(x)|$ and $\|h(x)\|_{L_1} = \int |h(x)|dx$. Since $\|F_n - F\|_\infty = O_p(n^{-1/2})$ as noted earlier in Remark 6.2.2B(ii), we can obtain $|\Delta_{1n}| = o_p(n^{-1/2})$ by showing

$$\text{(3A)} \qquad \|W_{F_n, F}\|_{L_1} \xrightarrow{p} 0.$$

To this effect, following Boos (1977, 1979), we introduce

Assumption A. J *is bounded and continuous a.e. Lebesgue and a.e.* F^{-1}, and

Assumption B. $\mathrm{J(u)}$ *vanishes for* $\mathrm{u} < \alpha$ *and* $\mathrm{u} > \beta$, *where* $0 < \alpha < \beta < 1$, and prove

Lemma A. *Under Assumptions A and B,*

$$\lim_{\|G-F\|_\infty \to 0} \|W_{G,F}\|_{L_1} = 0.$$

PROOF. First we utilize Assumption B. Let $0 < \varepsilon < \min\{\alpha, 1 - \beta\}$. Check that there exist a and b such that

$$-\infty < a < F^{-1}(\alpha - \varepsilon) < F^{-1}(\beta + \varepsilon) < b < \infty.$$

Then, for $\|G = F\|_\infty < \varepsilon$, $x < a$ implies $F(x) \le \alpha - \varepsilon < \alpha$ and thus $G(x) < F(x) + \varepsilon \le \alpha$, in which case (justify) $W_{G,F}(x) = 0$. Similarly, for $\|G - F\|_\infty < \varepsilon$, $x > b$ implies $W_{G,F}(x) = 0$. Therefore, for $\|G - F\|_\infty < \varepsilon$,

$$(*) \qquad \|W_{G,F}\|_{L_1} = \int_a^b |W_{G,F}(x)|\,dx.$$

Also, keeping a and b fixed, this identity continues to hold as $\varepsilon \to 0$.

Next we utilize Assumption A and apply dominated convergence (Theorem 1.3.7). For all x, we have

$$\begin{aligned} |W_{G,F}(x)| &\le |G(x) - F(x)|^{-1} \int_{F(x)}^{G(x)} |J(u)|\,du + |J(F(x))| \\ &\le 2\|J\|_\infty < \infty. \end{aligned}$$

Let $D = \{x\colon J$ is discontinuous at $F(x)\}$. For $x \notin D$, we have $W_{G,F}(x) \to 0$ as $G(x) \to F(x)$. But D is a Lebesgue-null set (why?). Hence

$$\lim_{\|G-F\|_\infty \to 0} \int_a^b |W_{G,F}(x)|\,dx = 0. \quad \blacksquare$$

Therefore, under Assumptions A and B, we have

$$(4) \qquad n^{1/2}\Delta_{1n} \xrightarrow{p} 0.$$

Indeed (justify), these assumptions imply

$$(5) \qquad n^{1/2}\Delta_{1n} \overset{wp1}{=} o((\log\log n)^{1/2}).$$

Further, from Example 6.5D, we have (justify)

$$n^{1/2}\Delta_{2n} = o((\log\log n)^{1/2})$$

provided that F is twice differentiable at the points $F^{-1}(p_j)$, $1 \le j \le m$. Therefore, we have proved

Theorem A. *Consider the* L-*estimate* $T(F_n) = T_1(F_n) + T_2(F_n)$. *Suppose that Assumptions A and B hold, and that* F *has positive derivatives at its* p_j-*quantiles,* $1 \le j \le m$. *Assume* $0 < \sigma^2(T, F) < \infty$. *Then*

$$n^{1/2}(T(F_n) - T(F)) \xrightarrow{d} N(0, \sigma^2(T, F)).$$

If, further, F *is twice differentiable at its* p_j-*quantiles,* $1 \le j \le m$, *then the corresponding LIL holds.*

Examples A. (i) *The trimmed mean.* Consider $T_1(F)$ based on $J(t) = I(\alpha \le t \le \beta)/(\beta - \alpha)$ and $T_2(\cdot) \equiv 0$. The conditions of the theorem are satisfied if the α- and β-quantiles of F are unique.

(ii) *The Winsorized mean.* (Problem 8.P.13). ■

It is desirable also to deal with untrimmed J functions. To this effect, Boos (1977, 1979) uses the following implication of (1):

(2B) $$|\Delta_{1n}| \le \|(q \circ F)W_{F_n, F}\|_{L_1} \cdot \left\| \frac{(F_n - F)}{q \circ F} \right\|_\infty,$$

where q can be any strategically selected function satisfying

Assumption B*. $\int_{-\infty}^{\infty} q(F(x))dx < \infty$.

In this case the role of Lemma A is given to the following analogue.

Lemma B. *Under Assumptions A and B*,*

$$\lim_{\|G-F\|_\infty \to 0} \|(q \circ F)W_{G,F}\|_{L_1} = 0.$$

PROOF. analogous to that of Lemma A (Problem 8.P.14). ■

In order to exploit Lemma B to establish (4) and (5), we require that $\|(F_n - F)/q \circ F\|_\infty$ satisfy analogues of the properties $O_p(n^{-1/2})$ and

$$O_{wp1}(n^{-1/2}(\log \log n)^{1/2})$$

known for $\|F_n - F\|_\infty$. O'Reilly (1974) gives weak convergence results which yield the first property for a class of q functions containing in particular

$$Q = \{q: q(t) = [t(1 - t)]^{(1/2)-\delta}, \qquad 0 < t < 1; 0 < \delta < \tfrac{1}{2}\}.$$

James (1975) gives functional LIL results which yield the second property for a class of q functions also containing Q.

On the other hand, Gaenssler and Stute (1976) note that the $O_p(n^{-1/2})$ property fails for $q(t) = [t(1-t)]^{1/2}$. For this q, the other property also fails, by results of James (1975). Although some of the aforementioned results are established only for *uniform* (0, 1) variates, the conclusions we are drawing are valid for general F. We assert:

Lemma C. *For* $q \in Q$,

$$\|(F_n - F)/q \circ F\|_\infty = O_p(n^{-1/2})$$

and

$$\overset{\text{wp1}}{=} O(n^{-1/2}(\log \log n)^{1/2}).$$

Consequently, we have (4) and (5) under Assumptions A and B*. That is,

Theorem B. *Assume the conditions of Theorem A, with Assumption B replaced by Assumption B* for some* $q \in Q$. *Then the assertions of Theorem A remain valid.*

Examples B. Let F satisfy $\int [F(x)(1 - F(x))]^{(1/2)-\delta}\, dx < \infty$ for some $\delta > 0$. Then Theorem B is applicable to

(i) The mean: $J(u) \equiv 1$;
(ii) Gini's mean difference: $J(u) = 4u - 2$;
(iii) The asymptotically efficient L-estimator for location for the logistic family: $J(u) = 6u(1 - u)$. ■

Remark. Boos (1977, 1979) actually establishes that

$$T(F; \Delta) = -\int \Delta(x) J(F(x))dx$$

is a *differential* of $T(\cdot)$ at F w.r.t. suitable $\|\cdot\|$'s. ■

Still more can be extracted from (1), via the implication

(2C) $$|\Delta_{1n}| \le \|W_{F_n,F}\|_\infty \cdot \|F_n - F\|_{L_1}.$$

Thus one approach toward obtaining (4) is to establish $\|F_n - F\|_{L_1} = O_p(n^{-1/2})$ under suitable restrictions on F, and to establish $\|W_{F_n,F}\|_\infty \to 0$ under suitable restrictions on J. We start with $\|F_n - F\|_{L_1}$.

Lemma D. *Let* F *satisfy* $\int [F(x)[1 - F(x)]]^{1/2}\, dx < \infty$. *Then*

$$E\{\|F_n - F\|_{L_1}\} = O(n^{-1/2}).$$

PROOF. Write $F_n(x) - F(x) = n^{-1}\sum_{i=1}^n Y_i(x)$, where $Y_i(x) = I(X_i \le x) - F(x)$. Then

$$\|F_n - F\|_{L_1} = \int \left| n^{-1} \sum_{i=1}^n Y_i(x) \right| dx.$$

By Tonelli's Theorem (Royden (1968), p. 270),

$$E\{\|F_n - F\|_{L_1}\} = \int E\left\{ \left| n^{-1} \sum_{i=1}^n Y_i(x) \right| \right\} dx.$$

Now check that

$$E\left\{ \left| n^{-1} \sum_{i=1}^n Y_i(x) \right| \right\} \le n^{-1/2}[F(x)(1 - F(x))]^{1/2}. \quad \blacksquare$$

Now we turn to $\|W_{F_n, F}\|_\infty$ and adopt

Assumption A*. *J is continuous on* [0, 1].

Lemma E. *Under Assumption A*,*

$$\lim_{\|G - F\|_\infty \to 0} \|W_{G,F}\|_\infty = 0.$$

(Prove as an exercise.) We thus have arrived at

Theorem C. *Let* F *satisfy* $\int [F(x)(1 - F(x))]^{1/2}\,dx < \infty$ *and have positive derivatives at its* p_j*-quantiles,* $1 \le j \le m$. *Let* J *be continuous on* [0, 1]. *Assume* $0 < \sigma^2(T, F) < \infty$. *Then*

$$n^{1/2}(T(F_n) - T(F)) \xrightarrow{d} N(0, \sigma^2(T, F)).$$

Compared with Theorem B, this theorem requires slightly less on F and slightly more on J. Examples B are covered by the present theorem also.

Note that Theorem C remains true if $T(F_n)$ is replaced by $T_n = T_{n1} + T_2(F_n)$, where

$$T_{n1} = n^{-1} \sum_{i=1}^n J\left(\frac{i}{n}\right) X_{ni}.$$

Show (Problem 8.P.17) that this assertion follows from

Lemma F. *Under Assumption A*,*

$$\max_{1 \le i \le n} \left| K\left(\frac{i}{n}\right) - K\left(\frac{i-1}{n}\right) - \frac{1}{n} J\left(\frac{i}{n}\right) \right| = o(n^{-1}).$$

Prove Lemma F as an exercise.

8.2.5 Berry–Esséen Rates

For L-estimates in the case of zero weight given to extreme order statistics, Rosenkrantz and O'Reilly (1972) derived the Berry–Esséen rate $O(n^{-1/4})$. However, as we saw in Theorem 2.3.3C, for sample quantiles the rate $O(n^{-1/2})$ applies. Thus it is not surprising that the rate $O(n^{-1/4})$ can be improved to $O(n^{-1/2})$. We shall give three such results. Theorem A, due to Bjerve (1977), is obtained by a refinement of the approach of CGJ (1967) discussed in **8.2.1**. The result permits quite general weights on the observations between the αth and βth quantiles, where $0 < \alpha < \beta < 1$, but requires zero weights on the remaining observations. Thus the distribution F need not satisfy any moment condition. However, strong smoothness is required. Theorem B, due to Helmers (1977a, b), allows weights to be put on all the observations, under sufficient smoothness of the weight function and under moment restrictions on F. However, F need not be continuous. Helmers' methods, as well as Bjerve's, incorporate Fourier techniques. Theorem C, due to Boos and Serfling (1979), applies the method developed in **6.4.3** and thus implicitly uses the Berry–Esséen theorem for U-statistics (Theorem 5.5.1B) due to Callaert and Janssen (1978). Thus Fourier techniques are bypassed, being subsumed into the U-statistic result. Theorem C is close to Theorem B in character. It should be noted that a major influence underlying all of these developments was provided by ideas in Bickel (1974).

Bjerve treats L-estimates in the form

$$T_n = n^{-1} \sum_{i=1}^{n} c_{ni} h(X_{ni})$$

and utilizes the function $\tilde{H} = h \circ F^{-1} \circ G$ and the notation μ_n and σ_n defined in **8.2.1**. He confines attention to the case that

$$c_{ni} = 0 \quad \text{for} \quad i \le \alpha n \quad \text{or} \quad i > \beta n, \qquad \text{where} \quad 0 < \alpha < \beta < 1,$$

and introduces constants a and b satisfying $0 < a < -\log(1 - \alpha)$ and $-\log(1 - \beta) < b < \infty$. His theorem imposes further conditions on the c_{ni}'s as well as severe regularity conditions on $\tilde{H}$. Namely, Bjerve proves

Theorem A. *Let* $\tilde{H}$ *satisfy a first order Lipschitz condition on* $[a, b]$ *and assume for some constants* $c > 0$ *and* $d < \infty$ *that*

(i) $\sigma_n^2 > c$, *all* n,

and

(ii) $n^{-1} \sum_{i=1}^{n} |c_{ni}| < d$, *all* n.

Then

$$(*) \qquad \sup_t \left| P\left(\frac{n^{1/2}(T_n - \mu_n)}{\sigma_n} \le t\right) - \Phi(t) \right| = O(n^{-1/2}).$$

PROOF (Sketch). The representation $T_n = \mu_n + Q_n + R_n$ of **8.2.1** is refined by writing $R_n = M_n + \Delta_n$, where

$$M_n = n^{-1} \sum_{i=1}^{n} \frac{1}{2} c_{ni} \tilde{H}''(\tilde{v}_{ni})(V_{ni} - \tilde{v}_{ni})^2$$

and

$$\Delta_n = n^{-1} \sum_{i=1}^{n} c_{ni}(V_{ni} - \tilde{v}_{ni})^3 \tilde{G}_{ni}(V_{ni}),$$

with

$$\tilde{G}_{ni}(v) = \frac{\left\{\dfrac{G_{ni}(v)}{v - \tilde{v}_{ni}} - \dfrac{1}{2}\tilde{H}''(\tilde{v}_{ni})\right\}}{(v - \tilde{v}_{ni})}, \qquad v \ne \tilde{v}_{ni},$$

$$= 0, \qquad v = \tilde{v}_{ni}.$$

It can be shown that

$$P(|\Delta_n| > n^{-1/2}) = O(n^{-1/2})$$

and, by a characteristic function approach, that

$$\sup_t \left| P\left(\frac{n^{1/2}(Q_n + M_n)}{\sigma_n} \le t\right) - \Phi(t) \right| = O(n^{-1/2}).$$

(See Bjerve (1977) for details.) The result then follows by Lemma 6.4.3. ■

For the special case

$$T_n = n^{-1} \sum_{i=1}^{n} J\left(\frac{i}{n+1}\right) h(X_{ni}) + \sum_{j=1}^{m} a_j h(X_{n,[np_j]}),$$

considered also by CGJ (1967), Theorem A yields (*) with μ_n and σ_n replaced by the constants μ and σ defined at the end of **8.2.1**; that is, with $H = h \circ F^{-1}$, we have

Corollary. *Let* J *and* H″ *satisfy a first order Lipschitz condition on an open interval containing* $[\alpha, \beta]$, $0 < \alpha < \beta < 1$, *and let* J *vanish outside* $[\alpha, \beta]$. *Let* $p_1, \ldots p_m \in [\alpha, \beta]$. *Then* (*) *holds with* μ_n *and* σ_n *replaced by* μ *and* σ.

We next give Helmers' result, which pertains to L-estimates in the form

$$T_n = n^{-1} \sum_{i=1}^{n} J\left(\frac{i}{n+1}\right) X_{ni}.$$

Theorem B. *Suppose that*

(i) $E_F|X|^3 < \infty$;

(ii) J *is bounded and continuous on* (0, 1);

(iiia) J′ *exists except possibly at finitely many points*;

(iiib) J′ *is Lipschitz of order* $> \frac{1}{2}$ *on the open intervals where it exists*;

(iv) F^{-1} *is Lipschitz of order* $> \frac{1}{2}$ *on neighborhoods of the points where* J′ *does not exist*;

(v) $0 < \sigma^2(J, F) = \iint J(F(x))J(F(y))[F(\min(x, y)) - F(x)F(y)]dx\, dy < \infty$.

Then

$$(**) \qquad \sup_t \left| P\left(\frac{T_n - E\{T_n\}}{\sigma(T_n)} \le t\right) - \Phi(t) \right| = O(n^{-1/2}).$$

This theorem is proved in Helmers (1977a) under the additional restriction $\int |J'| dF^{-1} < \infty$, which is eliminated in Helmers (1977b).

We now establish a closely parallel result for L-estimates of the form $T(F_n)$ based on the functional $T(F) = \int_0^1 F^{-1}(u)J(u)du$.

Theorem C. *Assume conditions* (i), (ii) *and* (v) *of Theorem B. Replace* (iii) *and* (iv) *by*

(iii′) J' *exists and is Lipschitz of order* $\delta > \frac{1}{3}$ *on* (0, 1).

Then

$$(***) \quad \sup_t \left| P\left(\frac{n^{1/2}(T(F_n) - T(F))}{\sigma(J, F)} \le t\right) - \Phi(t) \right| = O(n^{-1/2}).$$

Remark A. Compared to Theorem B, Theorem C requires existence of J' at *all* points but permits a *lower order* Lipschitz condition. Also, from the proof it will be evident that under a higher order moment assumption $E|X_1|^\nu < \infty$ for integer $\nu > 3$, the Lipschitz order may be relaxed to $\delta > 1/\nu$. ∎

The proof of Theorem C will require the following lemmas, the second of which is a parallel of Lemma 8.2.4D. Here $\|h\|_{L_2} = [\int h^2(x)dx]^{1/2}$.

Lemma A. *Let the random variable* X *have distribution* F *and satisfy* $E|X|^k < \infty$, *where* k *is a positive integer. Let* g *be a bounded function. Then*

(i) $E\{\int [I(X \le y) - F(y)]g(y)dy\} = 0$

and

(ii) $E\{[\int |I(X \le y) - F(y)]g(y)|dy]^k\} < \infty$.

PROOF. Since $E|X| < \infty$, we have (why?) $y[F(-y) + 1 - F(y)] \to 0$ as $y \to \infty$. Thus $y|I(X \le y) - F(y)| \to 0$ as $y \to \pm\infty$ and hence, by integration by parts,

$$\int |I(X \le y) - F(y)|dy \le |X| + E|X|.$$

Thus (ii) readily follows. Also, by Fubini's theorem, this justifies an interchange of $E\{\cdot\}$ and $\int$ in (i). ■

Lemma B. *Let* $E_F|X|^k < \infty$, *where* k *is a positive integer. Then*

$$E\{\|F_n - F\|_{L_2}^{2k}\} = O(n^{-k}).$$

PROOF. Put $Y_i(t) = I(X_i \le t) - F(t)$, $1 \le i \le n$. Then

$$\text{(a)} \qquad E\{\|F_n - F\|_{L_2}^{2k}\} = n^{-2k} \sum_{i_1=1}^{n} \sum_{j_1=1}^{n} \cdots \sum_{i_k=1}^{n} \sum_{j_k=1}^{n} E\left\{\prod_{l=1}^{k} \int Y_{i_l}(t) Y_{j_l}(t) dt\right\}.$$

By the use of Lemma A and Fubini's Theorem (check), we have

$$\text{(b)} \qquad E\left\{\prod_{l=1}^{k} \int Y_{i_l}(t) Y_{j_l}(t) dt\right\} = \int \cdots \int E\left\{\prod_{l=1}^{k} Y_{i_l}(t_l) Y_{j_l}(t_l)\right\} dt_1 \cdots dt_k.$$

Check that we have $E\{Y_{i_1}(t_1)Y_{j_1}(t_1) \cdots Y_{i_k}(t_k)Y_{j_k}(t_k)\} = 0$ *except* possibly in the case that each index in the list $i_1, j_1, \ldots, i_k, j_k$ appears at least twice. In this case the number of distinct elements in the set $\{i_1, j_1, \ldots, i_k, j_k\}$ is $\le k$. It follows that the number of ways to choose $i_1, j_1, \ldots, i_k, j_k$ such that the expectation in (b) is nonzero is $O(n^k)$. Thus the number of nonzero terms in the summation in (a) is $O(n^k)$. ■

PROOF OF THEOREM C. We apply Theorem 6.4.3. Thus we express $T(F_n) - T(F)$ as $V_{2n} + R_{2n}$, where

$$\begin{aligned} V_{2n} &= d_1 T(F; F_n - F) + \tfrac{1}{2} d_2 T(F; F_n - F) \\ &= n^{-2} \sum_{i=1}^{n} \sum_{j=1}^{n} h(F; X_i, X_j). \end{aligned}$$

By Problem 8.P.5,

$$d_1 T(F; F_n - F) = -\int_{-\infty}^{\infty} [F_n(t) - F(t)]J \circ F(t)dt$$

and

$$d_2 T(F; F_n - F) = -\int_{-\infty}^{\infty} [F_n(t) - F(t)]^2 J' \circ F(t)dt.$$

Thus (check) the desired $h(F; x, y) = \frac{1}{2}[\alpha(x) + \alpha(y) + \beta(x, y)]$, where

$$\alpha(x) = -\int_{-\infty}^{\infty} [I(x \le t) - F(t)]J \circ F(t)dt$$

and

$$\beta(x, y) = -\int_{-\infty}^{\infty} [I(x \le t) - F(t)][I(y \le t) - F(t)]J' \circ F(t)dt.$$

Therefore (check), R_{2n} is given by

$$-\int_{-\infty}^{\infty} \{K \circ F_n(t) - K \circ F(t) - J \circ F(t)[F_n(t) - F(t)] - \tfrac{1}{2}J' \circ F(t)[F_n(t) - F(t)]^2\}dt,$$

where $K(u) = \int_0^u J(v)dv$. By the Lip condition on J', we obtain

$$|R_{2n}| \le \tfrac{1}{2}C \int |F_n(t) - F(t)|^{2+\delta}\, dt \le \tfrac{1}{2}C\|F_n - F\|_\infty^\delta \|F_n - F\|_{L_2}^2. \tag{1}$$

Let $A > 0$ be given. By (1),

$$P(|R_{2n}| > An^{-1}) \le P\left(n\|F_n - F\|_\infty^\delta \|F_n - F\|_{L_2}^2 > \frac{2A}{C}\right) \le P\left(n^{1/6}\|F_n - F\|_\infty^\delta > \frac{2A}{C}\right) + P(n^{5/6}\|F_n - F\|_{L_2}^2 > 1). \tag{2}$$

For $\delta > \frac{1}{3}$, the first right-hand term is (check) $O(n^{-1/2})$ by an application of Theorem 2.1.3A. The second term is $O(n^{-1/2})$ by Lemma B above. Therefore,

$$P(|R_{2n}| > An^{-1}) = O(n^{-1/2}),$$

as required in Theorem 6.4.3.

The required properties of $h(F; x, y)$ are obtained by use of Lemma A above (check). ■

Remark B (Problem 8.P.21). Under the same conditions on F and J, (***) holds also with $T(F_n)$ replaced by

$$T_n = n^{-1}\sum_{i=1}^{n} J\left(\frac{i}{n+1}\right)X_{ni}.$$

(Hint: Show that $|T(F_n) - T(F)| \le Mn^{-2}\sum_1^n |X_i|$ for a constant M. Thus show that $P(|T(F_n) - T(F)| > 2ME|X_1|n^{-1}) = O(n^{-1})$, using Chebyshev's inequality.) ■

8.P PROBLEMS

Section 8.1

1. Complete details for Example 8.1.1A (Gini's mean difference).

2. For the functional $T(F) = F^{-1}(p)$, show that the Gâteaux derivative of T at F in the direction of G is

$$d_1T(F; G - F) = \frac{p - G(F^{-1}(p))}{f(F^{-1}(p))},$$

in the case that F has a positive density f at $F^{-1}(p)$. (Hint: following Huber (1977), put $F_\lambda = F + \lambda(G - F)$ and differentiate implicitly with respect to λ in the equation $F_\lambda(F_\lambda^{-1}(p)) = p$.)

3. Prove Lemma 8.1.1A. (Hint: Let D be the discontinuity set of F and put $A = [0, 1] - D$. Deal with $\int_A F^{-1}(t)dK_1(t)$ by a general change of variables lemma (e.g., Dunford and Schwartz (1963), p. 182).)

4. Prove Lemma 8.1.1B. (Hint: Apply Lemma 8.1.1A and integrate by parts.)

5. For the functional $T_1(F) = \int_0^1 F^{-1}(t)J(t)dt$, put $F_\lambda = F + \lambda(G - F)$ and show

$$\frac{d^kT(F_\lambda)}{d_\lambda^k} = -\int_{-\infty}^{\infty}[G(x) - F(x)]^k J^{(k-1)}(F_\lambda(x))dx.$$

(Hint: Apply Lemma 8.1.1B.)

6. (Continuation). Show that the influence curve of $T_1(\cdot)$ is differentiable, with derivative $J(F(x))$.

7. (Complement to Problem 5). For the functional $T(F) = F^{-1}(p)$, find $d_2T(F; G - F)$ for arbitrary G.

8. Derive the influence curve of the α-Winsorized mean (Example 8.1.3C).

Section 8.2

9. Prove Lemma 8.2.1B.

10. (a) Let $\{\alpha_{ni}\}$ be arbitrary constants and put $\sigma_n^2 = n^{-1} \sum_{i=1}^n \alpha_{ni}^2$. Let $\{Z_i\}$ be IID negative exponential variates. Put $X_i = Z_i - 1$, $i = 1, 2, \ldots$, and $W_n = n^{-1} \sum_{i=1}^n \alpha_{ni} X_i$. Show that W_n is $AN(0, n^{-1}\sigma_n^2)$ if and only if

(*) $$\max_{1 \le i \le n} |\alpha_{ni}| = o(n^{1/2}\sigma_n).$$

(Hint: use characteristic functions.)

(b) Now let $\{X_i\}$ be IID F, where F has mean 0 and finite variance. Show that (*) suffices for W_n to be $AN(0, n^{-1}\sigma_n^2)$. (Hint: apply Theorem 1.9.3.)

11. Show that Example 8.2.3A1 is a special case of Example 8.2.3A.

12. Complete the details of proof of Lemma 8.2.4A.

13. Details for Example 8.2.4A(ii).

14. Verify Lemma 8.2.4B.

15. Minor details for proof of Lemma 8.2.4D.

16. Prove Lemma 8.2.4E.

17. Prove the assertion preceding Lemma 8.2.4F.

18. Prove Lemma 8.2.4F.

19. Details for proof of Lemmas 8.2.5A, B.

20. Details for proof of Theorem 8.2.5C.

21. Verify Remark 8.2.5B.

CHAPTER 9

R-Estimates

Consider a sample of independent observations $X_1, \ldots, X_N$ having respective distribution functions $F_1, \ldots, F_N$ *not necessarily identical.* For example, the X_i's may correspond to a combined sample formed from samples from several different populations. It is often desired to base inference purely on the *ranks* $R_1, \ldots, R_N$ of $X_1, \ldots, X_N$. This may be due to invariance considerations, or to gain the mathematical simplicity of having a finite sample space, or because rank procedures are convenient to apply. Section 9.1 provides a basic formulation and some examples. We shall confine attention primarily to *simple linear rank statistics* and present in Section 9.2 several methodologies for treating asymptotic normality. Some complements are provided in Section 9.3, including, in particular, the connections between *R*-estimates and the *M*- and *L*-estimates of Chapters 7 and 8.

9.1 BASIC FORMULATION AND EXAMPLES

A motivating example is provided in **9.1.1**, and the class of *linear* rank statistics is examined in **9.1.2**. Our treatment in this chapter emphasizes *test* statistics. However, in **9.1.3** the role of rank-type statistics in *estimation* is noted, and here the connection with the "statistical function" approach of Chapter 6 is seen.

9.1.1 A Motivating Example: Testing Homogeneity of Two Samples

Consider mutually independent observations $X_1, \ldots, X_N$, where $X_1, \ldots, X_m$ have continuous distribution F and $X_{m+1}, \ldots, X_N$ have continuous distribution function G. The problem is to test the hypothesis $H_0\colon F = G$.

An instructive treatment of the problem is provided by Fraser (1957), §5.3, to which the reader is referred for details. By invariance considerations, the data vector $\mathbf{X} = (X_1, \ldots, X_N)$ is reduced to the vector of ranks $\mathbf{R} = (R_1, \ldots, R_N)$. By sufficiency considerations, a further reduction is made, to the vector $(R_{m1}, \ldots, R_{mm})$ of ordered values of the ranks $R_1, \ldots, R_m$ of the first sample. Hence we consider basing a test of H_0 upon the statistic

$$T(\mathbf{X}) = \mathbf{R}_{(m)} = (R_{m1}, \ldots, R_{mm}).$$

The "best" test statistic based on $T(\mathbf{X})$ depends on the particular class of alternatives to H_0 against which protection is most desired. We shall consider three cases.

(i) $H_1: G = F^2$. For this alternative, the most powerful rank test is found to have test function of the form

$$P(\text{reject } H_0 \,|\, \mathbf{R}_{(m)}) = \begin{cases} 1 \\ \gamma \\ 0 \end{cases} \quad \text{if} \quad \sum_{i=1}^{m} (R_{mi} + i - 1) \gtreqless c.$$

Accordingly, an appropriate test statistic is

$$S_1 = \sum_{i=1}^{m} \log(R_{mi} + i - 1).$$

(ii) $H_2: G = qF + pF^2 (0 < p \leq 1, q = 1 - p)$. For this alternative, the locally most powerful rank test (for p in a neighborhood of the "null" value 0) is based on the test statistic

$$S_2 = \sum_{i=1}^{m} R_{mi}.$$

(iii) $H_3: F = N(\mu_1, \sigma^2), G = N(\mu_2, \sigma^2), \mu_1 < \mu_2$. For this alternative, the locally most powerful rank test (for $\mu_2 - \mu_1$ in a neighborhood of 0) is the "c_1-test," based on the statistic

$$S_3 = \sum_{i=1}^{m} E(Z_{N, R_i}),$$

where $(Z_{N1}, \ldots, Z_{NN})$ denotes the order statistic for a random sample of size N from $N(0, 1)$.

Observe that, even having reduced the data to $T(\mathbf{X}) = \mathbf{R}_{(m)}$, a variety of statistics based on T arise for consideration. The class of useful rank statistics is clearly very rich.

Note that in each of the three preceding cases, the relevant statistic is of the form

$$S = \sum_{i=1}^{m} a_N(i, R_{mi})$$

for some choice of constants $a_N(i, j)$, $1 \leq i, j \leq N$.

9.1.2 Linear Rank Statistics

In general, any statistic T which is a function of $\mathbf{R} = (R_1, \ldots, R_N)$ is called a *rank* statistic. An important class of rank statistics consists of the *linear* type, given by the form

$$T(R) = \sum_{i=1}^{N} a(i, R_i),$$

where $\{a(i, j)\}$ is an arbitrary $N \times N$ matrix. Any choice of the set of constants defines such a statistic. As will be discussed in **9.2.5**, an arbitrary rank statistic may often be suitably approximated by its projection into the family of linear rank statistics.

A useful subclass of the linear rank statistics consists of the *simple* type, given by the form

$$S(R) = \sum_{i=1}^{N} c_i a_N(R_i),$$

where $c_1, \ldots, c_N$ are arbitrary "regression" constants and $a_N(1), \ldots, a_N(N)$ are "scores." Typically, the scores are generated by a function $h(t), 0 < t < 1$, either by

(i) $a_N(i) = h(i/(N + 1)), 1 \leq i \leq N,$

or by

(ii) $a_N = Eh(U_{Ni}), 1 \leq i \leq N,$

where U_{Ni} denotes the ith order statistic in a random sample of size N from the uniform [0, 1] distribution. The scores given by (ii) occur in statistics yielding *locally most powerful* tests. Those given by (i) have the appeal of *simplicity*.

The special case of $c_i = 1$ for $1 \leq i \leq m$ and $c_i = 0$ for $m + 1 \leq i \leq N$ is called a *two-sample* simple linear rank statistic. Note that the statistics S_2 and S_3 mentioned in **9.1.1** are of this type, with scores generated by

$$h(t) = t, \qquad 0 \leq t \leq 1,$$

and

$$h(t) = \Phi^{-1}(t), \qquad 0 \leq t \leq 1,$$

respectively. The statistic S_1 of **9.1.1** is of linear form, but not of the simple type.

9.1.3 *R*-Estimates

Consider a two-sample simple linear rank statistic for *shift*. That is, the null hypothesis is $H_0: G(x) = F(x - \Delta)$, and the test statistic is of the form $S = \sum_{i=1}^m a_N(R_i)$. A related estimator $\hat{\Delta}_N$ of the shift parameter Δ may be developed as follows. Find the choice of d such that the statistic S, when recomputed using the values $X_{m+1} - d, \ldots, X_N - d$ in place of $X_{m+1}, \ldots, X_N$, comes as close as possible to its null hypothesis expected value, which is $mN^{-1} \sum_{i=1}^N a(i)$. This value $\hat{\Delta}_N$ makes the sample $X_{m+1} - \hat{\Delta}_N, \ldots, X_N - \hat{\Delta}_N$ appear to be distributed as a sample from the distribution F and thus serves as a natural estimator of Δ.

By a similar device, the *location* parameter of a *single* sample may be estimated. Let $X_1, \ldots, X_m$ be a sample from a distribution F symmetric about a location parameter θ. Construct (from the same observations) a "second sample"

$$2d - X_1, \ldots, 2d - X_m,$$

where d is chosen arbitrarily. Now find the value $d = \hat{\theta}_n$ such that the statistic S computed from the two samples comes as close as possible to its null value. For example, if S denotes the two-sample Wilcoxon statistic, based on the scores $a(i) = i$, then $\hat{\theta}_n$ turns out to be the Hodges–Lehmann estimate, median $\{\frac{1}{2}(X_i + X_j), 1 \le i < j \le m\}$.

Let the scores $a(i)$ be generated via

$$a(i) = \int_{(i-1)/m}^{i/m} h(t)dt.$$

Then the location estimator just discussed is given by $T(F_m)$, where F_m denotes the usual sample distribution function and $T(\cdot)$ is the functional defined by the implicit equation

$$\int h\{\tfrac{1}{2}[t + 1 - F(2T(F) - F^{-1}(t))]\}dt = 0.$$

See Huber (1977) for further details. Thus the methods of Chapter 6 may be applied.

9.2 ASYMPTOTIC NORMALITY OF SIMPLE LINEAR RANK STATISTICS

Several approaches to the problem of asymptotic normality will be described, broadly in **9.2.1** and more specifically in **9.2.2–4**. In **9.2.5** we examine in general form the important *projection* method introduced in **5.3.1** in dealing with U-statistics and further noted in **8.2.2** in dealing with L-estimates. In **9.2.6** we present *Berry–Esséen* rates, making use of the projection method.

9.2.1 Preliminary Discussion

The distribution theory of statistics of the form

$$S = \sum_{i=1}^{N} c_i a_N(R_i)$$

is determined by the following three entities:

(a) the regression constants $c_1, \ldots, c_N$;
(b) the scores generating function $h(\cdot)$;
(c) the distribution functions $F_1, \ldots, F_N$.

The conclusion of asymptotic normality of S, either with "natural" parameters $(E\{S\}, \text{Var}\{S\})$, or with other parameters (μ_N, σ_N^2) preferred for their simplicity, requires suitable regularity conditions to be imposed on these entities. Of course, less regularity in one entity may be balanced by strong regularity in another.

The most regular $c_1, \ldots, c_N$ are those generated by a linear function: $c_j = a + b_j$, $1 \le j \le N$. A typical relaxation of this degree of regularity is the condition that

$$\nu_N = \frac{\max_{1 \le i \le N} (c_i - \bar{c})^2}{1/N \sum_{i=1}^{N} (c_i - \bar{c})^2}$$

be bounded. The mildest condition yet used is that $\nu_N = o(N)$, $N \to \infty$.

The severest restriction on $F_1, \ldots, F_N$ corresponds to the "null" hypothesis $F_1 = \cdots = F_N$. Other conditions on $F_1, \ldots, F_N$ correspond to alternatives of the "local" type (i.e., converging to the null hypothesis in some sense as $N \to \infty$) or to fixed alternatives of special structure (as of the two-sample type).

The regularity conditions concerning the scores are expressed in terms of smoothness and boundedness of the scores generating function h. A linear h is ideal.

The asymptotic distribution theory for simple linear rank statistics falls roughly into three lines of development, each placing emphasis in a different way on the three entities involved. These approaches are described in **9.2.2–4.** Further background discussion is given in Hájek (1968).

9.2.2 Continuation: The Wald and Wolfowitz Approach

This line of development assumes the strongest regularity on $F_1, \ldots, F_N$, namely that $F_1 = \cdots = F_N$, and directs attention toward relaxation of restrictions on $c_1, \ldots, c_N$ and $a_N(1), \ldots, a_N(N)$. The series of results began with a result of Hotelling and Pabst (1936), discussed in the example following the theorem below. Their work was generalized by Wald and Wolfowitz (1943), (1944) in the following theorem.

Theorem (Wald and Wolfowitz). *Suppose that* $F_1 = \cdots = F_N$, *each* $N = 1, 2, \ldots$. *Suppose that the quantities*

$$\frac{\sum_{i=1}^N (c_{Ni} - \bar{c}_N)^r}{[\sum_{i=1}^N (c_{Ni} - \bar{c}_N)^2]^{r/2}}, \frac{\sum_{i=1}^N (a_{Ni} - \bar{a}_N)^r}{[\sum_{i=1}^N (a_{Ni} - \bar{a}_N)^2]^{r/2}}$$

are $O(1)$, $N \to \infty$, *for each* $r = 3, 4, \ldots$. *Then*

$$S_N = \sum_{i=1}^N c_{Ni} a_{NR_i} \quad \textit{is} \quad AN(\mu_N, \sigma_N^2)$$

and

$$E(S_N) = \mu_N, \qquad Var(S_N) = \sigma_N^2,$$

where $\mu_N = N\bar{c}_N\bar{a}_N$ *and* $\sigma_N^2 = (N - 1)\sigma_{Nc}^2\sigma_{Na}^2$, *with* $\sigma_{Nc}^2 = (N - 1)^{-1} \sum_1^N (c_{Ni} - \bar{c}_N)^2$ *and* $\sigma_{Na}^2 = (N - 1)^{-1} \sum_1^N (a_{Ni} - \bar{a}_N)^2$.

PROOF (Sketch). The moments of $(S_N - \mu_N)/\sigma_N$ are shown to converge to those of $N(0, 1)$. Then the Fréchet–Shohat Theorem (1.5.1B) is applied. For details, see Fraser (1957), Chapter 6, or Wilks (1962), §9.5. ■

Example. *Testing independence by the rank correlation coefficient.* A test may be based on

$$S_N = \sum_{i=1}^N iR_i,$$

which under the null hypothesis is found (check) by the preceding theorem to be $AN(\mu_N, \sigma_N^2)$, with

$$\mu_N = \frac{N(N + 1)^2}{4} \sim \frac{N^3}{4}$$

and

$$\sigma_N^2 = \frac{N^2(N^2 - 1)^2}{144(N - 1)} \sim \frac{N^5}{144}. \quad ■$$

A series of extensions of the preceding theorem culminated with necessary and sufficient conditions being provided by Hájek (1961). For detailed bibliographic discussion and further results, see Hájek and Šidák (1967), pp. 152–168 and 192–198.

9.2.3 Continuation: The Chernoff and Savage Approach

The line of development, initiated by Chernoff and Savage (1958), concerns the two-sample problem and allows broad assumptions regarding $F_1, \ldots, F_N$ but imposes stringent conditions on the regression constants and the scores generating function. The basic device introduced by Chernoff and Savage is the representation of a simple linear rank statistic as a function of the sample distribution function, in order to utilize theory for the latter. The representation is as follows.

Let $\{X_1, \ldots, X_m\}$ and $\{X_{m+1}, \ldots, X_N\}$ be independent random samples from (*not necessarily continuous*) distribution functions F and G, respectively. Put $\lambda_N = m/N$ and $n = N - m$. Then the distribution function for the combined sample is

$$H(t) = \lambda_N F(t) + (1 - \lambda_N)G(t), \qquad -\infty < t < \infty.$$

Likewise, if F_m^* and G_n^* denote the sample distribution functions of the subsamples

$$H_N(t) = \lambda_N F_m^*(t) + (1 - \lambda_N)G_n^*(t), \qquad -\infty < t < \infty.$$

The statistic of interest is

$$S_{mn} = \sum_{i=1}^{m} a_N(R_i).$$

Define

$$J_N\left(\frac{i}{N}\right) = a_N(i), \qquad 1 \leq i \leq N.$$

Then

$$S_{mn} = m \int_{-\infty}^{\infty} J_N(H_N(x)) dF_m^*(x),$$

since if X_i has rank R_i, then $H_N(X_i) = R_i/N$ and thus $J_N(H_N(X_i)) = J_N(R_i/N) = a_N(R_i)$.

The following regularity conditions are assumed for the scores $a_N(i)$, with respect to some nonconstant function h:

(1) $$\lim_{N\to\infty} a_N(1 + [uN]) = h(u), \qquad 0 < u < 1;$$

(2) $$\frac{1}{\sqrt{N}} \sum_{\substack{i=1 \\ R_i < N}}^{N} \left[a_N(R_i) - h\left(\frac{R_i}{N}\right)\right] \xrightarrow{p} 0;$$

(3) $$\lim_{N\to\infty} \frac{1}{\sqrt{N}} a_N(N) = 0;$$

(4)

$$\left|\frac{d^k h(t)}{dt^k}\right| \leq K[t(1-t)]^{-k-1/2+\delta} (k = 0, 1, \ldots) \qquad \text{for some} \quad \delta > 0,\ K < \infty.$$

Theorem (Chernoff and Savage). *Let* m, n → ∞ *such that* λ_N *is bounded away from* 0 *and* 1. *Assume conditions* (1)–(4). *Then*

$$S_{mn} \quad \textit{is} \quad AN(\mu_{mn}, \sigma^2_{mn}),$$

where

$$\mu_{mn} = m \int_{-\infty}^{\infty} h(H(x))dF(x)$$

and

$$\sigma^2_{mn} = \frac{mn}{N}\{(1 - \lambda_N)\text{Var}[B(X_1)] + \lambda_N \text{Var}[B^*(X_N)]\},$$

with

$$B(x) = \int_{x_0}^{x} h'(H(y))dG(y)$$

and

$$B^*(x) = \int_{x_0}^{x} h'(H(y))dF(y),$$

provided that $\sigma^2_{mn} > 0$. *Further, the asymptotic normality holds uniformly in* (F, G) *satisfying*

$$\inf_{(F,G)} \text{Var}[B(X_1)] > 0, \qquad \inf_{(F,G)} \text{Var}[B^*(X_N)] > 0.$$

For proof, see Chernoff and Savage (1958) or, for a somewhat more straightforward development utilizing stochastic process methods, see Pyke and Shorack (1968). For related results and extensions, see Hájek and Šidák (1967), pp. 233–237, Hájek (1968), Hoeffding (1973), and Lai (1975). In Hájek (1968) the method of projection is used, and a much broader class of regression constants is considered. (In **9.2.6** we follow up Hájek's treatment with corresponding Berry–Esséen rates.)

9.2.4 Continuation: The LeCam and Hájek Approach

This line of development was originated independently by Le Cam (1960) and Hájek (1962). As regards $F_1, \ldots, F_N$, this approach is intermediate between the two previously considered ones. It is assumed that the set of distributions $F_1, \ldots, F_N$ is "local" to the (composite) null hypothesis $F_1 = \cdots = F_N$ in a certain special sense called *contiguous*. However, the c_i's are allowed to satisfy merely the weakest restrictions on

$$\frac{\max_{1 \le i \le N} (c_i - \bar{c})^2}{1/N \sum_{i=1}^{N} (c_i - \bar{c})^2},$$

and the function h is allowed to be merely square integrable. For introductory discussion, see Hájek and Šidák (1967), pp. 201–210.

9.2.5 The Method of Projection

Here we introduce in general form the technique used in **5.3.1** with U-statistics and in **8.2.2** with L-estimates. Although the method goes back to Hoeffding (1948), its recent popularization is due to Hájek (1968), who gives the following result (Problem 9.P.2).

Lemma (Hájek). *Let* $\mathrm{Z}_1, \ldots, \mathrm{Z}_n$ *be indpendent random variables and* $\mathrm{S} = \mathrm{S}(\mathrm{Z}_1, \ldots, \mathrm{Z}_n)$ *any statistic satisfying* $\mathrm{E}(\mathrm{S}^2) < \infty$. *Then the random variable*

$$\hat{\mathrm{S}} = \sum_{i=1}^{n} \mathrm{E}(\mathrm{S} \mid \mathrm{Z}_i) - (n - 1)\mathrm{E}(\mathrm{S})$$

satisfies

$$\mathrm{E}(\hat{\mathrm{S}}) = \mathrm{E}(\mathrm{S})$$

and

$$\mathrm{E}(\hat{\mathrm{S}} - \mathrm{S})^2 = \mathrm{Var}(\mathrm{S}) - \mathrm{Var}(\hat{\mathrm{S}}).$$

The random variable $\hat{S}$ is called the *projection* of S on $X_1, \ldots, X_n$. Note that it is conveniently a sum of *independent* random variables. In cases that $E(S - \hat{S})^2 \to 0$ at a suitable rate as $n \to \infty$, the asymptotic normality of S may be established by applying classical theory to $\hat{S}$. For example, Hájek (1968) uses this approach in treating simple linear rank statistics.

It is also possible to apply the technique to project a statistic onto *dependent* random variables. For example, Hájek and Šidák (1967), p. 59, associate with an *arbitrary* rank statistic T a *linear* rank statistic

$$\hat{T} = \frac{N - 1}{N} \sum_{i=1}^{N} \hat{a}(i, R_i) - (n - 2)E(T),$$

where

$$\hat{a}(i, j) = E(T \mid R_i = j), \qquad 1 \le i, j \le N.$$

This random variable is shown to be the projection of T upon the family of linear rank statistics. In this fashion, Hájek and Šidák derive properties of the rank correlation measure known as Kendall's tau,

$$\tau = \frac{1}{N(N - 1)} \sum_{1 \neq j}^{N} \mathrm{sign}(i - j)\mathrm{sign}(R_i - R_j),$$

which is a nonlinear rank statistic, by considering the *linear* rank statistic

$$\hat{\tau} = \frac{8}{N^2(N - 1)} \sum_{i=1}^{N} \left(i - \frac{N + 1}{2}\right)\left(R_i - \frac{N + 1}{2}\right)$$

and showing that $\mathrm{Var}(\hat{\tau})/\mathrm{Var}(\tau) \to 1$. (Note that, up to a multiplication constant, $\hat{\tau}$ is the rank correlation coefficient known as Spearman's rho.)

9.2.6 Berry–Esséen Rates for Simple Linear Rank Statistics

The rate of convergence $O(N^{-1/2+\delta})$ for any $\delta > 0$ is established for two theorems of Hájek (1968) on asymptotic normality of simple linear rank statistics. These pertain to smooth and bounded scores, arbitrary regression constants, and broad conditions on the distributions of individual observations. The results parallel those of Bergström and Puri (1977). Whereas Bergström and Puri provide explicit constants of proportionality in the $O(\cdot)$ terms, the present development is in closer touch with Hájek (1968), provides some alternative arguments of proof, and provides explicit application to relax the conditions of a theorem of Jurečková and Puri (1975) giving the above rate for the case of location-shift alternatives.

Generalizing the line of development of Chernoff and Savage (see **9.2.3**), Hájek (1968) established the asymptotic normality of simple linear rank statistics under broad conditions. Corresponding to his asymptotic normality theorems for the case of *smooth and bounded scores*, rates of convergence are obtained in Theorems B and C below. The method of proof consists in approximating the simple linear rank statistic by a sum of independent random variables and establishing, for arbitrary ν, a suitable bound on the νth moment of the error of approximation (Theorem A).

Let $X_{N1}, \ldots, X_{NN}$ be independent random variables with ranks $R_{N1}, \ldots, R_{NN}$. The simple linear rank statistic to be considered is

$$S_N = \sum_{i=1}^{N} c_{Ni} a_N(R_{Ni}),$$

where $c_{N1}, \ldots, c_{NN}$ are arbitrary "regression constants" and $a_N(1), \ldots, a_N(N)$ are "scores." Throughout, the following condition will be assumed.

Condition A. (i) The scores are generated by a function $\phi(t)$, $0 < t < 1$, in either of the following ways:

$$\text{(A1)} \qquad a_N(i) = \phi\left(\frac{i}{N+1}\right), \qquad 1 \le i \le N,$$

$$\text{(A2)} \qquad a_N(i) = E\phi(U_N^{(i)}), \qquad 1 \le i \le N,$$

where $U_N^{(i)}$ denotes the ith order statistic in a sample of size N from the uniform distribution on (0, 1).

(ii) ϕ has a bounded second derivative.

(iii) The regression constants satisfy

$$\text{(A3)} \qquad \sum_{i=1}^{N} c_{Ni} = 0, \qquad \sum_{i=1}^{N} c_{Ni}^2 = 1,$$

$$\text{(A4)} \qquad \max_{1 \le i \le N} c_{Ni}^2 = O(N^{-1} \log N), \qquad N \to \infty. \quad \blacksquare$$

Note that (A3) may be assumed without loss of generality.

The X_{Ni}'s are assumed to have continuous distribution functions F_{Ni}, $1 \le i \le N$. Put $H_N(x) = N^{-1} \sum_{i=1}^{N} F_{Ni}(x)$. The derivatives of ϕ will be denoted by ϕ', ϕ'', etc. Also, put $\mu_\phi = \int_0^1 \phi(t)dt$ and $\sigma_\phi^2 = \int_0^1 [\phi(t) - \mu_\phi]^2\,dt$. As usual, denote by Φ the standard normal cdf. Hereafter the suffix N will be omitted from X_{Ni}, R_{Ni}, c_{Ni}, S_N, F_{Ni}, H_N and other notation.

The statistic S will be approximated by the same sum of independent random variables introduced by Hájek (1968), namely

$$T = \sum_{i=1}^{N} l_i(X_i),$$

where

$$l_i(x) = N^{-1} \sum_{j=1}^{N} (c_j - c_i) \int [u(y - x) - F_i(y)]\phi'(H(y))dF_j(y),$$

with

$$u(x) = 1, \qquad x \ge 0; \qquad u(x) = 0, \qquad x < 0.$$

Theorem A. *Assume Condition A. Then, for every integer* r, *there exists a constant* M = M(ϕ, r) *such that*

$$E(S - ES - T)^{2r} \le MN^{-r}, \qquad \textit{all } N.$$

The case $r = 1$ was proved by Hájek (1968). The extension to higher order is needed for the present purposes.

Theorem B. *Assume Condition A.* (i) *If* Var S > B > 0, N → ∞, *then for every* δ > 0.

$$\sup_x |P(S - ES < x(\mathrm{Var}\, S)^{1/2}) - \Phi(x)| = O(N^{-1/2+\delta}), \qquad N \to \infty.$$

(ii) *The assertion remains true with* Var S *replaced by* Var T.
(iii) *Both assertions remain true with* ES *replaced by*

$$\mu = \sum_{i=1}^{N} c_i \int \phi(H(x))dF_i(x).$$

Compare Theorem 2.1 of Hájek (1968) and Theorem 1.2 of Bergström and Puri (1977).

Theorem C. *Assume Condition A and that*

$$\sup_{i,j,x}|F_i(x) - F_j(x)| = O(N^{-1/2}\log N), \qquad N \to \infty.$$

Then for every $\delta > 0$

$$\sup_x |P(S - ES < x\sigma_\Phi) - \Phi(x)| = O(N^{-1/2+\delta}), \qquad N \to \infty.$$

The assertion remains true with σ_Φ^2 *replaced by either* Var S *or* Var T, *and/or* ES *replaced by* μ.

Compare Theorem 2.2 of Hájek (1968). As a corollary of Theorem C, the case of local location-shift alternatives will be treated. The following condition will be assumed.

Condition B. (i) The cdf's F_i are generated by a cdf F as follows: $F_i(x) = F(x - \Delta d_i)$, $1 \leq i \leq N$, with $\Delta \neq 0$.

(ii) F has a density f with bounded derivative f'.

(iii) The shift coefficients satisfy

$$\text{(B1)} \qquad \sum_{i=1}^N d_i = 0, \qquad \sum_{i=1}^N d_i^2 = 1,$$

$$\text{(B2)} \qquad \max_{1\leq i\leq N} d_i^2 = O(N^{-1}\log N), \qquad N \to \infty. \quad \blacksquare$$

Note that (B1) may be assumed without loss of generality.

Corollary. *Assume Conditions A and B and that*

$$\text{(C)} \qquad \sum_{i=1}^N c_i^2 d_i^2 = O(N^{-1}\log N), \qquad N \to \infty.$$

Then for every $\delta > 0$

$$\sup_x |P(S - \tilde{\mu} < x\sigma_\Phi) - \Phi(x)| = O(N^{-1/2+\delta}), \qquad N \to \infty.$$

where

$$\tilde{\mu} = \Delta\left(\sum_{i=1}^N c_i d_i\right)\int \phi'(F(x))f^2(x)dx.$$

(The corresponding result of Jurečková and Puri (1975) requires ϕ to have four bounded derivatives and requires further conditions on the c_i's and d_i's. On the other hand, their result for the case of all F_i's identical requires only a single bounded derivative for ϕ.)

In proving these results, the main development will be carried out for the case of scores given by (A1). In Lemma G it will be shown that the case of scores given by (A2) may be reduced to this case.

Assuming ϕ'' bounded, put $K_1 = \sup_{0<t<1} |\phi'(t)|$ and $K_2 = \sup_{0<t<1} |\phi''(t)|$. By Taylor expansion the statistic S may be written as

$$S = U + V + W,$$

where, with $\rho_i = R_i/(N+1)$, $1 \le i \le N$,

$$U = \sum_{i=1}^{N} c_i \phi(E(\rho_i | X_i)),$$

$$V = \sum_{i=1}^{N} c_i \phi'(E(\rho_i | X_i))[\rho_i - E(\rho_i | X_i)]$$

and

$$W = \sum_{i=1}^{N} c_i K_2 \xi_i [\rho_i - E(\rho_i | X_i)]^2,$$

the random variables ξ_i satisfying $|\xi_i| \le 1$, $1 \le i \le N$. It will first be shown that W may be neglected. To see this, note that, with $u(\cdot)$ as above,

$$R_i = \sum_{j=1}^{N} u(X_i - X_j), \qquad 1 \le i \le N.$$

Thus

$$E(\rho_i | X_i) = \left[\sum_{\substack{j=1 \\ j \ne i}}^{N} F_j(X_i) + 1\right] \Big/ (N+1)$$

and

$$\rho_i - E(\rho_i | X_i) = \frac{1}{N+1} \sum_{\substack{j=1 \\ j \ne i}}^{N} [u(X_i - X_j) - F_j(X_i)]. \tag{1}$$

Observe that, given X_i, the summands in (1) are conditionally independent random variables centered at means. Hence the following classical result, due to Marcinkiewicz and Zygmund (1937), is applicable. (Note that it contains Lemma 2.2.2B, which we have used several times in previous chapters.)

Lemma A. *Let* $Y_1, Y_2, \ldots$ *be independent random variables with mean* 0. *Let* ν *be an even integer. Then*

$$E\left|\sum_{i=1}^{N} Y_i\right|^{\nu} \le A_\nu n^{(1/2)\nu - 1} \sum_{i=1}^{n} E|Y_i|^{\nu},$$

where A_ν *is a universal constant depending only on* ν.

Lemma B. *Assume (A3). For each positive integer* r,

(2) $$EW^{2r} \le K_2^{2r} A_{4r} N^{-r}, \qquad \textit{all } N.$$

PROOF. Write W in the form $W = K_2 \sum_{i=1}^{N} c_i W_i$. Apply the Cauchy–Schwarz inequality, (A3), Minkowski's inequality, and Lemma A to obtain

$$EW_i^{2r} \le E[\rho_i - E(\rho_i | X_i)]^{4r} \le A_{4r} N^{-2r}.$$

Thus (2) follows. ■

Thus S may be replaced by $Z = U + V$, in the sense that $E(S - Z)^{2r} = O(N^{-r})$, $N \to \infty$, each r. It will next be shown that, in turn, Z may be replaced in the same sense by a sum of independent random variables, namely by its *projection*

$$\hat{Z} = \sum_{i=1}^{N} E(Z | X_i) - (N - 1)E(Z).$$

Clearly, $\hat{Z} = \hat{U} + \hat{V}$ and $\hat{U} = U$. Thus $Z - \hat{Z} = V - \hat{V}$.

Lemma C. *The projection of* V *is*

$$V = \frac{1}{N+1} \sum_{i=1}^{N} \sum_{\substack{j=1 \\ j \ne i}}^{N} c_j l_{ji}(X_i),$$

where

(4) $$l_{ji}(x) = \int [u(y - x) - F_i(y)] \phi'(E(\rho_j | X_j = y)) dF_j(y).$$

PROOF. Put

$$Y_{ij} = \phi'(E(\rho_i | X_i))[u(X_i - X_j) - F_j(X_i)].$$

Show that the projection of Y_{ij}, for $i \ne j$, is $\hat{Y}_{ij} = l_{ij}(X_j)$. Since, by (1) and the definition of V,

(5) $$V = \frac{1}{N+1} \sum_{i=1}^{N} \sum_{\substack{j=1 \\ j \ne i}}^{N} c_j Y_{ji},$$

the projection $\hat{V}$ is thus given by (3). ■

Lemma D. *Assume* (A3). *For each positive integer* r, *there exists a constant* B_r *such that*

(6) $$E(V - \hat{V})^{2r} \le K_1^{2r} B_r N^{-r}, \qquad \textit{all } N.$$

PROOF. By (3) and (5),

$$E(V - V)^{2r} = (N+1)^{-2r} \sum_{i_1=1}^{N} \cdots \sum_{i_{2r}=1}^{N} c_{i_1} \cdots c_{i_{2r}} \sum_{\substack{j_1=1 \\ j_1 \neq i_1}}^{N} \cdots \sum_{\substack{j_{2r}=1 \\ j_{2r} \neq i_{2r}}}^{N} \delta_{i_1, j_1, \ldots, i_{2r}, j_{2r}}, \tag{7}$$

where

$$\delta_{i_1, j_1, \ldots, j_{2r}} = E \prod_{k=1}^{2r} [Y_{i_k j_k} - l_{i_k j_k}(X_{j_k})]. \tag{8}$$

Consider a typical term of the form (8). Argue that the expectation in (8) is possibly nonzero only if each factor has both indices repeated in other factors. Among such cases, consider now only those terms corresponding to a given pattern of the possible identities $i_a = i_b$, $i_a = j_b$, $j_a = j_b$ for $1 \leq a \leq 2r$, $1 \leq b \leq 2r$. For example, for $r = 3$, one such specific pattern is: $i_2 = i_1$, $i_3 \neq i_1$, $i_4 = i_1$, $i_5 = i_3$, $i_6 \neq i_1$, $i_6 \neq i_3$, $j_2 = j_1$, $j_3 = j_1$, $j_4 \neq j_1$, $j_5 = j_4$, $j_6 = j_4, j_1 = i_3, j_4 \neq i_1$. In general, there are at most 2^{6r} such patterns. For such a pattern, let q denote the number of distinct values among $i_1, \ldots, i_{2r}$ and p the number of distinct values among $j_1, \ldots, j_{2r}$. Let p_1 denote the number of distinct values among $j_1, \ldots, j_{2r}$ *not* appearing among $i_1, \ldots, i_{2r}$ and put $p_2 = p - p_1$. Within the given constraints, and after selection of $i_1, \ldots, i_{2r}$, the number of choices for $j_1, \ldots, j_{2r}$ clearly is of order $O(N^{p_1})$. Now clearly $2p_1 \leq 2r - p_2$, i.e., $p_1 \leq r - \frac{1}{2}p_2$. Now let q_1 denote the number of $i_1, \ldots, i_{2r}$ used only *once* among $i_1, \ldots, i_{2r}$. Then obviously $q_i \leq p_2$. It is thus seen that the contribution to (7) from summation over $j_1, \ldots, j_{2r}$ is of order at most $O(N^{r-(1/2)q_1})$, since the quantity in (8) is of magnitude $\leq K_1^{2r}$. It follows that

$$E(V - \hat{V})^{2r} \leq (N+1)^{-2r} K_2^{2r} [O(N^{r-(1/2)q_1})] \sum_{l_1=1}^{N} \cdots \sum_{l_q=1}^{N} |c_{l_1}^{a_1} \cdots c_{l_q}^{a_q}|,$$

where $a_1, \ldots, a_q$ are integers satisfying $a_i \geq 1$, $a_1 + \cdots + a_q = 2r$, and exactly q_1 of the a_i's are equal to 1. Now, for $a \geq 2$,

$$\sum_{i=1}^{N} |c_i|^a \leq \left(\sum_{i=1}^{N} c_i^2 \right)^{(1/2)a} = 1,$$

by (A3). Further,

$$\sum_{i=1}^{N} |c_i| \leq N \left(\frac{1}{N} \sum_{i=1}^{N} c_i^2 \right)^{1/2} = N^{1/2}. \tag{9}$$

Thus

$$\sum_{l_1=1}^{N} \cdots \sum_{l_q=1}^{N} |c_{l_1}^{a_1} \cdots c_{l_q}^{a_q}| \leq N^{(1/2)q_1},$$

and we obtain (6). ■

Next it is shown that $\hat{Z}$ may be replaced by $\tilde{Z} = \tilde{U} + \tilde{V}$, where

$$\tilde{U} = \sum_{i=1}^{N} c_i \phi(H(X_i))$$

and

$$\tilde{V} = \frac{1}{N} \sum_{i=1}^{N} \sum_{\substack{j=1 \\ j \neq i}}^{N} c_j \tilde{l}_{ji}(X_i),$$

with

$$\tilde{l}_{ji}(x) = \int [u(y - x) - F_i(y)] \phi'(H(y)) dF_j(y).$$

Lemma E. *Assume* (A3). *Then* $|\hat{Z} - \tilde{Z}| \leq (K_2 + 3K_1)N^{-1/2}$.

PROOF. Check that

$$E(\rho_i | X_i) = H(X_i) + \frac{1 - F_i(X_i) - H(X_i)}{N + 1}.$$

And hence

$$|\phi(E(\rho_i | X_i)) - \phi(H(X_i))| \leq K_1 N^{-1}.$$

Therefore, by (9), $|U - \tilde{U}| \leq K_1 N^{-1/2}$, Now

$$\hat{V} - \tilde{V} = \sum_{i=1}^{N} \sum_{j=1}^{N} \frac{1}{N} c_j [l_{ji}(X_i) - \tilde{l}_{ji}(X_i)] - \frac{1}{N} \hat{V} - \frac{1}{N} \sum_{i=1}^{N} c_i l_{ii}(X_i).$$

But $|\hat{V}| \leq K_1 \sum_{i=1}^{N} |c_i| \leq K_1 N^{1/2}$, i.e., $N^{-1}|\hat{V}| \leq K_1 N^{-1/2}$. Similarly, $|N^{-1} \sum_{i=1}^{N} c_i l_{ii}(X_i)| \leq K_1 N^{-1/2}$. Finally, $|l_{ji}(X_i) - \tilde{l}_{ji}(X_i)| \leq K_2 N^{-1}$ so that

$$\left| \sum_{i=1}^{N} \sum_{j=1}^{N} \frac{1}{N} c_j [l_{ji}(X_i) - \tilde{l}_{ji}(X_i)] \right| \leq K_2 N^{-1/2}. \quad ■$$

Now we connect with the random variable T of Theorem A.

Lemma F. *We have* $\tilde{Z} - \mu = T$ *and there exists a constant* $K_4 = K_4(\phi)$ *such that* $|ES - \mu| \leq K_4 N^{-1/2}$.

PROOF. The second assertion is shown by Hájek (1968), p. 340. To obtain the first, check that

$$\tilde{Z} - \mu - T = \sum_{i=1}^{N} c_i\{\phi(H(X_i)) - E\phi(H(X_i)) + \int [u(x - X_i) - F_i(x)]\phi'(H(x))dH(x)\}.$$

Now, by integration by parts, for any distribution function G we have

$$\int \phi'(H(x))G(x)dH(x) = -\int \phi(H(x))dG(x) + \text{constant},$$

where the constant may depend on ϕ and $H(\cdot)$ but not on $G(\cdot)$. Thus the above sum reduces to 0. ■

Up to this point, only the scores given by (A1) have been considered. The next result provides the basis for interchanging with the scores given by (A2).

Lemma G. *Denote* $\sum_{i=1}^{N} c_i a_N(R_i)$ *by* S *in the case corresponding to* (A1) *and by* S′ *in the case corresponding to* (A2). *Assume* (A3). *Then there exists* $K_5 = K_5(\phi)$ *such that*

$$|S - ES - (S' - ES')| \le K_5 N^{-1/2}.$$

PROOF. It is easily found (see Hájek (1968), p. 341) that

$$\left|\phi\left(\frac{i}{N+1}\right) - E\phi(U_N^{(i)})\right| \le K_0 N^{-1},$$

where K_0 does not depend on i or N. Thus, by (9), $|S - S'| \le K_0 N^{-1/2}$ and hence also $|ES - ES'| \le K_0 N^{-1/2}$. Thus the desired assertion follows with $K_5 = 2K_0$. ■

PROOF OF THEOREM A. Consider first the case (A1). By Minkowski's inequality,

$$[E(S - ES - T)^{2r}]^{1/2r} \le [E(S - Z)^{2r}]^{1/2r} + [E(Z - \hat{Z})^{2r}]^{1/2r} + [E(\hat{Z} - \tilde{Z})^{2r}]^{1/2r} + [E(\tilde{Z} - \mu - T)^{2r}]^{1/2r} + |ES - \mu|. \tag{10}$$

By Lemmas B, D, E and F, each term on the right-hand side of (10) may be bounded by $KN^{-1/2}$ for a constant $K = K(\phi, r)$ depending only on ϕ and r. Thus follows the assertion of the theorem. In the case of scores given by (A2), we combine Lemma G with the preceding argument. ■

PROOF OF THEOREM B. First assertion (i) will be proved. Put

$$\alpha_N = \sup_x |P(S - ES < x(\text{Var } S)^{1/2}) - \Phi(x)|,$$

$$\beta_N = \sup_x |P(T < x(\text{Var } S)^{1/2}) - \Phi(x)|,$$

and

$$\gamma_N = \sup_x |P(T < x(\text{Var } T)^{1/2}) - \Phi(x)|.$$

By Lemma 6.4.3, if

$$\beta_N = O(a_N), \qquad N \to \infty, \tag{11}$$

for a sequence of constants $\{a_N\}$, then

$$\alpha_N = O(a_N) + P(|S - ES - T|/(\text{Var } S)^{1/2} > a_N), \qquad N \to \infty. \tag{12}$$

We shall obtain a condition of form (11) by first considering γ_N. By the classical Berry–Esséen theorem (1.9.5),

$$\gamma_N \le C(\text{Var } T)^{-3/2} \sum_{i=1}^{N} E|l_i(X_i)|^3,$$

where C is a universal constant. Clearly,

$$|l_i(X_i)| \le K_1 N^{-1} \sum_{j=1}^{N} |c_j - c_i|.$$

Now

$$\left(\sum_{j=1}^{N} |c_j - c_i|\right)^2 \le N\left[\sum_{j=1}^{N} (c_j - c_i)^2\right] = N[1 + Nc_i^2].$$

By the elementary inequality (Loève (1977), p. 157)

$$|x + y|^m \le \theta_m |x|^m + \theta_m |y|^m,$$

where $m > 0$ and $\theta_m = 1$ or 2^{m-1} according as $m \le 1$ or $m \ge 1$, we thus have

$$\left(\sum_{j=1}^{N} |c_j - c_i|\right)^3 \le N^{3/2} 2^{1/2} (1 + N^{3/2} |c_i|^3)$$

and hence

$$\sum_{i=1}^{N} E|l_i(X_i)|^3 \le 2^{1/2} K_1^3 \left[N^{-1/2} + \sum_{i=1}^{N} |c_i|^3\right]. \tag{13}$$

Check (Problem 9.P.8) that

$$|(\text{Var } S)^{1/2} - (\text{Var } T)^{1/2}| \le (\text{Var}\{S - T\})^{1/2},$$

so that by Theorem A we have

$$|(\text{Var } S)^{1/2} - (\text{Var } T)^{1/2}| \leq M_0 N^{-1/2}, \tag{14}$$

where the constant M_0 depends only on ϕ. It follows that if Var S is bounded away from 0, then the same holds for Var T, and conversely. Consequently, by the hypothesis of the theorem, and by (12), (13) and (14), we have

$$\gamma_N = O(N^{-1/2}) + O\left(\sum_{i=1}^{N} |c_i|^3\right), \qquad N \to \infty.$$

Therefore, by (A3) and (A4), $\gamma_N = O(N^{-1/2} \log N)$, $N \to \infty$. Now it is easily seen that

$$\beta_N \leq \gamma_N + O\left(\left|\frac{(\text{Var } S)^{1/2}}{(\text{Var } T)^{1/2}} - 1\right|\right).$$

By (14) the right-most term is $O(N^{-1/2})$. Hence $\beta_N = O(N^{-1/2} \log N)$. Therefore, for any sequence of constants a_N satisfying $N^{-1/2} \log N = O(a_N)$, we have (11) and thus (12). A further application of Theorem A, with Markov's inequality, yields for arbitrary r

$$P\left\{\frac{|S - ES - T|}{(\text{Var } S)^{1/2}} > a_N\right\} \leq a_N^{-2r}(\text{Var } S)^{-r} M N^{-r}.$$

Hence (12) becomes

$$\alpha_N = O(a_N) + O(a_N^{-2r} N^{-r}).$$

Choosing $a_N = O(N^{-r/(2r+1)})$, we obtain

$$\alpha_N = O(N^{-r/(2r+1)}), \qquad N \to \infty.$$

Since this holds for arbitrarily large r, the first assertion of Theorem B is established.

Assertions (ii) and (iii) are obtained easily from the foregoing arguments. ■

PROOF OF THEOREM C. It is shown by Hájek (1968), p. 342, that

$$|(\text{Var } T)^{1/2} - \sigma_\phi| \leq 2^{1/2}(K_1 + K_2) \sup_{i, j, x} |F_i(x) - F_j(x)|.$$

The proof is now straightforward using the arguments of the preceding proof. ■

PROOF OF THE COROLLARY. By Taylor expansion,

$$|F_i(x) - F(x) - (-\Delta\, d_i f(x))| \leq A\Delta^2\, d_i^2, \tag{15}$$

where A is a constant depending only on F. Hence, by (B1) and (B2),

$$\sup_{i,j,x} |F_i(x) - F_j(x)| = O\left(\max_i |d_i|\right) = O(N^{-1/2} \log N),$$

so that the hypothesis of Theorem C is satisfied. It remains to show that ES may be replaced by the more convenient parameter $\tilde{\mu}$. A further application of (15), with (B1), yields $|H(x) - F(x)| \le A\Delta^2 N^{-1}$, so that $|\phi(H(x)) - \phi(F(x))| \le K_1 A\Delta^2 N^{-1}$. Hence, by (9).

$$\left| \mu - \sum_{i=1}^{N} c_i \int \phi(F(x)) dF_i(x) \right| \le K_1 A \Delta^2 N^{-1/2}.$$

By integration by parts, along with (A3) and (15),

$$\begin{aligned}\sum_{i=1}^{N} c_i \int \phi(F(x)) dF_i(x) &= -\sum_{i=1}^{N} c_i \int F_i(x) \phi'(F(x)) dF(x) \\ &= -\sum_{i=1}^{N} c_i \int (-\Delta\, d_i) f(x) \phi'(F(x)) dF(x) \\ &\quad + \eta A \Delta^2 \sum_{i=1}^{N} c_i\, d_i^2 \\ &= \tilde{\mu} + \eta A \Delta^2 \sum_{i=1}^{N} c_i\, d_i^2,\end{aligned}$$

where $|\eta| \le 1$. Now, by (B1), $\sum_i |c_i| d_i^2 \le (\sum_i c_i^2\, d_i^2)^{1/2}$. Therefore, by (C) and the above steps,

$$|\mu - \tilde{\mu}| = O(N^{-1/2} \log N), \qquad N \to \infty.$$

Thus μ may be replaced by $\tilde{\mu}$ in Theorem C. ■

9.3 COMPLEMENTS

(i) **Deviation theory for linear rank statistics.** Consider a linear rank statistic T_N which is $AN(\mu_N, \sigma_N^2)$. In various efficiency applications (as we will study in Chapter 10), it is of interest to approximate a probability of the type

$$P_N(x_N) = P\left(\frac{T_N - \mu_N}{\sigma_N} \ge x_N\right)$$

for $x_N \to \infty$. For application to the computation of Bahadur efficiencies, the case $x_N \sim cN^{1/2}$ is treated by Woodworth (1970). For application to the computation of Bayes risk efficiencies, the case $x_N \sim c(\log N)^{1/2}$ is treated by Clickner and Sethuraman (1971) and Clickner (1972).

(ii) **Connection with sampling of finite populations.** Note that a two-sample simple linear rank statistic may be regarded, *under the null-hypothesis* $F_1 = \cdots = F_N$, as the mean of a sample of size m drawn without replacement from the population $\{a_N(1), \ldots, a_N(N)\}$.

(iii) **Probability inequalities for two-sample linear rank statistics.** In view of (ii) just discussed, see Serfling (1974).

(iv) **Further general reading.** See Savage (1969).

(v) **Connections between *M*-, *L*-, and *R*-statistics.** See Jaeckel (1971) for initial discussion of these interconnections. One such relation, as discussed in Huber (1972, 1977), is as follows. For estimation of the location parameter θ of a location family based on a distribution F with density f, by an estimate $\hat{\theta}$ given as a statistical function $T(F_n)$, where F_n is the sample distribution function, we have: an M-estimate of $T(G)$ is defined by solving

$$\int \psi(x - T(G))dG(x) = 0;$$

an L-estimate of $T(G)$ is defined by

$$T(G) = \int J(t)F^{-1}(t)dt;$$

an R-estimate of $T(G)$ is defined by solving

$$\int J\left(\frac{G(x) + 1 - G(2T(G) - x)}{2}\right)dG(x) = 0.$$

It turns out that the M-estimate for $\psi_0 = -f'/f$, the L-estimate for $J(t) = \psi_0'(F^{-1}(t))/I_F$, where $I_F = \int [f'/f]^2 \, dF$, and the R-estimate for $J(t) = \psi_0(F^{-1}(t))$, are all asymptotically equivalent in distribution and, moreover, *asymptotically efficient*. For general comparison of M-, L- and R-estimates, see Bickel and Lehmann (1975).

9.P PROBLEMS

Section 9.2

1. Complete the details for Example 9.2.2.
2. Prove Lemma 9.2.5 (Projection Lemma).
3. Complete details for proof of Lemma 9.2.6B.
4. Details for proof of Lemma 9.2.6C.
5. Details for proof of Lemma 9.2.6D.
6. Details for proof of Lemma 9.2.6E.
7. Details for proof of Lemma 9.2.6F.

8. Provide the step required in the proof of Theorem 9.2.6B. That is, show that for any random variables S and T,

$$|(\text{Var } S)^{1/2} - (\text{Var } T)^{1/2}| \leq (\text{Var}\{S - T\})^{1/2}.$$

(Hint: Apply the property $|\text{Cov}(S, T)| \leq (\text{Var } S)^{1/2}(\text{Var } T)^{1/2}$.)

9. Let $\Pi_N = \{x_{N1}, \ldots, x_{NN}\}$, $N = 1, 2, \ldots$, be a sequence of finite populations such that Π_N has mean μ_N and variance σ_N^2. Let $\bar{X}_{N,n}$ denote the mean of a random sample of size n drawn without replacement from the population Π_N. State a central limit theorem for $\bar{X}_{N,n}$ as $n, N \to \infty$. (Hint: note Section 9.3 (ii).)

CHAPTER 10

Asymptotic Relative Efficiency

Here we consider a variety of approaches toward assessment of the relative efficiency of two test procedures in the case of large sample size. The various methods of comparison differ with respect to the manner in which the Type I and Type II error probabilities vary with increasing sample size, and also with respect to the manner in which the alternatives under consideration are required to behave. Section 10.1 provides a general discussion of six contributions, due to Pitman, Chernoff, Bahadur, Hodges and Lehman, Hoeffding, and Rubin and Sethuraman. Detailed examination of their work is provided in Sections 10.2–7, respectively. The roles of central limit theory, Berry–Esséen theorems, and general deviation theory will be viewed.

10.1 APPROACHES TOWARD COMPARISON OF TEST PROCEDURES

Let H_0 denote a null hypothesis to be tested. Typically, we may represent H_0 as a specified family $\mathscr{F}_0$ of distributions for the data. For any test procedure T, we shall denote by T_n the version based on a sample of size n. The function

$$\gamma_n(T, F) = P_F(T_n \text{ rejects } H_0),$$

defined for distribution functions F, is called the *power function* of T_n (or of T). For $F \in \mathscr{F}_0$, $\gamma_n(T, F)$ represents the probability of a *Type I* error. The quantity

$$\alpha_n(T, \mathscr{F}_0) = \sup_{F \in \mathscr{F}_0} \gamma_n(T, F)$$

is called the *size* of the test. For $F \notin \mathscr{F}_0$, the quantity

$$\beta_n(T, F) = 1 - \gamma_n(T, F)$$

represents the probability of a *Type II* error. Usually, attention is confined to *consistent* tests: for fixed $F \notin \mathscr{F}_0$, $\beta_n(T, F) \to 0$ as $n \to \infty$. Also, usually attention is confined to *unbiased* tests: for $F \notin \mathscr{F}_0$, $\gamma_n(T, F) \geq \alpha_n(T, \mathscr{F}_0)$.

A general way to compare two such test procedures is through their power functions. In this regard we shall use the concept of *asymptotic relative efficiency* (ARE) given in **1.15.4**. For two test procedures T_A and T_B, suppose that a performance criterion is tightened in such a way that the respective sample sizes n_1 and n_2 for T_A and T_B to perform "equivalently" tend to ∞ but have ratio n_1/n_2 tending to some limit. Then this limit represents the ARE of procedure T_B relative to procedure T_A and is denoted by $e(T_B, T_A)$.

We shall consider several performance criteria. Each entails specifications regarding

(a) $\alpha = \lim_n \alpha_n(T, \mathscr{F}_0)$,

(b) an alternative distribution $F^{(n)}$ allowed to depend on n,
and

(c) $\beta = \lim_n \beta_n(T, F^{(n)})$.

With respect to (a), the cases $\alpha = 0$ and $\alpha > 0$ are distinguished. With respect to (c), the cases $\beta = 0$ and $\beta > 0$ are distinguished. With respect to (b), the cases $F^{(n)} \equiv F$ (fixed), and $F^{(n)} \to \mathscr{F}_0$ in some sense, are distinguished.

The following table gives relevant details and notation regarding the methods we shall examine in Sections 10.2–7.

Names of Contributors	Behavior of Type I Error Probability α_n	Behavior of Type II Error Probability β_n	Behavior of Alternatives	Notation for ARE	Section
Pitman	$\alpha_n \to \alpha > 0$	$\beta_n \to \beta > 0$	$F^{(n)} \to \mathscr{F}_0$	$e_P(\cdot, \cdot)$	10.2
Chernoff	$\alpha_n \to 0$	$\beta_n \to 0$	$F^{(n)} = F$ fixed	$e_C(\cdot, \cdot)$	10.3
Bahadur	$\alpha_n \to 0$	$\beta_n \to \beta > 0$	$F^{(n)} = F$ fixed	$e_B(\cdot, \cdot)$	10.4
Hodges & Lehmann	$\alpha_n \to \alpha > 0$	$\beta_n \to 0$	$F^{(n)} = F$ fixed	$e_{HL}(\cdot, \cdot)$	10.5
Hoeffding	$\alpha_n \to 0$	$\beta_n \to 0$	$F^{(n)} = F$ fixed	$e_H(\cdot, \cdot)$	10.6
Rubin & Sethuraman	$\alpha_n \to 0$	$\beta_n \to 0$	$F^{(n)} \to \mathscr{F}_0$	$e_{RS}(\cdot, \cdot)$	10.7

Each of the approaches has its own special motivation and appeal, as we shall see. However, it should be noted also that each method, apart from intuitive and philosophical appeal, is in part motivated by the availability of convenient mathematical tools suitable for theoretical derivation of the relevant quantities $e(\cdot, \cdot)$. In the Pitman approach, the key tool is *central limit theory*. In the Rubin–Sethuraman approach, the theory of *moderate deviations* is used. In the other approaches, the theory of *large deviations* is employed. The technique of application of these ARE approaches in any actual statistical problem thus involves a trade-off between relevant intuitive considerations and relevant technical issues.

10.2 THE PITMAN APPROACH

The earliest approach to asymptotic relative efficiency was introduced by Pitman (1949). For exposition, see Noether (1955).

In this approach, two tests sequences $T = \{T_n\}$ and $U = \{U_n\}$ are compared as the Type I and Type II error probabilities tend to positive limits α and β, respectively. In order that $\alpha_n \to \alpha > 0$ and simultaneously $\beta_n \to \beta > 0$, it is necessary to consider $\beta_n(\cdot)$ evaluated at an alternative $F^{(n)}$ converging at a suitable rate to the null hypothesis $\mathscr{F}_0$. (Why?)

In justification of this approach, we might argue that large sample sizes would be relevant in practice only if the alternative of interest were close to the null hypothesis and thus hard to distinguish with only a small sample .On the other hand, a practical objection to the Pitman approach is that the measure of ARE obtained does not depend upon a particular alternative. In any case, the approach is very easily carried out, requiring mainly just a knowledge of the asymptotic distribution theory of the relevant test statistics. As we have seen in previous chapters, such theorems are readily available under mild restrictions. Thus the Pitman approach turns out to be widely applicable.

In **10.2.1** we develop the basic theorem on Pitman ARE and in **10.2.2** exemplify it for the problem of testing location. The relationships between Pitman ARE and the asymptotic correlation of test statistics is examined in **10.2.3**. Some complements are noted in **10.2.4**.

10.2.1 The Basic Theorem

Suppose that the distributions F under consideration may be indexed by a set $\Theta \subset R$, and consider a simple null hypothesis

$$H_0: \theta = \theta_0$$

to be tested against alternatives

$$\theta > \theta_0.$$

Consider the comparison of test sequences $T = \{T_n\}$ satisfying the following conditions, relative to a neighborhood $\theta_0 \le \theta \le \theta_0 + \delta$ of the null hypothesis.

Pitman Conditions

(P1) For some continuous strictly increasing distribution function G, and functions $\mu_n(\theta)$ and $\sigma_n(\theta)$, the F_θ-distribution of $(T_n - \mu_n(\theta))/\sigma_n(\theta)$ converges to G uniformly in $[\theta_0, \theta_0 + \delta]$:

$$\sup_{\theta_0 \le \theta \le \theta_0 + \delta} \sup_{-\infty < t < \infty} \left| P\left(\frac{T_n - \mu_n(\theta)}{\sigma_n(\theta)} \le t\right) - G(t) \right| \to 0, \qquad n \to \infty.$$

(P2) For $\theta \in [\theta_0, \theta_0 + \delta]$, $\mu_n(\theta)$ is k times differentiable, with $\mu_n^{(1)}(\theta_0) = \cdots = \mu_n^{(k-1)}(\theta_0) = 0 < \mu_n^{(k)}(\theta_0)$.

(P3) For some function $d(n) \to \infty$ and some constant $c > 0$,

$$\sigma_n(\theta_0) \sim c\, \frac{\mu_n^{(k)}(\theta_0)}{d(n)}, \qquad n \to \infty.$$

(P4) For $\theta_n = \theta_0 + O([d(n)]^{-1/k})$,

$$\mu_n^{(k)}(\theta_n) \sim \mu_n^{(k)}(\theta_0), \qquad n \to \infty.$$

(P5) For $\theta_n = \theta_0 + O([d(n)]^{-1/k})$,

$$\sigma_n(\theta_n) \sim \sigma_n(\theta_0), \qquad n \to \infty. \quad \blacksquare$$

Remarks A. (i) Note that the constant c in (P3) satisfies

$$c = \lim_n \frac{d(n)\sigma_n(\theta_0)}{\mu_n^{(k)}(\theta_0)}.$$

(ii) In typical cases, the test statistics under consideration will satisfy (P1)–(P5) with $G = \Phi$ in (P1), $k = 1$ in (P2), and $d(n) = n^{1/2}$ in (P3). In this case

$$c = \lim_n \frac{n^{1/2}\sigma_n(\theta_0)}{\mu_n'(\theta_0)}. \quad \blacksquare$$

Theorem (Pitman–Noether). (i) *Let* $T = \{T_n\}$ *satisfy* (P1)–(P5). *Consider testing* H_0 *by critical regions* $\{T_n > u_{\alpha_n}\}$ *with*

$$\alpha_n = P_{\theta_0}(T_n > u_{\alpha n}) \to \alpha, \tag{1}$$

where $0 < \alpha < 1$. *For* $0 < \beta < 1 - \alpha$, *and* $\theta_n = \theta_0 + O([d(n)]^{-1/k})$, *we have*

$$\beta_n(\theta_n) = P_{\theta_n}(T_n \le u_{\alpha n}) \to \beta \tag{2}$$

if and only if

$$\frac{(\theta_n - \theta)^k}{k!} \cdot \frac{d(n)}{c} \to G^{-1}(1 - \alpha) - G^{-1}(\beta). \tag{3}$$

(ii) *Let* $\mathrm{T_A} = \{\mathrm{T_{An}}\}$ *and* $\mathrm{T_B} = \{\mathrm{T_{Bn}}\}$ *each satisfy* (P1)–(P5) *with common* G, k *and* d(n) *in* (P1)–(P3). *Let* $\mathrm{d(n)} = \mathrm{n^q}$, $\mathrm{q} > 0$. *Then the Pitman ARE of* $\mathrm{T_A}$ *relative to* $\mathrm{T_B}$ *is given by*

$$e_P(T_A, T_B) = \left(\frac{c_B}{c_A}\right)^{1/q}.$$

PROOF. Check that, by (P1),

$$\left|\beta_n(\theta_n) - G\left(\frac{u_{\alpha n} - \mu_n(\theta_n)}{\sigma_n(\theta_n)}\right)\right| \to 0, \qquad n \to \infty.$$

Thus $\beta_n(\theta_n) \to \beta$ if and only if

$$\frac{u_{\alpha n} - \mu_n(\theta_n)}{\sigma_n(\theta_n)} \to G^{-1}(\beta). \tag{4}$$

Likewise (check), $\alpha_n \to \alpha$ if and only if

$$\frac{u_{\alpha n} - \mu_n(\theta_n)}{\sigma_n(\theta_0)} \to G^{-1}(1 - \alpha). \tag{5}$$

It follows (check, utilizing (P5)) that (4) and (5) together are equivalent to (5) and

$$\frac{\mu_n(\theta_n) - \mu_n(\theta_0)}{\sigma_n(\theta_0)} \to G^{-1}(1 - \alpha) - G^{-1}(\beta) \tag{6}$$

together. By (P2) and (P3),

$$\frac{\mu_n(\theta_n) - \mu_n(\theta_0)}{\sigma_n(\theta_0)} \sim \frac{\mu_n^{(k)}(\tilde{\theta}_n)}{\mu_n^{(k)}(\theta_0)} \cdot \frac{(\theta_n - \theta_0)^k}{k!} \cdot \frac{d(n)}{c}, \qquad n \to \infty,$$

where $\theta_0 \le \tilde{\theta}_n \le \theta_n$. Thus, by (P4), (6) is equivalent to (3). This completes the proof of (i).

Now consider tests based on T_A and T_B, having sizes $\alpha_{An} \to \alpha$ and $\alpha_{Bn} \to \alpha$. Let $0 < \beta < 1 - \alpha$. Let $\{\theta_n\}$ be a sequence of alternatives of the form

$$\theta_n = \theta_0 + A[d(n)]^{-1/k}.$$

It follows by (i) that if $h(n)$ is the sample size at which T_N performs "equivalently" to T_A with sample size n, that is, at which T_B and T_A have the same limiting power $1 - \beta$ for the given sequence of alternatives, so that

$$\beta_{An}(\theta_n) \to \beta, \; \beta_{Bh(n)}(\theta_n) \to \beta,$$

then we must have $d(h(n))$ proportional to $d(n)$ and

$$\frac{(\theta_n - \theta_0)^k}{k!} \frac{d(n)}{c_A} \sim \frac{(\theta_n - \theta_0)^k}{k!} \frac{d(h(n))}{c_B},$$

or

$$\frac{d(h(n))}{d(n)} \to \frac{c_B}{c_A}.$$

For $d(n) = n^q$, this yields $(h(n)/n)^q \to (c_B/c_A)$, proving (ii). ∎

Remarks B. (i) By Remarks A we see that for $d(n) = n^q$,

$$e_P(T_A, T_B) = \lim_n \left[\frac{\sigma_{Bn}(\theta_0)\mu_{An}^{(k)}(\theta_0)}{\sigma_{An}(\theta_0)\mu_{Bn}^{(k)}(\theta_0)}\right]^{1/q}.$$

For the typical case $k = 1$ and $d(n) = n^{1/2}$, we have

$$e_P(T_A, T_B) = \lim_n \left[\frac{\sigma_{Bn}(\theta_0)\mu'_{An}(\theta_0)}{\sigma_{An}(\theta_0)\mu'_{Bn}(\theta_0)}\right]^2.$$

(ii) For a test T satisfying the Pitman conditions, the limiting power against local alternatives is given by part (i) of the theorem: for

$$\theta_n = \theta_0 + A[d(n)]^{-1/k} + o([d(n)]^{-1/k}),$$

we have from (3) that

$$\frac{A^k}{k!\,c} = G^{-1}(1 - \alpha) - G^{-1}(\beta),$$

yielding as limiting power

$$1 - \beta = 1 - G\left[G^{-1}(1 - \alpha) - \frac{A^k}{k!\,c}\right].$$

In particular, for $G = \Phi$, $k = 1$, $d(n) = n^{1/2}$, we have simply

$$1 - \beta = 1 - \Phi\left(\Phi^{-1}(1 - \alpha) - \frac{A}{c}\right). \quad ∎$$

10.2.2 Example: Testing for Location

Let $X_1, \ldots, X_n$ be independent observations having distribution $F(x - \theta)$, for an unknown value $\theta \in R$, where F is a distribution having density f symmetric about 0 and continuous at 0, and having variance $\sigma_F^2 < \infty$. Consider testing

$$H_0: \theta = 0$$

versus alternatives

$$\theta > 0.$$

For several test statistics, we shall derive the Pitman ARE's as functions of F, and then consider particular cases of F.

The test statistics to be considered are the "mean statistic"

$$T_{1n} = \frac{1}{n}\sum_{i=1}^{n} X_i = \bar{X},$$

the "t-statistic"

$$T_{2n} = \frac{\bar{X}}{s}$$

(where $s^2 = (n-1)^{-1}\sum_1^n (X_i - \bar{X})^2$), the "sign test" statistic

$$T_{3n} = \frac{1}{n}\sum_{i=1}^{n} I(X_i > 0),$$

and the "Wilcoxon test" statistic

$$T_{4n} = \binom{n}{2}^{-1} \sum\sum_{1 \le i < j \le n} I(X_i + X_j > 0).$$

The statistics based on $\bar{X}$ have optimal features when F is a normal distribution. The sign test has optimal features for F double exponential and for a nonparametric formulation of the location problem (see Fraser (1957), p. 274, for discussion). The Wilcoxon statistic has optimal features in the case of F logistic.

We begin by showing that T_{1n} satisfies the Pitman conditions with $\mu_{1n}(\theta) = \theta$, $\sigma_{1n}^2(\theta) = \sigma_F^2/n$, $G = \Phi$, $k = 1$ and $d(n) = n^{1/2}$. Firstly (justify),

$$P_\theta\left(\frac{T_{1n} - \mu_{1n}(\theta)}{\sigma_{1n}(\theta)} \le t\right) = P_\theta\left(n^{1/2}\frac{\bar{X} - \theta}{\sigma_F} \le t\right) = P_0\left(n^{1/2}\frac{\bar{X}}{\sigma_F} \le t\right).$$

Also,

$$\sup_t \left| P_0\left[n^{1/2}\frac{\bar{X}}{\sigma_F} \le t\right] - \Phi(t)\right| \to 0, \qquad n \to \infty,$$

by now-familiar results. Thus (P1) is satisfied with $G = \Phi$. Also, $\mu'_{1n}(\theta) = 1$, so that (P2) holds with $k = 1$, and we see that (P3) holds with $c_1 = \sigma_F$ and $d(n) = n^{1/2}$. Finally, clearly (P4) and (P5) hold.

We next consider the statistic T_{2n}, and find that it satisfies the Pitman conditions with $G = \Phi$, $k = 1$, $d(n) = n^{1/2}$, and $c_2 = c_1 = \sigma_F$. We take $\mu_{2n}(\theta) = \theta/\sigma_F$ and $\sigma_{2n}^2(\theta) = 1/n$. Then

$$P_\theta\left[\frac{T_{2n} - \mu_{2n}(\theta)}{\sigma_{2n}(\theta)} \le t\right] = P_\theta\left[n^{1/2}\left(\frac{\bar{X}}{s} - \frac{\theta}{\sigma}\right) \le t\right]$$
$$= P_0\left[n^{1/2}\frac{\bar{X}}{s} \le t\right].$$

Further, by Problem 2.P.10,

$$\sup_t \left| P_0\left[n^{1/2}\frac{\overline{X}}{s} \le t\right] - \Phi(t)\right| \to 0, \qquad n \to \infty.$$

Thus (P1) is satisfied with $G = \Phi$. Also, $\mu'_{2n}(\theta) = 1/\sigma_F$ and we find easily that (P2)–(P4) are satisfied with $k = 1$, $d(n) = n^{1/2}$ and, in (P3), $c_2 = \sigma_F$.

At this point we may see from Theorem 10.2.1 that the mean statistic and the t-statistic are *equivalent* test statistics from the standpoint of Pitman ARE:

$$e_P(T_1, T_2) = 1.$$

Considering now T_{3n}, take

$$\mu_{3n}(\theta) = E_\theta T_{3n} = E_\theta I(X_1 > 0) = P_\theta(X_1 > 0) = 1 - F(-\theta) = F(\theta)$$

and

$$\sigma_{3n}^2(\theta) = \mathrm{Var}_\theta\, T_{3n} = \frac{1}{n}\mu_{3n}(\theta)[1 - \mu_{3n}(\theta)] = \frac{F(\theta)[1 - F(\theta)]}{n}.$$

Then

$$\frac{T_{3n} - \mu_{3n}(\theta)}{\sigma_{3n}(\theta)}$$

is a standardized binomial $(n, F(\theta))$ random variable. Since $F(\theta)$ lies in a neighborhood of $\frac{1}{2}$ for θ in a neighborhood of 0, it follows by an application of the Berry–Esséen Theorem (1.9.5), as in the proof of Theorem 2.3.3A, that (P1) holds with $G = \Phi$. Also, $\mu'_{3n}(\theta) = f(\theta)$ and it is readily found that conditions (P2)–(P5) hold with $k = 1$, $d(n) = n^{1/2}$ and $c_3 = 1/2f(0)$.

The treatment of T_{4n} is left as an exercise. By considering T_{4n} as a U-statistic, show that the Pitman conditions are satisfied with $G = \Phi$, $k = 1$, $d(n) = n^{1/2}$ and $c_4 = 1/(12)^{1/2} \int f^2(x)dx$.

Now denote by M the "mean" test T_1, by t the "t-test" T_2, by S the "sign" test T_3, and by W the "Wilcoxon" test T_4. It now follows from Theorem 10.2.1 that

$$e_P(M, t) = 1,$$

$$e_P(S, M) = 4\sigma_F^2 f^2(0),$$

and

$$e_P(W, M) = 12\sigma_F^2\left[\int f^2(x)dx\right]^2.$$

(Of course, $e_P(S, W)$ is thus determined also.) Note that these give the same measures of asymptotic relative efficiency as obtained in **2.6.7** for the associated *confidence interval* procedures.

We now examine these measures for some particular choices of F.

Examples. For each choice of F below, the values of $e_P(S, M)$ and $e_P(W, M)$ will not otherwise depend upon σ_F^2. Hence for each F we shall take a "conventional" representative.

(i) *F normal*: $F = \Phi$. In this case,

$$e_P(S, M) = \frac{2}{\pi} = 0.637$$

and

$$e_P(W, M) = \frac{3}{\pi} = 0.955.$$

It is of interest that in this instance the limiting value $e_P(S, M)$ represents the *worst* efficiency of the sign test relative to the mean (or t-) test. The *exact* relative efficiency is 0.95 for $n = 5$, 0.80 for $n = 10$, 0.70 for $n = 20$, decreasing to $2/\pi = 0.64$ as $n \to \infty$. For details, see Dixon (1953).

We note that the Wilcoxon test is a very good competitor of the mean test even in the present case of optimality of the latter. See also the remarks following these examples.

(ii) *F double exponential*: $f(x) = \frac{1}{2}e^{-|x|}$, $-\infty < x < \infty$. In this case (check)

$$e_P(S, M) = 2$$

and

$$e_P(W, M) = \tfrac{3}{2}.$$

(iii) *F uniform*: $f(x) = 1$, $|x| < \frac{1}{2}$. In this case (check)

$$e_P(S, M) = \tfrac{1}{3}$$

and

$$e_P(W, M) = 1.$$

(iv) *F logistic*: $f(x) = e^{-x}(1 + e^{-x})^{-2}$, $-\infty < x < \infty$. Explore as an exercise. ■

Remark. Note in the preceding examples that $e_P(W, M)$ is quite high for a variety of F's. In fact, the inequality

$$e_P(W, M) \geq \tfrac{108}{125} = 0.864$$

is shown to hold for *all* continuous F, with the equality attained for a particular F, by Hodges and Lehmann (1956).

For consideration of the "normal scores" statistic (recall Chapter 9) in this regard, see Hodges and Lehmann (1961). ■

10.2.3 Relationship between Pitman ARE and Correlation

Note that if a test sequence $T = \{T_n\}$ satisfies the Pitman conditions, then also the "standardized" (in the null hypothesis sense) test sequence $T^* = \{T_n^*\}$, where

$$T_n^* = \frac{T_n - \mu_n(\theta_0)}{\sigma_n(\theta_0)},$$

satisfies the conditions with

$$\mu_n^*(\theta) = \frac{\mu_n(\theta) - \mu_n(\theta_0)}{\sigma_n(\theta_0)}, \qquad \sigma_n^*(\theta) = \frac{\sigma_n(\theta)}{\sigma_n(\theta_0)}$$

and with the same G, k, $d(n)$ and c. Thus, it is equivalent to deal with T^* in place of T.

In what follows, we consider two standardized test sequences $T_0 = \{T_{0n}\}$ and $T_1 = \{T_{1n}\}$ satisfying the Pitman conditions with $G = \Phi$, $k = 1$, $d(n) = n^{1/2}$, and with constants $c_0 \le c_1$. Thus $e_P(T_1, T_0) = (c_0/c_1)^2 \le 1$, so T_0 is as good as T_1, if not better. We also assume condition

(P6) T_{0n} and T_{1n} are asymptotically bivariate normal uniformly in θ in a neighborhood of θ_0.

We shall denote by $\rho(\theta)$ the asymptotic correlation of T_{0n} and T_{1n} under the θ-distribution.

We now consider some results of van Eeden (1963).

Theorem. *Let* $T_0 = \{T_{0n}\}$ *and* $T_1 = \{T_{1n}\}$ *satisfy the conditions* (P1)–(P6) *in standardized form and suppose that*

$$\rho(\theta_n) \to \rho(\theta_0) = \rho, \qquad \text{as } \theta_n \to \theta_0.$$

(i) *For* $0 \le \lambda \le 1$, *tests of the form*

$$T_{\lambda n} = (1 - \lambda)T_{0n} + \lambda T_{1n}$$

satisfy the Pitman conditions.

(ii) *The "best" such test, that is, the one which maximizes* $e_P(T_\lambda, T_0)$, *is* T_γ *for*

$$\gamma = \frac{c_0 - \rho c_1}{(1 - \rho)(c_0 + c_1)} = \frac{e_P^{1/2}(T_1, T_0) - \rho}{(1 - \rho)[1 + e_P^{1/2}(T_1, T_0)]}$$

if $\rho \ne 1$ *and for* γ *taking any value if* $\rho = 1$.

(iii) *For this "best" test,*

$$e_P(T_\gamma, T_0) = 1 + \frac{[e_P^{1/2}(T_1, T_0) - \rho]^2}{1 - \rho^2}.$$

PROOF. (i) Put $\mu_{\lambda n}(\theta) = (1 - \lambda)\mu_{0n}(\theta) + \lambda\mu_{1n}(\theta)$ and $\sigma_{\lambda n}^2(\theta) = (1 - \lambda)^2\sigma_{0n}^2(\theta) + \lambda^2\sigma_{1n}^2(\theta) + 2\lambda(1 - \lambda)\sigma_{0n}(\theta)\sigma_{1n}(\theta)\rho(\theta)$. Then

$$\mu'_{\lambda n}(\theta_0) = (1 - \lambda)\mu'_{0n}(\theta_0) + \lambda\mu'_{1n}(\theta_0) \sim n^{1/2}\left(\frac{1 - \lambda}{c_0} + \frac{\lambda}{c_1}\right)$$

and

$$\sigma_{\lambda n}^2(\theta_0) = (1 - \lambda)^2 + \lambda^2 + 2\lambda(1 - \lambda)\rho.$$

Thus (P1)–(P5) are satisfied with $G = \Phi$, $k = 1$, $d(n) = n^{1/2}$ and

$$c_\lambda = \frac{[(1 - \lambda)^2 + \lambda^2 + 2\lambda(1 - \lambda)\rho]^{1/2}}{\left(\frac{1 - \lambda}{c_0}\right) + \left(\frac{\lambda}{c_1}\right)}.$$

To prove (ii), note that

$$e_P(T_\lambda, T_0) = \left(\frac{c_0}{c_\lambda}\right)^2,$$

so it suffices to minimize c_λ as a function of λ. This is left as an exercise. Finally, (iii) follows by substitution of γ for λ in the formula for $e_P(T_\lambda, T_0)$. ■

Corollary A. *If* T_0 *is a best test satisfying* (P1)–(P5), *then the Pitman ARE of any other test* T *satisfying* (P1)–(P5) *is given by the square of the "correlation" between* T *and* T_0, *i.e.,*

$$e_P(T, T_0) = \rho^2.$$

PROOF. Put $T = T_1$ in the theorem. Then, by (iii), since $e_P(T_\gamma, T_0) = 1$, we have $e_P(T_1, T_0) = \rho^2$. ■

Corollary B. *If* T_0 *and* T_1 *have* $\rho = 1$, *then* $e_P(T_1, T_0) = 1$ *and* $e_P(T_\lambda, T_0) = 1$, *all* λ. *If* $\rho \neq 1$, *but* $e_P(T_1, T_0) = 1$, *then* $\gamma = \frac{1}{2}$ *and* $e_P(T_{1/2}, T_0) = 2/(1 + \rho)$.

Thus no improvement in Pitman ARE can result by taking a linear combination of T_0 and T_1 having $\rho = 1$. However, if $\rho \neq 1$, some improvement is possible.

Remark. Under certain regularity conditions, the result of Corollary A holds also in the fixed sample size sense. ■

10.2.4 Complements

(i) *Efficacy.* We note that the strength of a given test, from the standpoint of Pitman ARE, is an increasing function of $1/c$, where c is the constant appearing in the Pitman condition (P3). The quantity $1/c$ is called the "efficacy" of the test. Thus the Pitman ARE of one test relative to another is given by the corresponding ratio of their respective efficacies.

(ii) *Contiguity.* A broader approach toward asymptotic power against alternatives local to a null hypothesis involves the notion of "contiguity." See Hájek and Šidák (1967) for exposition in the context of rank tests and Roussas (1972) for general exposition.

10.3 THE CHERNOFF INDEX

One might argue that error probabilities (of both types) of a test ought to decrease to 0 as the sample size tends to ∞, in order that the increasing expense be justified. Accordingly, one might compare two tests asymptotically by comparing the rate of convergence to 0 of the relevant error probabilities. Chernoff (1952) introduced a method of comparison which falls within such a context.

Specifically, consider testing a *simple* hypothesis H_0 versus a *simple* alternative H_1, on the basis of a test statistic which is a *sum of I.I.D.*'s,

$$S_n = \sum_{i=1}^{n} Y_i,$$

whereby H_0 is rejected if $S_n > c_n$, where c_n is a selected constant. For example, the likelihood ratio test for fixed sample size may be reduced to this form (exercise).

For such sums S_n, Chernoff establishes a useful large deviation probability result: for $t \geq E\{Y\}$, $P(S_n > nt)$ behaves roughly like m^n, where m is the minimum value of the moment generating function of $Y - t$. (Thus note that $P(S_n > nt)$ decreases at an *exponential* rate.) This result is applied to establish the following: if c_n is chosen to minimize $\beta_n + \lambda\alpha_n$ (where $\lambda > 0$), then the minimum value of $\beta_n + \lambda\alpha_n$ behaves roughly like ρ^n, where ρ does not depend upon λ. In effect, the critical point c_n is selected so that the Type I and Type II error probabilities tend to 0 at the same rate. The value ρ is called the *index* of the test. In this spirit we may compare two tests A and B by comparing sample sizes at which the tests perform equivalently with respect to the criterion $\beta_n + \lambda\alpha_n$. The corresponding ARE turns out to be $(\log \rho_A)/(\log \rho_B)$.

These remarks will now be precise. In **10.3.1** we present Chernoff's general large deviation theorem, which is of interest in itself and has found wide application. In **10.3.2** we utilize the result to develop Chernoff's ARE.

10.3.1 A Large Deviation Theorem

Let $Y_1, \ldots, Y_n$ be I.I.D. with distribution F and put $S_n = Y_1 + \cdots + Y_n$. Assume existence of the moment generating function $M(z) = E_F\{e^{zY}\}$, z real, and put

$$m(t) = \inf_z E\{e^{z(Y-t)}\} = \inf_z e^{-zt}M(z).$$

The behavior of large deviation probabilities $P(S_n \geq t_n)$, where $t_n \to \infty$ at rates slower than $O(n)$, has been discussed in **1.9.5**. The case $t_n = tn$ is covered in the following result.

Theorem (Chernoff). *If* $-\infty < t \leq E\{Y\}$, then

(1) $$P(S_n \leq nt) \leq [m(t)]^n.$$

If $E\{Y\} \leq t < +\infty$, *then*

(2) $$P(S_n \geq nt) \leq [m(t)]^n.$$

If $0 < \varepsilon < m(t)$, *then for the given cases of* t, *respectively,*

(3) $$\lim_n \frac{(m(t) - \varepsilon)^n}{P(S_n \leq nt)} = \lim_n \frac{(m(t) - \varepsilon)^n}{P(S_n \geq nt)} = 0.$$

Remark. Thus $P(S_n \geq nt)$ is bounded above by $[m(t)]^n$, yet for any small $\varepsilon > 0$ greatly exceeds, for all large n, $[m(t) - \varepsilon]^n$. ■

PROOF. To establish (1) we use two simple inequalities. First, *check* that for any $z \leq 0$,

$$P(S_n \leq nt) \leq [e^{-tz}M(z)]^n.$$

Then *check* that for $t \leq E\{Y\}$ and for any $z \geq 0$, $e^{-tz}M(z) \geq 1$. Thus deduce (1). In similar fashion (2) is obtained.

We now establish (3). First, *check* that it suffices to treat the case $t = 0$, to which we confine attention for convenience. Now *check* that if $P(Y > 0) = 0$ or $P(Y < 0) = 0$, then $m(0) = P(Y = 0)$ and (3) readily follows. Hereafter we assume that $P(Y > 0) > 0$ and $P(Y < 0) > 0$. We next show that the general case may be reduced to the discrete case, by defining

$$Y^{(s)} = \frac{i}{s} \quad \text{if} \quad \frac{i-1}{s} < Y \leq \frac{i}{s}, \qquad i = -1, 0, 1, \ldots, s = 1, 2, \ldots.$$

Letting $S_n^{(s)}$ = the sum of the $Y_i^{(s)}$ corresponding to $Y_1, \ldots, Y_n$, we have

$$P(S_n \leq 0) \geq P(S_n^{(s)} \leq 0)$$

and

$$M^{(s)}(z) = E\{e^{zY^{(s)}}\} \geq e^{-|z|/s}M(z).$$

Since $P(Y > 0) > 0$ and $P(Y < 0) < 0$, $M(z)$ attains its minimum value for a finite value of z (*check*) and hence there exists an s sufficiently large so that

$$\inf_z M^{(s)}(z) \geq \inf_z M(z) - \tfrac{1}{2}\varepsilon.$$

Thus (*check*) (3) follows for the general case if already established for the discrete case.

Finally, we attack (3) for the discrete case that $P(Y = y_i) = p_i > 0$, $i = 1, 2, \ldots$. Given $\varepsilon > 0$, select an integer r such that

$$\min(y_1, \ldots, y_r) < 0 < \max(y_1, \ldots, y_r)$$

and

$$\inf_z \sum_{i=1}^{r} e^{zy_i} p_i > \inf_z \sum_{i=1}^{\infty} e^{zy_i} p_i - \tfrac{1}{2}\varepsilon.$$

Put

$$m^* = \sum_{i=1}^{r} e^{z^* y_i} p_i = \inf_z \sum_{i=1}^{r} e^{zy_i} p_i.$$

It now suffices (*justify*) to show that for sufficiently large n there exist r positive integers $n_1, \ldots, n_r$ such that

(1) $$\sum_{i=1}^{r} n_i = n,$$

(2) $$\sum_{i=1}^{r} n_i y_i \leq 0,$$

and

(3) $$P(n_1, \ldots, n_r) = \frac{n!\, p_1^{n_1} \cdots p_r^{n_r}}{n_1! \cdots n_r!} > (m^* - \tfrac{1}{2}\varepsilon)^n.$$

For large $n_1, \ldots, n_r$ (not necessarily integers) Stirling's formula gives

(4) $$P(n_1, \ldots, n_r) \geq \left\{ \prod_{i=1}^{r} \left(\frac{np_i}{n_i} \right)^{n_i} \right\} n^{-(1/2)r}.$$

Now apply the method of Lagrange multipliers to show that the factor

$$Q(n_1, \ldots, n_r) = \prod_{i=1}^{r} \left(\frac{np_i}{n_i} \right)^{n_i}$$

attains a maximum of $(m^*)^n$ subject to the restrictions $\sum_{i=1}^{r} n_i = n$, $\sum_{i=1}^{r} n_i y_i = 0$, $n_1 > 0, \ldots, n_r > 0$, and the maximizing values of $n_1, \ldots, n_r$ are

$$n_i^{(0)} = \frac{np_i e^{z^* y_i}}{m^*}, \qquad 1 \leq i \leq r.$$

Assume that $y_1 \le y_i$ for $i \le r$, and put

$$n_i^{(1)} = [n_i^{(0)}], \qquad 2 \le i \le r,$$

$$n_1^{(1)} = n - \sum_{i=2}^{r} n_i^{(1)},$$

where $[\cdot]$ denotes greatest integer part. For large n, the $n_i^{(1)}$ are positive integers satisfying (1), (2) and

$$Q(n_1^{(1)}, \ldots, n_r^{(1)}) \ge \left(\frac{p_1}{n}\right)^r (m^*)^n,$$

and thus (3) by virtue of (4). This completes the proof. ■

The foregoing proof adheres to Chernoff (1952). For another approach, using "*exponential centering*," see the development in Bahadur (1971).

We have previously examined large deviation probabilities,

$$P(S_n - ES_n \ge nt), t > 0,$$

in **5.6.1**, in the more general context of U-statistics. There we derived exponential-rate *exact upper bounds* for such probabilities. Here, on the other hand, we have obtained exponential rate *asymptotic approximations*, but only for the narrower context of sums S_n. Specifically, from the above theorem we have the following useful

Corollary. *For* $t > 0$,

$$\lim_n n^{-1} \log P(S_n - ES_n \ge nt) = \log m(t + E\{Y\}).$$

10.3.2 A Measure of Asymptotic Relative Efficiency

Let H_0 and H_1 be two hypotheses which determine the distribution of Y so that $\mu_0 = E\{Y | H_0\} \le \mu_1 = E\{Y | H_1\}$. For each value of t, we consider a test which rejects H_0 if $S_n > nt$. Let $\alpha_n = P(S_n > nt | H_0)$, $\beta_n = P(S_n \le nt | H_1)$ and λ be any positive number. Put

$$m_i(t) = \inf_z E\{e^{z(Y-t)} | H_i\}, \qquad i = 0, 1,$$

and

$$\rho(t) = \max\{m_0(t), m_1(t)\}.$$

The *index* of the test determined by Y is defined as

$$\rho = \inf_{\mu_0 \le t \le \mu_1} \rho(t).$$

The role of this index is as follows. Let Q_n be the minimum value attained by $\beta_n + \lambda\alpha_n$ as the number t varies. Then the rate of exponential decrease of Q_n to 0 is characterized by the index ρ. That is, Chernoff (1952) proves

Theorem. *For* $\varepsilon > 0$,

(A1)
$$\lim_n \frac{Q_n}{(\rho + \varepsilon)^n} = 0.$$

For $0 < \varepsilon < \rho$,

(A2)
$$\lim_n \frac{Q_n}{(\rho - \varepsilon)^n} = \infty.$$

PROOF. For any t in $[\mu_0, \mu_1]$, we immediately have using Theorem 10.3.1 that

$$\begin{aligned} Q_n &\le P(S_n \le nt \mid H_1) + \lambda P(S_n \ge nt \mid H_0) \\ &\le [m_1(t)]^n + \lambda[m_0(t)]^n \\ &\le (1 + \lambda)[\rho(t)]^n. \end{aligned}$$

Let $\varepsilon > 0$ be given. By the definition of ρ, there exists t_1 in $[\mu_0, \mu_1]$ such that $\rho(t_1) \le \rho + \frac{1}{2}\varepsilon$. Thus (A1) follows.

On the other hand, for any t in $[\mu_0, \mu_1]$, we have

$$P(S_n \le nt \mid H_1) \le P(S_n \le nt' \mid H_1), \qquad \text{all } t' \ge t,$$

and

$$P(S_n \ge nt \mid H_0) \le P(S_n > nt' \mid H_0), \qquad \text{all } t' \le t,$$

yielding (*check*)

$$Q_n \ge \min\{P(S_n \le nt \mid H_1), \lambda P(S_n \ge nt \mid H_0)\}.$$

For $0 < \varepsilon < \min\{m_0(t), m_1(t)\}$, we thus have by the second part of Theorem 10.3.1 that

$$\lim_n \frac{[\min\{m_0(t), m_1(t)\} - \varepsilon]^n}{Q_n} = 0.$$

Thus, in order to obtain (A2), it suffices to find t_2 in $[\mu_0, \mu_1]$ such that $\rho \le \min\{m_0(t_2), m_1(t_2)\}$. Indeed, such a value is given by

$$t_2 = \inf\{t : m_1(t) \ge \rho, \mu_0 \le t \le \mu_1\}.$$

First of all, the value t_2 is well-defined since $m_1(\mu_1) = 1 \ge \rho$. Next we need to use the following continuity properties of the function $m(t)$. Let y_0 satisfy

$P(Y < y_0) = 0 < P(Y < y_0 + \varepsilon)$, all $\varepsilon > 0$. Then (*check*) $m(t)$ is right-continuous for $t < E\{Y\}$ and left-continuous for $y_0 < t \leq E\{Y\}$. Consequently, $m_1(t_2) \geq \rho$ and $m_1(t) < \rho$ for $t < t_2$. But then, by definition of ρ, $m_0(t) \geq \rho$ for $\mu_0 \leq t \leq t_2$. Then, by left-continuity of $m_0(t)$ for $t > \mu_0$ (*justify*), $m_0(t_2) \geq \rho$ if $t_2 > \mu$. Finally, if $t_2 = \mu_0$, $m_0(t_2) = 1 \geq \rho$. ■

We note that the theorem immediately yields

$$(*) \qquad \lim_n n^{-1} \log Q_n = \log \rho.$$

Accordingly, we may introduce a measure of asymptotic relative efficiency based on the criterion of minimizing $\beta_n + \lambda\alpha_n$ for any specified value of λ. Consider two tests T_A and T_B based on sums as above and having respective indices ρ_A and ρ_B. The *Chernoff ARE* of T_A relative to T_B is given by

$$e_C(T_A, T_B) = \frac{(\log \rho_A)}{(\log \rho_B)}.$$

Therefore, if $h(n)$ denotes the sample size at which T_B performs "equivalently" to T_A with sample size n, that is, at which

$$(1) \qquad Q^B_{h(n)} \sim Q^A_n, \qquad n \to \infty,$$

or merely

$$(2) \qquad \log Q^B_{h(n)} \sim \log Q^A_n, \qquad n \to \infty,$$

then

$$\lim_n \frac{h(n)}{n} = \frac{(\log \rho_A)}{(\log \rho_B)} = e_C(T_A, T_B).$$

(Note that (1) implies (2)—see Problem 10.P.10.)

Example A. *The index of a normal test statistic.* Let Y be $N(\mu_i, \sigma_i^2)$ under hypothesis H_i, $i = 0, 1$ $(\mu_0 < \mu_1)$. Then

$$e^{-tz}M_i(z) = \exp[(\mu_i - t)z + \tfrac{1}{2}\sigma_i^2 z^2],$$

so that (check)

$$m_i(t) = \exp\left[-\frac{\frac{1}{2}(\mu_i - t)^2}{\sigma_i^2}\right]$$

and thus (check)

$$\rho = \rho\left(\frac{\sigma_1\mu_0 + \sigma_0\mu_1}{\sigma_1 + \sigma_0}\right) = \exp\left[-\frac{\frac{1}{2}(\mu_1 - \mu_0)^2}{(\sigma_1 + \sigma_0)^2}\right]. \quad ■$$

Example B. The index of a binomial test statistic. Let Y be $B(r, p_i)$ under hypothesis H_i, $i = 0, 1(p_0 < p_1)$. Put $q_i = 1 - p_i$, $i = 0, 1$. Then

$$e^{-tz}M_i(z) = e^{-tz}(p_i e^z + q_i)^r,$$

so that (check)

$$\log m_i(t) = (r - t)\log\left[\frac{rq_i}{(r - t)}\right] + t \log\left[\frac{rp_i}{t}\right]$$

and (check)

$$\log \rho = r\left\{(1 - c)\log\left[\frac{q_0}{(1 - c)}\right] + c \log\left[\frac{p_0}{c}\right]\right\},$$

where

$$c = \frac{\log(q_0/q_1)}{\log(q_0/q_1) + \log(p_1/p_0)}, \qquad p_0 < c < p_1.$$

Show that $\log \rho \sim -r(p_1 - p_0)^2/8p_0q_0$ as $p_1 \to p_0$. ∎

Example C. *Comparison of Pitman ARE and Chernoff ARE.* To illustrate the differences between the Pitman ARE and the Chernoff ARE, we will consider the *normal location problem* and compare the *mean test* and the *sign test*. As in **10.2.2**, let $X_1, \ldots, X_n$ be independent observations having distribution $F(x - \theta)$, for an unknown value $\theta \in R$, but here confine attention to the case $F = \Phi = N(0, 1)$. Let us test the hypotheses.

$$H_0: \theta = 0 \quad \text{versus} \quad H_1: \theta = \theta_1,$$

where $\theta_1 > 0$. Let T_A denote the mean test, based on $\bar{X}$, and let T_B denote the sign test, based on $n^{-1}\sum_{i=1}^{n} I(X_i > 0)$. From Examples A and B we obtain (check)

$$e_C(T_A, T_B) = \frac{\theta_1^2/8}{\log\{2a(\theta_1)^{a(\theta_1)}[1 - a(\theta_1)]^{1-a(\theta_1)}\}},$$

where $a(\theta) = \{\log[1 - \Phi(\theta)]\}/\log\{[1 - \Phi(\theta)]/\Phi(\theta)\}$. By comparison, we have from **10.2.2**

$$e_P(T_A, T_B) = \tfrac{1}{2}\pi.$$

We note that the measure $e_C(T_A, T_B)$ depends on the particular alternative under consideration, whereas $e_P(T_A, T_B)$ does not. We also note the computational difficulties with the e_C measure. As an exercise, numerically evaluate the above quantity for a range of values of θ_1 near and far from the null value. Show also that $e_C(T_A, T_B) \to \frac{1}{2}\pi = e_P(T_A, T_B)$ as $\theta_1 \to 0$. ∎

10.4 BAHADUR'S "STOCHASTIC COMPARISON"

A popular procedure in statistical hypothesis testing is to compute the *significance level* of the observed value of the test statistic. This is interpreted as a measure of the strength of the observed sample as evidence *against* the null hypothesis. This concept provides another way to compare two test procedures, the better procedure being the one which, when the alternative is true, on the average yields stronger evidence against the null hypothesis. Bahadur (1960a) introduced a formal notion of such "stochastic comparison" and developed a corresponding measure of asymptotic relative efficiency. We present this method in **10.4.1**. The relationship between this "stochastic comparison" and methods given in terms of Type I and Type II error probabilities is examined in **10.4.2**. Here also the connection with *large deviation probabilities* is seen. In **10.4.4** a general theorem on the evaluation of Bahadur ARE is given. Various examples are provided in **10.4.5**.

10.4.1 "Stochastic Comparison" and a Measure of ARE

We consider I.I.D. observations $X_1, \ldots, X_n$ in a general sample space, having a distribution indexed by an *abstract* parameter θ taking values in a set Θ. We consider testing the hypothesis

$$H_0: \theta \in \Theta_0$$

by a real-valued test statistic T_n, whereby H_0 becomes rejected for sufficiently large values of T_n. Let $G_{\theta n}$ denote the distribution function of T_n under the θ-distribution of $X_1, \ldots, X_n$.

A natural indicator of the significance of the observed data *against* the null hypothesis is given by the "level attained," defined as

$$L_n = L_n(X_1, \ldots, X_n) = \sup_{\theta \in \Theta}[1 - G_{\theta n}(T_n)].$$

The quantity $\sup_{\theta \in \Theta_0}[1 - G_{\theta n}(t)]$ represents the maximum probability, under any one of the null hypothesis models, that the experiment will lead to a test statistic exceeding the value t. It is a decreasing function of t. Evaluated at the observed T_n, it represents the largest probability, under the possible null distributions, that a more extreme value than T_n would be observed in a repetition of the experiment. Thus the "level attained" is a random variable representing the degree to which the test statistic T_n tends to reject H_0. The lower the value of the level attained, the greater the evidence against H_0.

Bahadur (1960) suggests comparison of two test sequences $T_A = \{T_{An}\}$ and $T_B = \{T_{Bn}\}$ in terms of their performances with respect to "level attained," arguing as follows. Under a nonnull θ-distribution, the test T_{An} is "more successful" than the test T_{Bn} at the observed sample $X_1, \ldots, X_n$ if

$$L_{An}(X_1, \ldots, X_n) < L_{Bn}(X_1, \ldots, X_n).$$

Equivalently, defining

$$K_n = -2 \log L_n,$$

T_{An} is more successful than T_{Bn} at the observed sample if $K_{An} > K_{Bn}$. Note that this approach is a *stochastic* comparison of T_A and T_B.

In typical cases the behavior of L_n is as follows. For $\theta \in \Theta_0$, L_n converges in θ-distribution to some nondegenerate random variable. On the other hand, under an alternative $\theta \notin \Theta_0$, $L_n \to 0$ at an exponential rate depending on θ.

Example. *The Location Problem.* Let the X_i's have distribution function $F(x - \theta)$, where F is continuous and $\theta \in \Theta = [0, \infty)$. Let $\Theta_0 = \{0\}$. Consider the mean test statistic,

$$T_n = n^{1/2} \frac{\bar{X}}{\sigma_F},$$

where $\sigma_F^2 = \mathrm{Var}_F\{X\}$. We have

$$L_n = 1 - G_{0n}(T_n)$$

and thus

$$\begin{aligned} P_0(L_n \le l) &= P_0(G_{0n}(T_n) \ge 1 - l) \\ &= P_0(T_n \ge G_{0n}^{-1}(1 - l)) \\ &= 1 - G_{0n}(G_{0n}^{-1}(1 - l)) = l, \end{aligned}$$

that is, under H_0 L_n has the distribution uniform (0, 1). Note also that

$$G_{0n} \Rightarrow \Phi = N(0, 1).$$

Now consider $\theta \notin \Theta_0$. We have, by the SLLN,

$$P_\theta(n^{-1/2} T_n \to \theta) = 1,$$

in which case L_n behaves approximately as

$$1 - \Phi(n^{1/2}\theta) \sim (2\pi n)^{-1/2}\theta^{-1} \exp(-\tfrac{1}{2}n\theta^2).$$

That is, in the nonnull case L_n behaves approximately as a quantity tending to 0 exponentially fast. Equivalently, K_n behaves approximately as a quantity tending to a finite positive limit (in this case θ^2). These considerations will be made more precise in what follows. ■

It is thus clear how the stochastic comparison of two test sequences is influenced in the case of nonnull θ by the respective indices of exponential convergence to 0 of the levels attained. A test sequence $T = \{T_n\}$ is said to have (*exact*) *slope* $c(\theta)$ when θ "obtains" (that is, when the X_i's have θ-distribution) if

$$(*) \qquad n^{-1}K_n \to c(\theta) \quad \text{a.s.} \quad (P_\theta).$$

In the nonnull case the limit $c(\theta)$ may be regarded as a measure of the performance of T_n; the larger the value of $c(\theta)$, the "faster" T_n tends to reject H_0. For two such test sequences T_A and T_B, the ratio

$$\frac{c_A(\theta)}{c_B(\theta)}$$

thus represents a measure of the asymptotic relative efficiency of T_A relative to T_B at the (fixed) alternative θ. Indeed, if $h(n)$ represents the sample size at which procedure T_B performs "equivalently" to T_A in the sense of being equally "successful" asymptotically, that is, $K_{Bh(n)}$ may replace K_{An} in relation (*), then we must have (check)

$$\frac{h(n)}{n} \to \frac{c_A(\theta)}{c_B(\theta)}.$$

Thus the (*exact*) *Bahadur ARE* of T_A relative to T_B is defined as $e_B(T_A, T_B) = c_A(\theta)/c_B(\theta)$.

The qualification "exact" in the preceding definitions is to distinguish from "approximate" versions of these concepts, also introduced by Bahadur (1960a), based on the substitution of G for $G_{\theta n}$ in the definition of L_n, where $G_{\theta n} \Rightarrow G$ for all $\theta \in \Theta_0$. We shall not pursue this modification.

The terminology "slope" for $c(\theta)$ is motivated by the fact that in the case of nonnull θ the random sequence of points $\{(n, K_n), n \geq 1\}$ moves out to infinity in the plane in the direction of a ray from the origin, with angle $\tan^{-1}c(\theta)$ between the ray and the n-axis.

A useful characterization of the slope is as follows. Given ε, $0 < \varepsilon < 1$, and the sequence $\{X_i\}$, denote by $N(\varepsilon)$ the random sample size required for the test sequence $\{T_n\}$ to become significant at the level ε and remain so. Thus

$$N(\varepsilon) = \inf\{m: L_n < \varepsilon, \text{ all } n \geq m\} (\leq \infty).$$

Bahadur (1967) gives

Theorem. *If* (*) *holds, with* $0 < c(\theta) < \infty$, *then*

$$\lim_{\varepsilon \to 0} \frac{-2 \log \varepsilon}{N(\varepsilon)} = c(\theta) \quad a.s. \quad (P_\theta).$$

PROOF. Let Ω denote the sample space in the P_θ-model. By (*), $\exists\Omega_0 \subset \Omega$ such that $P_\theta(\Omega_0) = 1$ and for $\omega \in \Omega_0$ the sequence $\{X_n(\omega)\}$ is such that $n^{-1}K_n(\omega) \to c(\theta)$. Now fix $\omega \in \Omega_0$. Since $c(\theta) > 0$, we have $L_n(\omega) > 0$ for all sufficiently large n and $L_n(\omega) \to 0$ as $n \to \infty$. Therefore (justify), $N(\varepsilon, \omega) < \infty$ for every $\varepsilon > 0$ and thus $N(\varepsilon, \omega) \to \infty$ through a subsequence of the integers as $\varepsilon \to 0$. Thus $2 \le N(\varepsilon, \omega) < \infty$ for all ε sufficiently small, say $<\varepsilon_1$. For all $\varepsilon < \varepsilon_1$, we thus may write

$$L_{N(\varepsilon, \omega)}(\omega) < \varepsilon \le L_{N(\varepsilon, \omega) - 1}(\omega).$$

The proof is readily completed (as an exercise). ■

It follows that the sample sizes $N_A(\varepsilon)$ and $N_B(\varepsilon)$ required for procedures T_A and T_B to perform "equivalently," in the sense of becoming and remaining significant at level ε, must satisfy

$$\lim_{\varepsilon \to 0} \frac{N_B(\varepsilon)}{N_A(\varepsilon)} = \frac{c_A(\theta)}{c_B(\theta)} = e_B(T_A, T_B) \quad \text{a.s.} \quad (P_\theta),$$

providing another interpretation of the Bahadur ARE.

Another important aspect of the Bahadur ARE is the connection with Type I and Type II error probabilities. Not only does this afford another way to interpret e_B, but also it supports comparison with other ARE measures such as e_P and e_C. These considerations will be developed in **10.4.2.** and will lead toward the issue of *computation* of e_B, treated in **10.4.3**.

Further important discussion of slopes and related matters, with references to other work also, is found in Bahadur (1960a, 1967, 1971).

10.4.2 Relationship between Stochastic Comparison and Error Probabilities

Consider testing H_0 by critical regions $\{T_n > t_n\}$ based on $\{T_n\}$. The relevant Type I and Type II error probabilities are

$$\alpha_n = \sup_{\theta \in \Theta_0} P_\theta(T_n > t_n)$$

and

$$\beta_n(\theta) = P_\theta(T_n \le t_n),$$

respectively.

Theorem (Bahadur). *Suppose that*

$$\frac{-2 \log \alpha_n}{n} \to d$$

and

$$\frac{K_n}{n} \xrightarrow{P_\theta} c(\theta).$$

Then

(i) $$d > c(\theta) \Rightarrow \beta_n(\theta) \to 1,$$

and

(ii) $$d < c(\theta) \Rightarrow \beta_n(\theta) \to 0.$$

PROOF. Write

$$\begin{aligned}\beta_n(\theta) &= P_\theta(L_n > \alpha_n) = P_\theta(K_n < -2\log\alpha_n)\\ &= P_\theta(n^{-1}K_n < n^{-1}(-2\log\alpha_n)).\end{aligned}$$

If $d > c(\theta) + \varepsilon$, then for n sufficiently large we have $n^{-1}(-2\log\alpha_n) > c(\theta) + \varepsilon$ and thus $\beta_n(\theta) \geq P_\theta(n^{-1}K_n < c(\theta) + \varepsilon) \to 1$, proving (i). Similarly, (ii) is proved. ∎

Corollary. *Suppose that*

$$\frac{-2\log\alpha_n}{n} \to d,$$

$$\beta_n(\theta) \to \beta(\theta), \qquad 0 < \beta(\theta) < 1,$$

and

$$\frac{K_n}{n} \xrightarrow{P_\theta} c(\theta).$$

Then $d = c(\theta)$.

By virtue of this result, we see that $e_B(T_A, T_B)$, although based on a concept of "stochastic comparison," may also be formulated as the measure of ARE obtained by comparing the rates at which the Type I error probabilities (of T_A and T_B) tend to 0 while the Type II error probabilities remain fixed at (or tend to) a value $\beta(\theta)$, $0 < \beta(\theta) < 1$, for fixed θ. That is, if $h(n)$ denotes the sample size at which T_B performs equivalently to T_A with sample size n, in the sense that

$$\frac{(\log\alpha_{B,h(n)})}{(\log\alpha_{A,n})} \to 1, \qquad \beta_{B,h(n)}(\theta) \to \beta(\theta), \qquad \beta_{A,n} \to \beta(\theta),$$

then

$$\frac{h(n)}{n} \sim \frac{[(\log\alpha_{A,n})/n]}{[(\log\alpha_{B,h(n)})/h(n)]} \to \frac{c_A(\theta)}{c_B(\theta)}.$$

Therefore, in effect, the Bahadur ARE relates to situations in which having small Type I error probability is of greater importance than having small Type II error probabilities. In Section 10.5 we consider a measure of similar nature but with the roles of Type I and Type II error reversed. In comparison, the Chernoff ARE relates to situations in which it is important to have both types of error probability small, on more or less an equal basis.

Like the Chernoff ARE, the Bahadur ARE depends upon a specific alternative and thus may pose more computational difficulty than the Pitman ARE. However, the Bahadur ARE is easier to evaluate than the Chernoff ARE, because it entails precise estimation only of α_n instead of *both* α_n and β_n. This is evident from the preceding corollary and will be further clarified from the theorem of **10.4.3**.

10.4.3 A Basic Theorem

We now develop a result which is of use in finding slopes in the Bahadur sense. The test statistics considered will be assumed to satisfy the following conditions. We put $\Theta_1 = \Theta - \Theta_0$.

Bahadur Conditions

(B1) For $\theta \in \Theta_1$,

$$n^{-1/2}T_n \to b(\theta) \quad \text{a.s.} \quad (P_\theta),$$

where $-\infty < b(\theta) < \infty$.

(B2) There exists an open interval I containing $\{b(\theta): \theta \in \Theta_1\}$, and a function g continuous on I, such that

$$\lim_n -2n^{-1} \log \sup_{\theta \in \Theta_0} [1 - G_{\theta n}(n^{1/2}t)] = g(t), \qquad t \in I. \quad ■$$

Theorem (Bahadur). *If* T_n *satisfies* (B1)–(B2), *then for* $\theta \in \Theta_1$

$$n^{-1}K_n \to g(b(\theta)) \quad a.s. \quad (P_\theta).$$

PROOF. Fix $\theta \in \Theta_1$, and let Ω denote the sample space in the P_θ-model. By (B1), $\exists \Omega_0 \subset \Omega$ such that $P_\theta(\Omega_0) = 1$ and for $\omega \in \Omega_0$ the sequence $\{X_n(\omega)\}$ is such that $n^{-1/2}T_n(\omega) \to b(\theta)$. Now fix $\omega \in \Omega_0$, For any $\varepsilon > 0$, we have

$$n^{1/2}(b(\theta) - \varepsilon) < T_n(\omega) < n^{1/2}(b(\theta) + \varepsilon)$$

for all sufficiently large n, and thus also

$$\frac{-2 \log \sup_{\theta \in \Theta_0} [1 - G_{\theta n}(n^{1/2}(b(\theta) + \varepsilon))]}{n} \le \frac{K_n(\omega)}{n}$$

$$\le \frac{-2 \log \sup_{\theta \in \Theta_0} [1 - G_{\theta n}(n^{1/2}(b(\theta) - \varepsilon))]}{n}$$

Therefore, for all ε sufficiently small that the interval I contains $b(\theta) \pm \varepsilon$, we have by condition (B2) that

$$g(b(\theta) + \varepsilon) \le \underline{\lim}_n \frac{K_n(\omega)}{n} \le \overline{\lim}_n \frac{K_n(\omega)}{n} \le g(b(\theta) - \varepsilon).$$

We apply continuity of g to complete the proof. ∎

Remarks. (i) Condition (B2) makes manifest the role of large deviation theory in evaluating Bahadur slopes. Only the *null hypothesis* large deviation probabilities for the test statistic are needed.

(ii) Variations of the theorem, based on other versions of (B2), have been established. See Bahadur (1960a, 1967, 1971) and references cited therein.

(iii) With (B1) relaxed to convergence in P_θ-probability, the conclusion of the theorem holds in the weak sense.

(iv) If a given $\{T_n\}$ fails to satisfy (B1)–(B2), it may well be the case that $T_n^* = h_n(T_n)$ does, where h_n is a strictly increasing function. In this case the slopes of $\{T_n\}$ and $\{T_n^*\}$ are identical. ∎

10.4.4 Examples

Example A. *The Normal Location Problem.* (Continuation of **10.2.2** and Examples 10.3.2C and 10.4.1). Here $\Theta_0 = \{0\}$ and $\Theta_1 = (0, \infty)$. The statistics to be compared are the "mean" test, "t-test," "sign" test, and "Wilcoxon" test (denoted T_{1n}, T_{2n}, T_{3n} and T_{4n}, respectively). In order to evaluate the Bahadur ARE's for these statistics, we seek their slopes (denoted $c_i(\theta)$, $i = 1, 2, 3, 4$).

For simplicity, we confine attention to the case that the θ-distribution of X is $N(\theta, 1)$.

We begin with the mean test statistic,

$$T_{1n} = n^{1/2}\overline{X},$$

which by Example 10.4.1 statisfies the Bahadur conditions with $b(\theta) = \theta$ in (B1) and $g(t) = t^2$ in (B2). Thus, by Theorem 10.4.3, T_{1n} has slope

$$c_1(\theta) = \theta^2.$$

The t-test statistic T_{2n}, $n^{1/2}\overline{X}/s$, has slope

$$c_2(\theta) = \log(1 + \theta^2).$$

For this computation, see Bahadur (1960b, or 1971). The interesting thing to observe is that the slope of the t-test is *not* the same as that of the mean. Thus

$$e_B(t, M) = \frac{\log(1 + \theta^2)}{\theta^2} < 1,$$

so that the t-test and mean test are *not* equivalent from the standpoint of Bahadur ARE, in contrast to the equivalence from the standpoint of Pitman ARE, as seen in **10.2.2**. (For further evidence against the t-test, see Example C below.)

The slope of the sign test statistic, $T_{3n} = n^{1/2}(2V_n - 1)$, where

$$V_n = n^{-1} \sum_{i=1}^{n} I(X_i > 0),$$

may be found by a direct handling of the level attained,

$$L_{3n} = \sum_{j=nV_n}^{n} \binom{n}{j} (\tfrac{1}{2})^n.$$

It is shown by Bahadur (1960b) that

$$\log L_{3n} = -\tfrac{1}{2}\left\{n[2 \log H(p)] + 2(npq)^{1/2}\xi_n \log\left(\frac{p}{q}\right) + \xi_n^2\right\}$$
$$- \tfrac{1}{2} \log n - \tfrac{1}{2} \log\left[\frac{2\pi q(2p - 1)^2}{p}\right] + o_{P_\theta}(1),$$

where $p = \Phi(\theta)$, $q = 1 - p$, $H(y) = 2y^y(1 - y)^{1-y}$ for $0 < y < 1$, and

$$\xi_n = (pq)^{-1/2} n^{1/2}(V_n - p).$$

Since $\xi_n \xrightarrow{d} N(0, 1)$, we have

$$\frac{K_{3n}}{n} = \frac{-2 \log L_{3n}}{n} \xrightarrow{P_\theta} 2 \log H(\Phi(\theta)).$$

Thus it is seen that T_{3n} has slope

$$c_3(\theta) = 2 \log\{2\Phi(\theta)^{\Phi(\theta)}[1 - \Phi(\theta)]^{1-\Phi(\theta)}\}.$$

We can also obtain this result by an application of Theorem 10.4.3, as follows. Check that condition (B1) holds with $b(\theta) = 2\Phi(\theta) - 1$. Next use Chernoff's Theorem (specifically, Corollary 10.3.1) in conjunction with Example 10.3.2B, to obtain condition (B2). That is, write $1 - G_{0n}(n^{1/2}t) = P(V_n > \frac{1}{2}(t + 1))$ and apply the Chernoff results to obtain (B2) with $g(t) = 2 \log H(\frac{1}{2}(1 + t))$, for $H(y)$ as above.

We thus have

$$e_B(S, M) = \frac{2 \log\{2\Phi(\theta)^{\Phi(\theta)}[1 - \Phi(\theta)]^{1-\Phi(\theta)}\}}{\theta^2}.$$

Show that $e_B(S, M) \to 2/\pi = e_P(S, M)$ as $\theta \to 0$. Some values of $e_B(S, M)$ are as follows.

θ	0	0.5	1.0	1.5	2.0	3.0	4.0	∞
$e_B(S, M)$	$2/\pi = 0.64$	0.60	0.51	0.40	0.29	0.15	0.09	0

The slope of the Wilcoxon test statistic has been found by Klotz (1965). It is

$$c_4(\theta) = \frac{1}{2}\left[\Delta\eta - \int_0^1 \log\cosh x\Delta\, dx\right],$$

where

$$\eta = P_\theta(X_1 + X_2 > 0) - \tfrac{1}{2}$$

and Δ is the solution of the equation

$$\int_0^\infty x \tanh x\Delta\, dx = \eta,$$

See also Bahadur (1971). ■

Example B. *The Kolmogorov–Smirnov Test.* (Abrahamson (1967)). Let Θ index the set of all continuous distribution functions $\theta(x)$ on the real line, and let H_0 be simple,

$$H_0: \theta = \theta_0,$$

where θ_0 denotes a specified continuous distribution function. Consider the Kolmogorov–Smirnov statistic

$$T_n = n^{1/2} \sup_x |F_n(x) - \theta_0(x)|,$$

where F_n denotes the sample distribution function. The slope is found by Theorem 10.4.3. First, check that condition (B1), with

$$b(\theta) = \sup_x |\theta(x) - \theta_0(x)|,$$

follows from the Glivenko-Cantelli Theorem (2.1.4A). Regarding condition (B2), the reader is referred to Abrahamson (1967) or Bahadur (1971) for derivations of (B2) with

$$g(t) = 2\inf\{h(t, p): 0 \le p \le 1\},$$

where for $0 \le p \le 1 - t$

$$h(t, p) = (t + p)\log\left(\frac{t + p}{p}\right) + (1 - t - p)\log\left(\frac{1 - t - p}{1 - p}\right)$$

and $h(t, p) = \infty$ for $p > 1 - t$. ■

Example C. *The t-Test for a Nonparametric Hypothesis.* Consider the *composite* null hypothesis H_0 that the data has distribution F belonging to the class $\mathscr{F}_0$ of all continuous distributions symmetric about 0. The slopes of various rank statistics such as the sign test, the Wilcoxon signed-rank test and the normal scores signed-rank test can be obtained by Theorem 10.4.3 in straightforward fashion because in each case the null distribution of the test statistic does not depend upon the particular $F \in \mathscr{F}_0$. For these slopes see Bahadur (1960b) and Klotz (1965). But how does the t-test perform in this context? The question of finding the slope of the t-test in this context leads to an *extremal* problem in large deviation theory, that of finding the rate of convergence to 0 of $\sup_{F \in \mathscr{F}_0} P(T_n \geq a)$, where $T_n = \bar{X}/s$. This problem is solved by Jones and Sethuraman (1978) and the result is applied via Bahadur's Theorem (10.4.3) to obtain the slope of the t-test at alternatives F_1 satisfying certain regularity conditions. It is found that for $F_1 = N(\theta, 1)$, $\theta \neq 0$, the t-test is somewhat inferior to the normal scores signed-rank test. ∎

10.5 THE HODGES–LEHMANN ASYMPTOTIC RELATIVE EFFICIENCY

How adequate is the Pitman efficiency? the Chernoff measure? the Bahadur approach? It should be clear by now that a *comprehensive* efficiency comparison of two tests cannot be summarized by a single number or measure. To further round out some comparisons, Hodges and Lehmann (1956) introduce an ARE measure which is pertinent when one is interested in "the region of high power." That is, two competing tests of size α are compared at fixed alternatives as the power tends to 1. In effect, the tests are compared with respect to the rate at which the Type II error probability tends to 0 at a fixed alternative while the Type I error probability is held fixed at a level α, $0 < \alpha < 1$. The resulting measure, $e_{HL}(T_A, T_B)$, which we call the *Hodges–Lehmann ARE*, is the dual of the Bahadur ARE. The relative importances of the Type I and Type II error probabilities are reversed.

Like the Bahadur ARE, the computation of the Hodges–Lehmann ARE is less formidable than the Chernoff index, because the exponential rate of convergence to 0 needs to be characterized for only one of the error probabilities instead of for both.

In the following example, we continue our study of selected statistics in the normal location problem and illustrate the computation of $e_{HL}(\cdot, \cdot)$.

Example. *The Normal Location Problem* (Continuation of **10.2.2** and Examples 10.3.2C, 10.4.1 and 10.4.4A) In general, the critical region $\{T_n > t_n\}$ is designed so that α_n tends to a limit α, $0 < \alpha < 1$, so that at alternatives θ we have $\beta_n(\theta) \to 0$ and typically in fact

$$(*) \qquad -2n^{-1} \log \beta_n(\theta) \to d(\theta),$$

for some value $0 < d(\theta) < \infty$. Let us now consider $\Theta_0 = \{0\}$ and $\Theta_1 = [0, \infty)$, and assume the θ-distribution of X to be $N(\theta, 1)$. For the *mean* test statistic, $T_n = n^{1/2}\overline{X}$, we must (why?) take t_n equal to a constant in order that α_n behave as desired. Thus $\beta_n(\theta)$ is of the form

$$\beta_n(\theta) = P_\theta(n^{1/2}\overline{X} \le c).$$

A straightforward application of Chernoff's results (see Corollary 10.3.1 and Example 10.3.2A) yields (*) with

$$d_M(\theta) = \theta^2.$$

Similarly, for the *sign* test, as considered in Example 10.4.4A, we obtain (*) with

$$d_S(\theta) = -\log\{4\Phi(\theta)[1 - \Phi(\theta)]\}.$$

Thus the Hodges–Lehmann ARE of the sign test relative to the mean test is

$$e_{HL}(S, M) = \frac{-\log\{4\Phi(\theta)[1 - \Phi(\theta)]\}}{\theta^2}.$$

Like the Bahadur ARE $e_B(S, M)$, this measure too converges to $2/\pi$ as $\theta \to 0$. Some values of $e_{HL}(S, M)$ are as follows.

θ	0	0.253	0.524	1.645	3.090	3.719	∞
$e_{HL}(S, M)$	$2/\pi = 0.64$	0.636	0.634	0.614	0.578	0.566	0.5

Interestingly, as $\theta \to \infty$, $e_{HL}(S, M) \to \frac{1}{2}$ whereas $e_B(S, M) \to 0$.

Hodges and Lehmann also evaluate the t-test and find that, like the Pitman ARE, $e_{HL}(t, M) = 1$. On the other hand, the Bahadur comparison gives (check)

$$e_B(t, M) = \frac{\log(1 + \theta^2)}{\theta^2} < 1. \quad \blacksquare$$

In **10.4.1** we mentioned an "approximate" version of the Bahadur slope. The analogous concept relative to the Hodges–Lehmann approach has been investigated by Hettmansperger (1973).

10.6 HOEFFDING'S INVESTIGATION (MULTINOMIAL DISTRIBUTIONS)

In the spirit of the Chernoff approach, Hoeffding (1965) considers the comparison of tests at fixed alternatives as both types of error probability tend to 0 with increasing sample size. He considers *multinomial data* and brings to light certain superior features of the *likelihood ratio* test, establishing the following

Proposition. *If a given test of size* α_n *is "sufficiently different" from a likelihood ratio test, then there is a likelihood ratio test of size* $\leq \alpha_n$ *which is considerably more powerful then the given test at "most" points in the set of alternatives when the sample size* n *is large enough, provided that* α_n *tends to* 0 *at a suitably fast rate.*

In particular, Hoeffding compares the chi-squared test to the likelihood ratio test and finds that, in the sense described, chi-square tests of simple hypotheses (and of some composite hypotheses) are *inferior* to the corresponding likelihood ratio tests.

In **10.6.1** we present a basic *large deviation* theorem for the multinomial distribution. This is applied in **10.6.2** to characterize *optimality of the likelihood ratio test*. Connections with *information numbers* are discussed in **10.6.3**. The chi-squared and likelihood ratio tests are compared in **10.6.4**, with discussion of the Pitman and Bahadur ARE's also.

10.6.1 A Large Deviation Theorem

Here we follow Hoeffding (1965), whose development is based essentially on work of Sanov (1957). A treatment is also available in Bahadur (1971).

Let $\mathbf{z}_n = (n_1/n, \ldots, n_k/n)$ denote the relative frequency vector associated with the point $(n_1, \ldots, n_k)$ in the sample space of the multinomial $(p_1, \ldots, p_k; n)$ distribution. Let Θ be the parameter space,

$$\Theta = \left\{\mathbf{p} = (p_1, \ldots, p_k): p_i \geq 0, \sum_{i=1}^{k} p_i = 1\right\}.$$

Let $P_n(\cdot \mid \mathbf{p})$ denote the probability function corresponding to the parameter $\mathbf{p}$. Thus

$$P_n(\{\mathbf{z}_n\} \mid \mathbf{p}) = n! \prod_{i=1}^{k} \frac{p_i^{n_i}}{n_i!}.$$

For any subset A of Θ, let $A^{(n)}$ denote the set of points of the form $\mathbf{z}_n$ which lie in A. We may extend the definition of $P_n(\cdot)$ to arbitrary sets A in Θ by defining

$$P_n(A \mid \mathbf{p}) = P_n(A^{(n)} \mid \mathbf{p}).$$

For points $\mathbf{x} = (x_1, \ldots, x_k)$ and $\mathbf{p} = (p_1, \ldots, p_k)$ in Θ, define

$$I(\mathbf{x}, \mathbf{p}) = \sum_{i=1}^{k} x_i \log\left(\frac{x_i}{p_i}\right).$$

As will be seen in **10.6.3**, this function may be thought of as a *distance* between the points $\mathbf{x}$ and $\mathbf{p}$ in Θ. For sets $A \subset \Theta$ and $\Lambda \subset \Theta$, define

$$I(A, \mathbf{p}) = \inf_{\mathbf{x} \in A} I(\mathbf{x}, \mathbf{p})$$

and

$$I(\mathbf{x}, \Lambda) = \inf_{\mathbf{p} \in \Lambda} I(\mathbf{x}, \mathbf{p}).$$

These extend $I(\mathbf{x}, \mathbf{p})$ to a distance between a point and a set. In this context, "large deviation" probability refers to a probability of the form $P_n(A|\mathbf{p})$ where the distance $I(A, \mathbf{p})$ is positive (and remains bounded away from 0 as $n \to \infty$).

Theorem. *For sets* $A \subset \Theta$ *and points* $\mathbf{p} \in \Theta$, *we have uniformly*

$$\frac{\log P_n(A|\mathbf{p})}{n} = -I(A^{(n)}, \mathbf{p}) + O\left(\frac{\log n}{n}\right), \qquad n \to \infty.$$

Remarks. (i) The qualification "uniformly" means that the $O(\cdot)$ function depends only on k and not on the choice of A, $\mathbf{p}$ and n.

(ii) Note that the above approximation is crude, in the sense of giving an asymptotic expression for $\log P_n(A|\mathbf{p})$ but not one for $P_n(A|\mathbf{p})$. However, as we have seen in Sections 10.3–10.5, this is strong enough for basic applications to asymptotic relative efficiency.

(iii) If the set A corresponds to the critical region of a test (of a hypothesis concerning $\mathbf{p}$), then the above result provides an approximation to the asymptotic behavior of the error probabilities. Clearly, we must confine attention to tests for which the size α_n tends to 0 faster than any power of n. The case where α_n tends to 0 more slowly is not resolved by the present development. (See Problem 10.P.19). ■

10.6.2 The Emergence of the Likelihood Ratio Test

It is quickly seen (exercise) that

$$(1) \qquad P_n(\mathbf{z}_n|\mathbf{p}) = P_n(\mathbf{z}_n|\mathbf{z}_n)e^{-nI(\mathbf{z}_n, \mathbf{p})}$$

Now consider the problem of testing the hypothesis

$$H_0: \mathbf{p} \in \Lambda (\Lambda \subset \Theta)$$

versus an alternative

$$H: \mathbf{p} \in \Lambda' = \Theta - \Lambda,$$

on the basis of an observation $\mathbf{Z}_n = \mathbf{Z}_n$.

The *likelihood ratio* (LR) test is based on the statistic

$$\frac{\sup_{\mathbf{p} \in \Lambda} P_n(\mathbf{z}_n|\mathbf{p})}{\sup_{\mathbf{p} \in \Lambda'} P_n(\mathbf{z}_n|\mathbf{p})} = e^{-nI(\mathbf{z}_n, \Lambda)}$$

(since $0 \leq I(\mathbf{x}, \mathbf{p}) \leq \infty$ and $I(\mathbf{x}, \mathbf{x}) = 0$). Thus the LR test rejects H_0 when

$$I(\mathbf{z}_n, \Lambda) > \text{constant}.$$

Now an *arbitrary* test rejects H_0 when $\mathbf{z}_n \in A_n$ where A_n is a specified subset of Θ. By the theorem, the size α_n of the test A_n satisfies

$$\alpha_n = \sup_{\mathbf{p} \in \Lambda} P_n(A_n | \mathbf{p}) = e^{-nI(A_n^{(n)}, \Lambda) + O(\log n)}.$$

Let us now compare the test A_n with the LR test which rejects H_0 when $\mathbf{z}_n \in B_n$, where

$$B_n = \{\mathbf{x}: I(\mathbf{x}, \Lambda) \geq c_n\}$$

and

$$c_n = I(A_n^{(n)}, \Lambda).$$

The critical region B_n contains the critical region A_n. In fact, B_n is the union of all critical regions of tests A_n' for which $I(A_n', \Lambda) \geq c_n$, that is, of all tests A_n' with size $\leq \alpha_n$ (approximately). Moreover, the size α_n^* of the test B_n satisfies

$$\alpha_n^* = e^{-nc_n + O(\log n)}$$

since $I(B_n^{(n)}, \Lambda) = c_n$. Hence we have

$$\log \alpha_n^* = \log \alpha_n + O(\log n).$$

Therefore, if the size $\alpha_n \to 0$ *faster than any power of n*, the right-hand side is dominated by the term $\log \alpha_n$, so that the sizes of the tests are approximately equal.

These considerations establish: Given any test A_n of size α_n, such that $\alpha_n \to 0$ faster than any power of n, there exists a LR test which is *uniformly at least as powerful* and asymptotically *of the same size.* (Why uniformly?)

Furthermore, at "most" points $\mathbf{p} \in \Theta - \Lambda$, the test B_n is *considerably more powerful* than A_n, in the sense that the ratio of Type II error probabilities at $\mathbf{p}$ tends to 0 more rapidly than any power of n. For we have

$$\frac{P_n(B_n | \mathbf{p})}{P_n(A_n | \mathbf{p})} = e^{-n[I(B_n^{(n)}, \mathbf{p}) - I(A_n^{(n)}, \mathbf{p})] + O(\log n)}.$$

At these points $\mathbf{p}$ for which

(i) $P_n(A_n | \mathbf{p}) \neq 0$

and

(ii) $$\frac{n[I(B_n^{(n)}, \mathbf{p}) - I(A_n^{(n)}, \mathbf{p})]}{\log n} \to \infty,$$

we have that the ratio of error probabilities $\to 0$ faster than any power of n.

10.6.3 The Function $I(x, p)$ as a Distance; Information

The function $I(\mathbf{x}, \mathbf{p})$ has a natural generalization including distributions other than multinomial. Suppose that a model is given by a family of distributions $\{F_\theta, \theta \in \Theta\}$. For any distributions F_{θ_0} and F_{θ_1}, suppose that F_{θ_0} and F_{θ_1} have densities f_{θ_i} with respect to some measure μ, for example the measure $d\mu = \frac{1}{2}(dF_{\theta_0} + dF_{\theta_1})$. Define

$$I(F_{\theta_0}, F_{\theta_1}) = \int f_{\theta_0} \log\left(\frac{f_{\theta_0}}{f_{\theta_1}}\right) d\mu,$$

with $f_{\theta_0} \log(f_{\theta_0}/f_{\theta_1})$ interpreted as 0 where $f_{\theta_0}(x) = 0$ and interpreted as ∞ where $f_{\theta_0}(x) > 0$, $f_{\theta_1}(x) = 0$. Note that this is a generalization of $I(\mathbf{x}, \mathbf{p})$. For example, let μ be *counting measure.*

$I(F, G)$ is an *asymmetric* measure of *distance* between F and G. We have

(i) $0 \le I(F, G) \le \infty$;
(ii) If $I(F, G) < \infty$, then $F \ll G$ and $I(F, G) = \int \log(dF/dG)dF$;
(iii) $I(F, G) = 0$ if and only if $F = G$.

$I(F, G)$ represents an *information measure.* It measures the ability to discriminate against G on the basis of observations taken from the distribution F. As such, this is asymmetric. To see what this means, consider the following example from Chernoff (1952).

Example Let $\Theta = \{\mathbf{p} = (p_1, p_2): p_i \ge 0, p_1 + p_2 = 1\}$. Consider the distributions $\mathbf{p}_0 = (1, 0)$ and $\mathbf{p}_1 = (0.9, 0.1)$. We have

$$I(\mathbf{p}_0, \mathbf{p}_1) < \infty, \qquad I(\mathbf{p}_1, \mathbf{p}_0) = \infty.$$

What does this mean? If $\mathbf{p}_1$ is the true distribution, only a finite number of observations will be needed to obtain an observation in the second cell, completely disproving the hypothesis $\mathbf{p}_0$. Thus the ability of $\mathbf{p}_1$ to discriminate against $\mathbf{p}_0$ is perfect, and this is measured by $I(\mathbf{p}_1, \mathbf{p}_0) = \infty$. On the other hand, if $\mathbf{p}_0$ is the true distribution, the fact that no observations ever occur in the second cell will build up evidence against $\mathbf{p}_1$ in only a gradual fashion. In general, points on the boundary of Θ are infinitely far from interiors points, but not vice versa. ■

Let us now interpret statistically the large deviation theorem of **10.6.1**, which may be expressed in the form

$$P_n(A \mid \mathbf{p}) = e^{-nI(A^{(n)}, \mathbf{p}) + O(\log n)}.$$

The quantity $I(A^{(n)}, \mathbf{p})$ represents the *shortest distance* from the point $\mathbf{p}$ to the set $A^{(n)}$. Suppose that A is the critical region of a test: "reject $H_0: \mathbf{p} = \mathbf{p}_0$ when the observed relative frequency vector falls in the region A." The above approximation tells us that the size α_n of the test is not much increased by adjoining to A all points whose "distance" from $\mathbf{p}_0$ is at least $I(A, \mathbf{p}_0)$. The test so obtained is at least as powerful, since the new critical region contains the first, and has approximately the same size. It turns out that the latter test is simply the LR test.

10.6.4 Comparison of χ^2 and *LR* Tests

From the considerations of **10.6.2**, we see the superiority of the LR test over the χ^2 test, with respect to power at "most" points in the parameter space, provided that the tests under consideration have size α_n tending to 0 at a suitably fast rate. The superiority of the LR test is also affirmed by the Bahadur approach. From Abrahamson (1965) (or see Bahadur (1971)), we have for the χ^2 test the slope

$$c_1(\boldsymbol{\theta}) = 2I(A(\boldsymbol{\theta}, \mathbf{p}_0), \mathbf{p}_0),$$

where $\mathbf{p}_0$ is the (simple) null hypothesis, and

$$A(\boldsymbol{\theta}, \mathbf{p}_0) = \left\{\mathbf{p}: \mathbf{p} \in \Theta, \sum_{i=1}^{k} \frac{(p_i - p_{0i})^2}{p_{0i}} \geq \sum_{i=1}^{k} \frac{(\theta_i - p_{0i})^2}{p_{0i}}\right\},$$

and we have for the *LR* test the slope

$$c_2(\boldsymbol{\theta}) = 2I(\boldsymbol{\theta}, \mathbf{p}_0).$$

It is readily checked that $c_1(\boldsymbol{\theta}) \leq c_2(\boldsymbol{\theta})$, that is,

$$e_B(\chi^2, LR) \leq 1.$$

As discussed in Bahadur (1971), the set E on which $c_1(\boldsymbol{\theta}) = c_2(\boldsymbol{\theta})$ is not yet known precisely, although some of its features are known.

With respect to Pitman ARE, however, the χ^2 and *LR* tests are equivalent. This follows from the equivalence in distribution under the null hypothesis, as we saw in Theorem 4.6.1.

10.7 THE RUBIN–SETHURAMAN "BAYES RISK" EFFICIENCY

Rubin and Sethuraman (1965b) consider efficiency of tests from a *Bayesian* point of view, and define the "*Bayes Risk*" *ARE* of two tests as the limit of the ratio of sample sizes needed to obtain equal Bayes risks. Namely, for a statistical procedure T_n, and for $\varepsilon > 0$, let $N(\varepsilon)$ denote the minimum n_0 such that for sample size $n > n_0$ the Bayes risk of T_n is $< \varepsilon$. Then the Rubin–Sethuraman ARE of a test T_A relative to a test T_B is given by

$$e_{RS}(T_A, T_B) = \lim_{\varepsilon \to 0} \frac{N_B(\varepsilon)}{N_A(\varepsilon)}.$$

By "Bayes risk of T_n" is meant the Bayes risk of the optimal critical region based on T_n.

Illustration. Consider the parameter space $\Theta = R$. Let the null hypothesis be $H_0: \theta = 0$. Let $l(\theta, i)$, $i = 1, 2$, denote the *losses* associated with accepting or rejecting H_0, respectively, when θ is true. Let $f(\theta)$, $\theta \in \Theta$, denote a *prior* distribution on Θ. Then the Bayes risk of a critical region C based on a statistic T_n is

$$B_n(C) = f(0)l(0, 2)P_0(T_n \in C) + \int_{\theta \neq 0} f(\theta)l(\theta, 1)P_\theta(T_n \notin C)d\theta$$

and the "Bayes risk of T_n" is $B_n^*(T) = \inf_C B_n(C)$. Typically, an asymptotically optimal critical region is given by

$$C_n = \{T_n > c(\log n)^{1/2}\},$$

where T_n is normalized to have a nondegenerate limit distribution under H_0. Thus *moderate deviation* probability approximations play a role in evaluating $B_n^*(T)$. Why do we wish to approximate the rate at which $B_n^*(T) \to 0$? Because these approximations enable us to compute $e_{RS}(.,.)$. In typical problems, it is found that $B_n^*(T)$ satisfies

$$B_n^*(T) \sim g(c_T^2 n^{-1}(\log n)), \qquad n \to \infty,$$

where g is a function depending on the problem but not upon the particular procedure T. For two such competing procedures, it thus follows by inverting $g(\cdot)$ that

$$e_{RS}(T_A, T_B) = \frac{c_B^2}{c_A^2}.$$

Moreover, in typical classical problems, this measure $e_{RS}(.,.)$ coincides with the Pitman ARE. Like the Pitman approach, the present approach is "local" in that the (optimal) Bayes procedure based on T places emphasis on "local" alternatives. However, the present approach differs from the Pitman approach in the important respect that the size of the test tends to 0 as $n \to \infty$. (Explain). ■

10.P PROBLEMS

Section 10.2

1. Complete details for proof of Theorem 10.2.1.

2. Provide details for the application of the Berry–Esséen theorem in showing that the sign test statistic considered as T_{3n} in **10.2.2** satisfies the Pitman condition (P1). Check the other conditions (P2)–(P5) also.

3. Show that the Wilcoxon statistic considered as T_{4n} in **10.2.2** satisfies the Pitman conditions.

4. Do the exercises assigned in Examples 10.2.2.
5. Complete details of proof of Theorem 10.2.3.

Section 10.3

6. Show that the likelihood ratio test may be represented as a test statistic of the form of a sum of I.I.D.'s.
7. Complete the details of proof of Theorem 10.3.1.
8. Complete the details of proof of Theorem 10.3.2.
9. Verify Corollary 10.3.1 and also relation (*) in **10.3.2**.
10. Let $\{a_n\}$ and $\{b_n\}$ be sequences of nonnegative constants such that $a_n \to 0$ and $a_n \sim b_n$. Show that $\log a_n \sim \log b_n$.
11. Let $\{a_n\}$ and $\{b_n\}$ be sequences of nonnegative constants such that $a_n \to 0$ and $\log a_n \sim \log b_n$. Does it follow that $a_n \sim b_n$?
12. Supply details for Example 10.3.2A.
13. Supply details for Example 10.3.2B.
14. Supply details for Example 10.3.2C.

Section 10.4

15. Justify that $e_B(,, .)$ is the limit of a ratio $h(n)/n$ of sample sizes for equivalent performance, as asserted in defining e_B.
16. Complete the details of proof of Theorem 10.4.1.
17. Complete details on computation of the slope of the sign test in Example 10.4.4A.

Section 10.5

18. Complete details for Example 10.5.

Section 10.6

19. Consider two tests $\{T_n\}$, $\{T_n^*\}$ having sizes α_n, α_n^* which satisfy

$$\log \alpha_n^* = \log \alpha_n + O(\log n), \qquad n \to \infty.$$

Show that if $\alpha_n \to 0$ faster than any power of n, then the right-hand side is dominated by the term $\log \alpha_n$. Thus, if $\alpha_n \to 0$ faster than any power of n, then $\log \alpha_n^* \sim \log \alpha_n$. Show, however, that this does not imply that $\alpha_n^* \sim \alpha_n$.

20. Verify relation (1) in **10.6.2**.
21. Check that the Bahadur slope of the χ^2 test does not exceed that of the *LR* test, as discussed in **10.6.4**.

Section 10.7

22. Justify the assertion at the conclusion of Example 10.7.

Appendix

1. CONTINUITY THEOREM FOR PROBABILITY FUNCTIONS

If events $\{B_n\}$ are monotone (either $B_1 \subset B_2 \subset \cdots$ or $B_1 \supset B_2 \supset \cdots$) with limit B, then

$$\lim_{n \to \infty} P(B_n) = P(B).$$

2. JENSEN'S INEQUALITY

If $g(\cdot)$ is a convex function on R, and X and $g(X)$ are integrable r.v.'s, then then

$$g(E\{X\}) \leq E(g(X)\}.$$

3. BOREL–CANTELLI LEMMA

(i) For arbitrary events $\{B_n\}$, if $\sum_n P(B_n) < \infty$, then $P(B_n$ infinitely often$) = 0$.
(ii) For independent events $\{B_n\}$, if $\sum_n P(B_n) = \infty$, then $P(B_n$ infinitely often$) = 1$.

4. MINKOWSKI'S INEQUALITY

For $p \geq 1$, and r.v.'s $X_1, \ldots, X_n$,

$$\left[E\left\{\left|\sum_{i=1}^{n} X_i\right|^p\right\}\right]^{1/p} \leq \sum_{i=1}^{n} [E\{|X_i|^p\}]^{1/p}.$$

5. FATOU'S LEMMA

If $X_n \geq 0$ $wp1$, then

$$E\left\{\varliminf_{n\to\infty} X_n\right\} \leq \varliminf_{n\to\infty} E\{X_n\}.$$

6. HELLY'S THEOREMS

(i) Any sequence of nondecreasing functions

$$F_1(x), F_2(x), \ldots, F_n(x), \ldots$$

which are uniformly bounded contains at least one subsequence

$$F_{n_1}(x), F_{n_2}(x), \ldots, F_{n_k}(x), \ldots$$

which converges weakly to some nondecreasing function $F(x)$.

(ii) Let $f(x)$ be a continuous function and let the sequence of nondecreasing uniformly bounded functions

$$F_1(x), F_2(x), \ldots, F_n(x), \ldots$$

converge weakly to the function $F(x)$ on some finite interval $a \leq x \leq b$, where a and b are points of continuity of the function $F(x)$; then

$$\lim_{n\to\infty} \int_a^b f(x)dF_n(x) = \int_a^b f(x)dF(x).$$

(iii) If the function $f(x)$ is continuous and bounded over the entire real line $-\infty < x < \infty$, the sequence of nondecreasing uniformly bounded functions $F_1(x)$, $F_2(x), \ldots$ converges weakly to the function $F(x)$, and $F_n(-\infty) \to F(-\infty)$ and $F_n(+\infty) \to F(+\infty)$, then

$$\lim_{n\to\infty} \int f(x)dF_n(x) = \int f(x)dF(x).$$

7. HOLDER'S INEQUALITY

For $p > 0$ and $q > 0$ such that $1/p + 1/q = 1$, and for random variables X and Y,

$$E|XY| \leq (E|X|^p)^{1/p}(E|Y|^q)^{1/q}.$$

References

Abrahamson, I. G. (1965), "On the stochastic comparison of tests of hypotheses," Ph.D. dissertation, University of Chicago.

Abrahamson, I. G. (1967), "The exact Bahadur efficiencies for the Kolmogorov–Smirnov and Kuiper one- and two-sample statistics," *Ann. Math. Statist.*, **38**, 1475–1490.

Abramowitz, M. and Stegun, I. A. (1965), eds. *Handbook of Mathematical Functions*, National Bureau of Standards, U.S. Government Printing Office, Washington, D.C.

Andrews, D. F., Bickel, P. J., Hampel, F. R., Huber, P. J., Rogers, W. H., and Tukey, J. W. (1972). *Robust Estimates of Location*. Princeton University Press, Princeton, N.J.

Apostol, T. M. (1957), *Mathematical Analysis*, Addison-Wesley, Reading, Mass.

Arvesen, J. N. (1969), "Jackknifing U-statistics," *Ann. Math. Statist.*, **40**, 2076–2100.

Bahadur, R. R. (1960a), "Stochastic comparison of tests," *Ann. Math. Statist.*, **31**, 276–295.

Bahadur, R. R. (1960b), "Simultaneous comparison of the optimum and sign tests of a normal mean," in *Contributions to Prob. and Statist.-Essays in Honor of Harold Hotelling*, Stanford University Press, 79–88.

Bahadur, R. R. (1966). "A note on quantiles in large samples," *Ann. Math. Statist.*, **37**, 577–580.

Bahadur, R. R. (1967), "Rates of convergence of estimates and test statistics," *Ann. Math. Statist.*, **38**, 303–324.

Bahadur, R. R. (1971), *Some Limit Theorems in Statistics*, SIAM, Philadelphia.

Bahadur, R. R. and Ranga Rao, R. (1960). "On deviations of the sample mean," *Ann. Math. Statist.*, **31**, 1015–1027.

Bahadur, R. R. and Zabell, S. L. (1979), "Large deviations of the sample mean in general vector spaces," *Ann. Prob.*, **7**, 587–621.

Basu, D. (1956), "On the concept of asymptotic relative efficiency," *Sankhyā*, **17**, 93–96.

Baum, L. E. and Katz, M. (1965), "Convergence rates in the law of large numbers," *Trans. Amer. Math. Soc.*, **120**, 108–123.

Bennett, C. A. (1952), "Asymptotic properties of ideal linear estimators," Ph.D. dissertation, University of Michigan.

Bennett, G. (1962), "Probability inequalities for the sum of independent random variables," *J. Amer. Statist. Assoc.*, **57**, 33–45.

Beran, R. (1977a), "Robust location estimates," *Ann. Statist.*, **5**, 431–444.

Beran, R. (1977b), "Minimum Hellinger distance estimates for parametric models," *Ann. Statist.*, **5**, 445–463.

Bergström, H. and Puri, M. L. (1977), "Convergence and remainder terms in linear rank statistics," *Ann. Statist.*, **5**, 671–680.

Berk, R. H. (1966), "Limiting behavior of posterior distributions when the model is incorrect," *Ann. Math. Statist.*, **37**, 51–58.

Berk, R. H. (1970), "Consistency a posteriori," *Ann. Math. Statist.*, **41**, 894–907.

Berman, S. M. (1963), "Limiting distribution of the studentized largest observation," *Skand. Aktuarietidskr.*, **46**, 154–161.

Berry, A. C. (1941), "The accuracy of the Gaussian approximation to the sum of independent variables," *Trans. Amer. Math. Soc.*, **49**, 122–136.

Bhapkar, V. P. (1961), "Some tests for categorical data," *Ann. Math. Statist.*, **32**, 72–83.

Bhapkar, V. P. (1966), "A note on the equivalence of two test criteria for hypotheses in categorical data," *J. Amer. Statist. Assoc.*, **61**, 228–236.

Bhattacharya, R. N. (1977), "Refinements of the multidimensional central limit theorem and applications," *Ann. Prob.*, **5**, 1–27.

Bhattacharya, R. N. and Rango Rao, R. (1976). *Normal Approximation and Asymptotic Expansions*, Wiley, New York.

Bickel, P. J. (1974), "Edgeworth expansions in nonparametric statistics," *Ann. Statist.*, **2**, 1–20.

Bickel, P. J., and Doksum, K. A. (1977), *Mathematical Statistics*, Holden-Day, San Francisco.

Bickel, P. J. and Lehmann, E. L. (1975), "Descriptive statistics for nonparametric models. I. Introduction," *Ann. Statist.*, **3**, 1038–1044; "II. Location," *Ann. Statist.*, **3**, 1045–1069.

Bickel, P. J. and Rosenblatt, M. (1973), "On some global measures of the deviations of density function estimates," *Ann. Statist.*, **1**, 1071–1096.

Billingsley, P. (1968), *Convergence of Probability Measures*, Wiley, New York.

Bjerve, S. (1977), "Error bounds for linear combinations of order statistics," *Ann. Statist.*, **5**, 357–369.

Blom, G. (1976), "Some properties of incomplete *U*-statistics," *Biometrika*, **63**, 573–580.

Bönner, N. and Kirschner, H.-P. (1977), "Note on conditions for weak convergence of von Mises' differentiable statistical functions," *Ann. Statist.*, **5**, 405–407.

Boos, D. D. (1977), "The differential approach in statistical theory and robust inference," Ph.D. dissertation, Florida State University.

Boos, D. D. (1979), "A differential for *L*-statistics," *Ann. Statist.*, **7**, 955–959.

Boos, D. D. and Serfling, R. J. (1979), "On Berry–Esséen rates for statistical functions, with application to *L*-estimates," Preprint.

Breiman, L. (1968), *Probability*, Addison-Wesley, Reading, Mass.

Brillinger, D. R. (1969), "An asymptotic representation of the sample distribution function," *Bull. Amer. Math. Soc.*, **75**, 545–547.

Brown, B. M. and Kildea, D. G. (1978), "Reduced *U*-statistics and the Hodges–Lehmann estimator," *Ann. Statist.*, **6**, 828–835.

Callaert, H. and Janssen, P. (1978), "The Berry–Esséen theorem for *U*-statistics," *Ann. Statist.*, **6**, 417–421.

Cantelli, F. P. (1933), "Sulla determinazione empirica delle leggi di probabilita," *Giorn. Inst. Ital. Attuari*, **4**, 421–424.

Carroll, R. J. (1978), "On almost sure expansions for *M*-estimates," *Ann. Statist.*, **6**, 314–318.

Chan, Y. K. and Wierman, J. (1977), "On the Berry–Esséen theorem for *U*-statistics," *Ann. Prob.*, **5**, 136–139.

Cheng, B. (1965), "The limiting distributions of order statistics," *Chinese Math.*, **6**, 84–104.

Chernoff, H. (1952), "A measure of asymptotic efficiency for tests of an hypothesis based on the sum of observations," *Ann. Math. Statist.*, **23**, 493–507.

Chernoff, H. (1954), "On the distribution of the likelihood ratio," *Ann. Math. Statist.*, **25**, 573–578.

Chernoff, H. (1956), "Large-sample theory: parametric case," *Ann. Math. Statist.*, **27**, 1–22.

Chernoff, H., Gastwirth, J. L., and Johns, M. V., Jr. (1967), "Asymptotic distribution of linear combinations of order statistics, with applications to estimation," *Ann. Math. Statist.*, **38**, 52–72.

Chernoff, H. and Savage, I. R. (1958), "Asymptotic normality and efficiency of certain non-parametric test statistics," *Ann. Math. Statist.*, **29**, 972–994.

Chibisov, D. M. (1965), "An investigation of the asymptotic power of the tests of fit," *Th. Prob. Applic.*, **10**, 421–437.

Chung, K. L. (1949), "An estimate concerning the Kolmogorov limit distribution," *Trans. Amer. Math. Soc.*, **67**, 36–50.

Chung, K. L. (1950), *Notes on Limit Theorems*, Columbia Univ. Graduate Mathematical Statistical Society (mimeo).

Chung, K. L. (1974), *A Course in Probability Theory*, 2nd ed., Academic Press, New York.

Clickner, R. P. (1972), "Contributions to rates of convergence and efficiencies in non-parametric statistics," Ph.D. dissertation, Florida State University.

Clickner, R. P. and Sethuraman, J. (1971), "Probabilities of excessive deviations of simple linear rank statistics—the two-sample case," Florida State University Statistics Report M226, Tallahassee.

Collins, J. R. (1976), "Robust estimation of a location parameter in the presence of asymmetry," *Ann. Statist.*, **4**, 68–85.

Cramér, H. (1938), "Sur un nouveau théorème-limite de la théorie des probabilités," *Act. Sci., et Ind.*, **736**, 5–23.

Cramér, H. (1946), *Mathematical Methods of Statistics*, Princeton Univ. Press, Princeton.

Cramér, H. (1970), *Random Variables and Probability Distributions*, 3rd ed., Cambridge Univ. Press, Cambridge.

Cramér, H. and Wold, H. (1936), "Some theorems on distribution functions," *J. London Math. Soc.*, **11**, 290–295.

Csáki, E. (1968), "An iterated logarithm law for semimartingales and its application to empirical distribution function," *Studia Sci. Math. Hung.*, **3**, 287–292.

David, H. A. (1970), *Order Statistics*, Wiley, New York.

Davidson, R. and Lever, W. (1970), "The limiting distribution of the likelihood ratio statistic under a class of local alternatives," *Sankhyā, Ser. A*, **32**, 209–224.

deHaan, L. (1974), "On sample quantities from a regularly varying distribution function," *Ann. Statist.*, **2**, 815–818.

Dieudonné, J. (1960), *Foundations of Modern Analysis*, Wiley, New York.

Dixon, W. J. (1953), "Power functions of the sign test and power efficiency for normal alternatives," *Ann. Math. Statist.*, **24**, 467–473.

Donsker, M. (1951), "An invariance principle for certain probability limit theorems," *Mem. Amer. Math. Soc.*, **6**.

Doob, J. L. (1953), *Stochastic Processes*, Wiley, New York.

Dudley, R. M. (1969), "The speed of mean Glivenko–Cantelli convergence," *Ann. Math. Statist.*, **40**, 40–50.

Dunford, N. and Schwartz, J. T. (1963), *Linear Operators*, Wiley, New York.

Durbin, J. (1973a), *Distribution Theory for Tests Based on the Sample Distribution Function*, SIAM, Philadelphia, PA.

Durbin, J. (1973b), "Weak convergence of the sample distribution function when parameters are estimated," *Ann. Statist.*, **1**, 279–290.

Duttweiler, D. L. (1973), The mean-square error of Bahadur's order-statistic approximation," *Ann. Statist.*, **1**, 446–453.

Dvoretzky, A. Kiefer, J., and Wolfowitz, J. (1956), "Asymptotic minimax character of the sample distribution function and of the classical multinomial estimator," *Ann. Math. Statist.*, **27**, 642–669.

Dwas, M. (1956), "The large-sample power of rank tests in the two-sample problem," *Ann. Math. Statist.*, **27**, 352–374.

Dwass, M. (1964), "Extremal processes," *Ann. Math. Statist.*, **35**, 1718–1725.

Eicker, F. (1966), "On the asymptotic representation of sample quantiles," (Abstract) *Ann. Math. Statist.*, **37**, 1424.

Esséen, C. G. (1945), "Fourier analysis of distribution functions," *Acta Math.*, **77**, 1–125.

Esséen, C. G. (1956), "A moment inequality with an application to the central limit theorem," *Skand. Aktuarietidskr.*, **39**, 160–170.

Feder, P. I. (1968), "On the distribution of the log likelihood ratio test statistic when the true parameter is 'near' the boundaries of the hypothesis regions," *Ann. Math. Statist.*, **39**, 2044–2055.

Feller, W. (1957, 1966), *An Introduction to Probability Theory and Its Applications*, Vol. I (2nd edit.) and Vol. II, Wiley, New York.

Ferguson, T. S. (1958), "A method of generating best asymptotically normal estimates with application to the estimation of bacterial densities," *Ann. Math. Statist.*, **29**, 1046–1062.

Ferguson, T. S. (1967), *Mathematical Statistics: A Decision Theoretic Approach*, Academic Press, New York.

Filippova, A. A. (1962), "Mises' theorem on the asymptotic behavior of functionals of empirical distribution functions and its statistical applications," *Th. Prob. Applic.*, **7**, 24–57.

Finkelstein, H. (1971), "The law of the iterated logarithm for empirical distributions," *Ann. Math. Statist.*, **42**, 607–615.

Fisher, R. A. (1912), "On an absolute criterion for fitting frequency curves," *Mess. of Math.*, **41**, 155.

Fraser, D. A. S. (1957), *Nonparametric Methods in Statistics*, Wiley, New York.

Fréchet, M. (1925), "La notion de differentielle dans l'analyse generale," *Ann. Ecole. Norm. Sup.*, **42**, 293.

Fréchet, M. and Shohat, J. (1931), "A proof of the generalized second limit theorem in the theory of probability," *Trans. Amer. Math. Soc.*, **33**.

Freedman, D. (1971), *Brownian Motion and Diffusion*, Holden-Day, San Francisco.

Funk, G. M. (1970), "The probabilities of moderate deviations of *U*-statistics and excessive deviations of Kolmogorov–Smirnov and Kuiper statistics," Ph.D. dissertation, Michigan State University.

Gaenssler, R. and Stute, W. (1979), "Empirical processes: a survey of results for independent and identically distributed random variables," *Ann. Prob.*, **7**, 193–243.

Galambos, J. (1978), *The Asymptotic Theory of Extreme Order Statistics*, Wiley, New York.

Gastwirth, J. L. (1966). "On robust procedures," *J. Amer. Statist. Assoc.*, **61**, 929–948.

Geertsema, J. C. (1970), "Sequential confidence intervals based on rank tests," *Ann. Math. Statist.*, **41**, 1016–1026.

Ghosh, J. K. (1971), "A new proof of the Bahadur representation of quantiles and an application," *Ann. Math. Statist.*, **42**, 1957–1961.

Glivenko, V. (1933), "Sulla determinazione empirica della legge di probabilita," *Giorn. Inst. Ital. Attuari*, **4**, 92–99.

Gnedenko, B. V. (1943), "Sur la distribution limite du terme maximum d'un serie aléatoire," *Ann. Math.*, **44**, 423–453.

Gnedenko, B. V. (1962), *The Theory of Probability*, 4th ed., Chelsea, New York.

Gnedenko, B. V. and Kolmogorov, A. N. (1954), *Limit Distribution for Sums of Independent Random Variables*, Addison–Wesley, Reading, Mass.

Grams, W. F. and Serfling, R. J. (1973), "Convergence rates for *U*-statistics and related statistics," *Ann. Statist.*, **1**, 153–160.

Gregory, G. G. (1977), "Large sample theory for *U*-statistics and test of fit," *Ann. Statist.*, **5**, 110–123.

Gupta, S. S. and Panchapakesan, S. (1974), "Inference for restricted families: (A) multiple decision procedures, (B) order statistics inequalities," *Proc. Conference on Reliability and Biometry*, Tallahassee (ed. by F. Proschan and R. J. Serfling), SIAM, 503–596.

Hájek, J. (1961), "Some extensions of the Wald–Wolfowitz–Noether theorem," *Ann. Math. Statist.*, **32**, 506–523.

Hájek, J. (1962), "Asymptotically most powerful rank-tests," *Ann. Math. Statist.*, **33**, 1124–1147.

Hájek, J. (1968), "Asymptotic normality of simple linear rank statistics under alternatives," *Ann. Math. Statist.*, **39**, 325–346.

Hájek, J. and Šidák, Z. (1967), *Theory of Rank Tests*, Academic Press, New York.

Hall, W. J., Kielson, J. and Simons, G. (1971), "Conversion of convergence in the mean to almost sure convergence by smoothing," Technical Report, Center for System Science, Univ. of Rochester, Rochester, N.Y.

Halmos, P. R. (1946), "The theory of unbiased estimation," *Ann. Math. Statist.*, **17**, 34–43.

Halmos, P. R. (1950), *Measure Theory*, Van Nostrand, Princeton.

Hampel, F. R. (1968), "Contributions to the theory of robust estimation," Ph.D. dissertation, Univ. of California.

Hampel, F. R. (1974), "The influence curve and ts role in robust estimation," *J. Amer. Statist. Assoc.*, **69**, 383–397.

Hardy, G. H. (1952), *A Course of Pure Mathematics*, 10th ed., Cambridge University Press, New York.

Hartman, P. and Wintner, A. (1941), "On the law of the iterated logarithm," *Amer. J. Math.*, **63**, 169–176.

Healy, M. J. R. (1968), "Disciplining of medical data," *British Med. Bull.*, **24**, 210–214.

Helmers, R. (1977a), "The order of the normal approximation for linear combinations of order statistics with smooth weight functions," *Ann. Prob.*, **5**, 940–953.

Helmers, R. (1977b), "A Berry–Esséen theorem for linear combinations of order statistics," Preprint.

Hettmansperger, T. P. (1973), "On the Hodges–Lehmann approximate efficiency," *Ann. Inst. Statist. Math.*, **25**, 279–286.

Hoadley, A. B. (1967), "On the probability of large deviations of functions of several empirical cdf's," *Ann. Math. Statist.*, **38**, 360–381.

Hodges, J. L., Jr. and Lehmann, E. L. (1956), "The efficiency of some nonparametric competitors of the t-test," *Ann. Math. Statist.*, **27**, 324–335.

Hodges, J. L., Jr., and Lehmann, E. L. (1961), "Comparison of the normal scores and Wilcoxon tests," *Proc. 4th Berk. Symp. on Math. Statist. and Prob.*, Vol. 1, 307–317.

Hoeffding, W. (1948), "A class of statistics with asymptotically normal distribution," *Ann. Math. Statist.*, **19**, 293–325.

Hoeffding, W. (1961), "The strong law of large numbers for U-statistics," Univ. of North Carolina Institute of Statistics Mimeo Series, No. 302.

Hoeffding, W. (1963), "Probability inequalities for sums of bounded random variables," *J. Amer. Statist. Assoc.*, **58**, 13–30.

Hoeffding, W. (1965), "Asymptotically optimal tests for multinomial distributions" (with discussion), *Ann. Math. Statist.*, **36**, 369–408.

Hoeffding, W. (1973), "On the centering of a simple linear rank statistic," *Ann. Statist.*, **1**, 54–66.

Hotelling, H. and Pabst, M. R. (1936), "Rank correlation and tests of significance involving no assumption of normality," *Ann. Math. Statist.*, **7**, 29–43.

Hsu, P. L. (1945), "The limiting distribution of functions of sample means and application to testing hypotheses," *Proc. First Berkeley Symp. on Math. Statist. and Prob.*, 359–402.

Hsu, P. L. and Robbins, H. (1947), "Complete convergence and the law of large numbers," *Proc. Nat. Acad. Sci. U.S.A.*, **33**, 25–31.

Huber, P. J. (1964), "Robust estimation of a location parameter," *Ann. Math. Statist.*, **35**, 73–101.

Huber, P. J. (1972), "Robust statistics: a review," *Ann. Math. Statist.*, **43**, 1041–1067.

Huber, P. J. (1973), "Robust regression: Asymptotics, conjectures and Monte Carlo," *Ann. Statist.*, **1**, 799–821.

Huber, P. J. (1977), *Robust Statistical Procedures*, SIAM, Philadelphia.

Ibragimov, I. A. and Linnik, Yu. V. (1971), *Independent and Stationary Sequences of Random Variables*, Wolters-Noordhoff, Groningen, Netherlands.

Jaeckel, L. A. (1971), "Robust estimates of location: symmetry and asymmetric contamination," *Ann. Math. Statist.*, **42**, 1020–1034.

James, B. R. (1975), "A functional law of the iterated logarithm for weighted empirical distributions," *Ann. Prob.*, **3**, 762–772.

Jones, D. H. and Sethuraman, J. (1978), "Bahadur efficiencies of the Student's t-tests," *Ann. Statist.*, **6**, 559–566.

Jung, J. (1955), "On linear estimates defined by a continuous weight function," *Arkiv. fur Matematik*, **3**, 199–209.

Jurečková, J. and Puri, M. L. (1975), "Order of normal approximation for rank test statistic distribution," *Ann. Prob.*, **3**, 526–533.

Kawata, T. (1951), "Limit distributions of single order statistics," *Rep. Stat. Appl. Res. Union of Jap. Scientists and Engineers*, **1**, 4–9.

Kiefer, J. (1961), "On large deviations of the empiric D. F. of vector chance variables and a law of iterated logarithm," *Pacific J. Math.*, **11**, 649–660.

Kiefer, J. (1967), "On Bahadur's representation of sample quantiles," *Ann. Math. Statist.*, **38**, 1323–1342.

Kiefer, J. (1970a), "Deviations between the sample quantile process and the sample df," *Proc. Conference on Nonparametric Techniques in Statistical Inference*, Bloomington (ed. by M. L. Puri), Cambridge University Press, 299–319.

Kiefer, J. (1970b), "Old and new methods for studying order statistics and sample quantiles," **ibid.**, 349–357.

Kiefer, J. and Wolfowitz, J. (1958), "On the deviations of the empiric distribution function of vector chance variables," *Trans. Amer. Math. Soc.*, **87**, 173–186.

Kingman, J. F. C. and Taylor, S. J. (1966), *Introduction to Measure and Probability*, Cambridge University Press, Cambridge.

Klotz, J. (1965), "Alternative efficiencies for signed rank tests," *Ann. Math. Statist.*, **36**, 1759–1766.

Kolmogorov, A. N. (1929), "Uber das Gesetz des iterierten Logarithmus," *Math. Annalen*, **101**, 126–136.

Kolmogorov, A. N. (1933), "Sulla determinazione empirica di una legge di distribuzione," *Giorn. Inst. Ital. Attuari*, **4**, 83–91.

Komlós, J., Major, P. and Tusnády, G. (1975), "An approximation of partial sums of independent rv's, and the sample df. I." *Z. Wahrscheinlichkeitstheorie und verw. Gebiete*, **32**, 111–131.

Lai, T. Z. (1975), "On Chernoff–Savage statistics and sequential rank tests," *Ann. Statist.*, **3**, 825–845.

Lai, T. L. (1977), "Power-one tests based on sample sums," *Ann. Statist.*, **5**, 866–880.

Lamperti, J. (1964), "On extreme order statistics," *Ann. Math. Statist.*, **35**, 1726–1737.

Lamperti, J. (1966), *Probability*, Benjamin, New York.

Landers, D. and Rogge, L. (1976), "The exact approximation order in the central limit theorem for random summation," *Z. Wahrscheinlichkeitstheorie und. Verw. Gebiete*, **36**, 269–283.

LaCam, L. (1960), "Locally asymptotically normal families of distributions," *Univ. of Calif. Public. in Statist.*, **3**, 37–98.

Lehmann, E. L. (1951), "Consistency and unbiasedness of certain nonparametric tests," *Ann. Math. Statist.*, **22**, 165–179.

Liapounoff, A. (1900), "Sur une proposition de la théorie des probabilités," *Bull. Acad. Sci. St. Pétersbourg*, **13**, 359–386.

Liapounoff, A. (1901), "Nouvelle forme du théorème sur la limite de probabilité," *Mém. Acad. Sci. St. Pétersbourg*, **12**, No. 5.

Lindgren, B. W. (1968), *Statistical Theory*, 2nd ed., Macmillan, Toronto.

Lloyd, E. H. (1952), "Least-squares estimation of location and scale parameters using order statistics," *Biometrika*, **34**, 41–67.

Loève, M. (1977), *Probability Theory I*, 4th ed., Springer-Verlag, New York.

Loève, M. (1978), *Probability Theory II*, 4th ed., Springer-Verlag, New York.

Loynes, R. M. (1970), "An invariance principle for reversed martingales," *Proc. Amer. Math. Soc.*, **25**, 56–64.

Luenberger, D. G. (1969), *Optimization by Vector Space Methods*, Wiley, New York.

Lukacs, E. (1970), *Characteristic Functions*, 2nd ed., Hafner, New York.

Mann, N. R., Schaefer, R. E., and Singpurwalla, N. D. (1974), *Methods for Statistical Analysis of Reliability and Life Data*, Wiley, New York.

Marcinkiewicz, J and Zygmund, A. (1937), "Sur les fonctions indépendantes," *Fund. Math.*, **29**, 60–90.

Mejzler, D. and Weissman, I. (1969), "On some results of N. V. Smirnov, concerning limit distributions for variational series," *Ann. Math. Statist.*, **40**, 480–491.

Miller, R. G., Jr. and Sen, P. K. (1972), "Weak convergence of U-statistics and von Mises' differentiable statistical functions," *Ann. Math. Statist.*, **43**, 31–41.

Nandi, H. K. and Sen, P. K. (1963), "On the properties of U-statistics when the observations are not independent. Part II. Unbiased estimation of the parameters of a finite population," *Cal. Statist. Assoc. Bull.*, **12**, 125–143.

Nashed, M. Z. (1971), "Differentiability and related properties of nonlinear operators: some aspects of the role of differentials in nonlinear functional analysis," in *Nonlinear Functional Analysis and Applications* (edit, by L. B. Rall), Academic Press, New York, 103–309.

Natanson, I. P. (1961), *Theory of Functions of a Real Variable*, Vol. I, rev. ed., Ungar, New York.

Neuhaus, G. (1976), "Weak convergence under contiguous alternatives of the empirical process when parameters are estimated: the D_k approach," *Proc. Conf. on Empirical Distributions and Processes*, Oberwolfach (ed. by P. Gaenssler and P. Révész), Springer-Verlag, 68–82.

Neyman, J. (1949), "Contributions to the theory of the χ^2 test," *First Berk. Symp. on Math. Statist. and Prob.*, 239–273.

Newman, J. and Pearson, E. S. (1928), "On the use and interpretation of certain test criteria for purposes of statistical inference," *Biometrika*, **20A**, 175–240 and 263–294.

Noether, G. E. (1955), "On a theorem of Pitman," *Ann. Math. Statist.*, **26**, 64–68.

Noether, G. E. (1967), *Elements of Nonparametric Statistics*, Wiley, New York.

Ogasawara, T. and Takehashi, M. (1951), "Independence of quadratic forms in normal system," *J. Sci. Hiroshima Univ.*, **15**, 1–9.

O'Reilly, N. E. (1974), "On the weak convergence of empirical processes in sup-norm metrics," *Ann. Prob.*, **4**, 642–651.

Parthasarathy, K. P. (1967), *Probability Measures on Metric Spaces*, Academic Press, New York.

Pearson, K. (1894), "Contributions to the mathematical theory of evolution," *Phil. Trans. Roy. Soc., London, A*, **185**, 71–78.

Petrov, V. V. (1966), "On the relation between an estimate of the remainder in the central limit theorem and the law of the iterated logarithm," *Th. Prob. Applic.*, **11**, 454–458.

Petrov, V. V. (1971), "A theorem on the law of the iterated logarithm," *Th. Prob. Applic.*, **16**, 700–702.

Pitman, E. J. G. (1949), "Lecture Notes on Nonparametric Statistical Inference," Columbia University.

Portnoy, S. L. (1977), "Robust estimation in dependent situations," *Ann. Statist.*, **5**, 22–43.

Puri, M. L. and Sen, P. K. (1971), *Nonparametric Methods in Multivariate Analysis*, Wiley, New York.

Pyke, R. (1965), "Spacings," *J. Roy Statist. Soc. (B)*, **27**, 395–436 (with discussion), 437–449.

Pyke, R. (1972), "Spacings revisited," *Proc. 6th Berk. Symp. on Math. Statist. and Prob.*, Vol. I, 417–427.

Pyke, R. and Shorack, G. (1968), "Weak convergence of a two-sample empirical process and a new approach to Chernoff-Savage theorems," *Ann. Math. Statist.*, **39**, 755–771.

Raghavachari, M. (1973), "Limiting distributions of Kolmogorov-Smirnov type statistics under the alternative," *Ann. Statist.*, **1**, 67–73.

Ranga Rao, R. (1962), "Relations between weak and uniform convergence of measures with applications," *Ann. Math. Statist.*, **33**, 659–681.

Rao, C. R. (1947), "Large sample tests of statistical hypotheses concerning several parameters with applications to problems of estimation," *Proc. Comb. Phil. Soc.*, **44**, 50–57.

Rao, C. R. (1973), *Linear Statistical Inference and Its Applications*, 2nd ed., Wiley, New York.

Rao, J. S. and Sethuraman, J. (1975), "Weak convergence of empirical distribution functions of random variables subject to perturbations and scale factors," *Ann. Statist.*, **3**, 299–313.

Reiss, R.-D. (1974), "On the accuracy of the normal approximation for quantiles," *Ann. Prob.*, **2**, 741–744.

Renyi, A. (1953), "On the theory of order statistics," *Acta Math. Acad. Sci. Hung.*, **4**, 191–231.

Révész, P. (1968), *The Laws of Large Numbers*, Academic Press, New York.

Richter, H. (1974), "Das Gesetz vom iterierten Logarithmus für empirische Verteilungsfunktionen im R^k," *Manuscripta math.*, **11**, 291–303.

Robbins, H. (1970), "Statistical methods related to the law of the iterated logarithm," *Ann. Math. Statist.*, **41**, 1397–1409.

Robbins, H. and Siegmund, D. (1973), "Statistical tests of power one and the integral representation of solutions of certain partial differential equations," *Bull. Inst. Math., Acad. Sinica*, **1**, 93–120.

Robbins, H. and Siegmund, D. (1974), "The expected sample size of some tests of power one," *Ann. Statist.*, **2**, 415–436.

Rosén, B. (1969), "A note on asymptotic normality of sums of higher-dimensionally indexed random variables," *Ark. for Mat.*, **8**, 33–43.

Rosenblatt, M. (1971), "Curve estimates," *Ann. Math. Statist.*, **42**, 1815–1842.

Roussas, G. G. (1972), *Contiguity of Probability Measures: Some Applications in Statistics*, Cambridge University Press, Cambridge.

Royden, H. L. (1968), *Real Analysis*, 2nd ed., Macmillan, New York.

Rubin, H. (1961), "The estimation of discontinuities in multivariate densities, and related problems in stochastic processes," *Proc. Fourth Berk. Symp. on Math. Statist. and Prob., Vol. 1*, 563–574.

Rubin, H. and Sethuraman, J. (1965a), "Probabilities of moderate deviations," *Sankhyā*, **27A**, 325–346.

Rubin, H. and Sethuraman, J. (1965b), "Bayes risk efficiency," *Sankhyā*, **27A**, 347–356.

Rubin, H. and Vitale, R. A. (1980), "Asymptotic distribution of symmetric statistics," *Ann. Statist.*, **8**, 165–170.

Sanov, I. N. (1957), "On the probability of large deviations of random variables," *Sel. Transl. Math. Statist. Prob.*, **1**, 213–244.

Sarhan, A. E. and Greenberg, E. G. (1962), eds., *Contributions to Order Statistics*, Wiley, New York.

Savage, I. R. (1969), "Nonparametric statistics: a personal review," *Sankhyā A*, **31**, 107–143.

Scheffé, H. (1947), "A useful convergence theorem for probability distributions," *Ann. Math. Statist.*, **18**, 434–438.

Schmid, P. (1958), "On the Kolmogorov and Smirnov limit theorems for discontinuous distribution functions," *Ann. Math. Statist.*, **29**, 1011–1027.

Sen, P. K. (1960), "On some convergence properties of U-statistics," *Cal. Statist. Assoc. Bull.*, **10**, 1–18.

Sen, P. K. (1963), "On the properties of U-statistics when the observations are not independent. Part one: Estimation of the non-serial parameters of a stationary process," *Cal. Statist. Assoc. Bull.*, **12**, 69–92.

Sen, P. K. (1965), "Some nonparametric tests for m-dependent time series," *J. Amer. Statist. Assoc.*, **60**, 134–147.

Sen, P. K. (1967), "U-statistics and combination of independent estimators of regular functionals," *Cal. Statist. Assoc. Bull.*, **16**, 1–14.

Sen, P. K. (1974), "Almost sure behavior of U-statistics and von Mises' differential statistical functions," *Ann. Statist.*, **2**, 387–395.

Sen, P. K. (1977), "Almost sure convergence of generalized U-statistics," *Ann. Prob.*, **5**, 287–290.

Sen, P. K. and Ghosh, M. (1971), "On bounded length sequential confidence intervals based on one-sample rank-order statistics," *Ann. Math. Statist.*, **42**, 189–203.

Serfling, R. J. (1968), "The Wilcoxon two-sample statistic on strongly mixing processes," *Ann. Math. Statist.*, **39**, 1202–1209.

Serfling, R. J. (1970), "Convergence properties of S_n under moment restrictions," *Ann. Math. Statist.*, **41**, 1235–1248.

Serfling, R. J. (1974), "Probability inequalities for the sum in sampling without replacement," *Ann. Statist.*, **2**, 39–48.

Serfling, R. J. and Wackerly, D. D. (1976), "Asymptotic theory of sequential fixed-width confidence intervals," *J. Amer. Statist. Assoc.*, **71**, 949–955.

Shapiro, C. P. and Hupert, L. (1979), "Asymptotic normality of permutation statistics derived from weighted sums of bivariate functions," *Ann. Statist.*, **7**, 788–794.

Shorack, G. R. (1969), "Asymptotic normality of linear combinations of functions of order statistics," *Ann. Math. Statist.*, **40**, 2041–2050.

Shorack, G. R. (1972), "Functions of order statistics," *Ann. Math. Statist.*, **43**, 412–427.

Sievers, G. L. (1978), "Weighted rank statistics for simple linear regression," *J. Amer. Statist. Assoc.*, **73**, 628–631.

Silverman, B. (1978), "Weak and strong uniform consistency of the kernel estimate of a density and its derivatives," *Ann. Statist.*, **6**, 177–184.

Simons, G. (1971), "Identifying probability limits," *Ann. Math. Statist.*, **42**, 1429–1433.

Skorokhod, A. V. (1956), "Limit theorems for stochastic processes," *Th. Prob. Applic.*, **1**, 261–290.

Slutsky, E. (1925), "Über stochastiche Asymptoter und Grenzwerte," *Math. Annalen*, **5**, 93.

Smirnov, N. V. (1944), "An approximation to the distribution laws of random quantities determined by empirical data," *Uspehi Mat. Nauk*, **10**, 179–206.

Smirnov, N. V. (1952), *Limit Distributions for the Terms of a Variational Series*, Amer. Math. Soc., Translation No. 67.

Soong, T. T. (1969), "An extension of the moment method in statistical estimation," *SIAM J. Applied Math.*, **17**, 560–568.

Sproule, R. N. (1969a), "A sequential fixed-width confidence interval for the mean of a U-statistic," Ph.D. dissertation, Univ. of North Carolina.

Sproule, R. N. (1969b), "Some asymptotic properties of U-statistics," (Abstract) *Ann. Math. Statist.*, **40**, 1879.

Stigler, S. M. (1969), "Linear functions of order statistics," *Ann. Math. Statist.*, **40**, 770–788.

Stigler, S. M. (1973), "The asymptotic distribution of the trimmed mean," *Ann. Statist.*, **1**, 472–477.

Stigler, S. M. (1974), "Linear functions of order statistics with smooth weight functions," *Ann. Statist.*, **2**, 676–693; (1979), "Correction note," *Ann. Statist.*, **7**, 466.

Stout, W. F. (1970a), "The Hartman–Wintner law of the iterated logarithm for martingales," *Ann. Math. Statist.*, **41**, 2158–2160.

Stout, W. F. (1970b), "A martingale analogue of Kolmogorov's law of the iterated logarithm," *Z. Wahrscheinlichkeitstheorie und. Verw. Gebiete*, **15**, 279–290.

Stout, W. F. (1974), *Almost Sure Convergence*, Academic Press, New York.

Uspensky, J. V. (1937), *Introduction to Mathematical Probability*, McGraw-Hill, New York.

van Beeck, P. (1972), "An application of Fourier methods to the problem of sharpening the Berry–Esséen inequality," *Z. Wahrscheinlichkeitstheorie und Verw. Gebiete*, **23**, 187–196.

van Eeden, C. (1963), "The relation between Pitman's asymptotic relative efficiency of two tests and the correlation coefficient between their test statistics," *Ann. Math. Statist.*, **34**, 1442–1451.

Varadarajan, V. S. (1958), "A useful convergence theorem," *Sankhyā*, **20**, 221–222.

von Mises, R. (1947), "On the asymptotic distribution of differentiable statistical functions," *Ann. Math. Statist.*, **18**, 309–348.

von Mises, R. (1964), *Mathematical Theory of Probability and Statistics* (edited and complemented by Hilda Geiringer), Academic Press, New York.

Wald, A. (1943), "Tests of statistical hypotheses concerning several parameters when the number of observations is large," *Trans. Amer. Math. Soc.*, **54**, 426–482.

Wald, A. (1949), "Note on the consistency of the maximum likelihood estimate," *Ann. Math. Statist.*, **20**, 595–601.

Wald, A. and Wolfowitz, J. (1943), "An exact test of randomness in the nonparametric case based on serial correlation," *Ann. Math. Statist.*, **14**, 378–388.

Wald, A. and Wolfowitz, J. (1944), "Statistical tests based on permutations of the observations," *Ann. Math. Statist.*, **15**, 358–372.

Watts, J. H. V. (1977), "Limit theorems and representations for order statistics from dependent sequences," Ph.D. dissertation, Univ. of North Carolina.

Wellner, J. A. (1977a), "A Glivenko–Cantelli theorem and strong laws of large numbers for functions of order statistics," *Ann. Statist.* **5**, 473–480 (correction note, *Ann. Statist.*, **6**, 1394).

Wellner, J. A. (1977b), "A law of the iterated logarithm for functions of order statistics," *Ann. Statist.*, **5**, 481–494.

Wilks, S. S. (1938), "The large-sample distribution of the likelihood ratio for testing composite hypothesis," *Ann. Math. Statist.*, **9**, 60–62.

Wilks, S. S. (1948), "Order statistics," *Bull. Amer. Math. Soc.*, **5**, 6–50.

Wilks, S. S. (1962), *Mathematical Statistics*, Wiley, New York.

Wood, C. L. (1975), "Weak convergence of a modified empirical stochastic process with applications to Kolmogorov–Smirnov statistics," Ph.D. dissertation, Florida State University.

Woodworth, G. C. (1970), "Large deviations and Bahadur efficiencies of linear rank statistics," *Ann. Math. Statist.*, **41**, 251–283.

Zolotarev, M. (1967), "A sharpening of the inequality of Berry–Esséen," *Z. Wahrscheinlichkeitstheorie und Verw. Gebiete*, **8**, 332–342.

Author Index

Note: Following each author's name are listed the subsections containing references to the author.

Subject Index

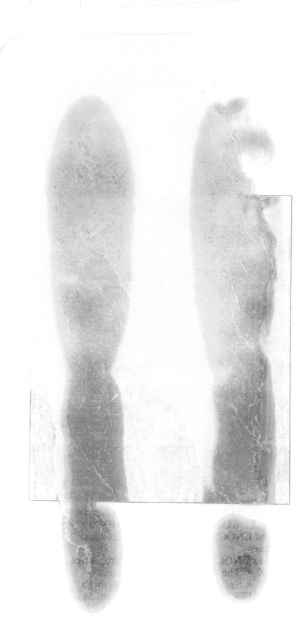

Applied Probability and Statistics (Continued)

FLEISS • Statistical Methods for Rates and Proportions, *Second Edition*
GALAMBOS • The Asymptotic Theory of Extreme Order Statistics
GIBBONS, OLKIN, and SOBEL • Selecting and Ordering Populations: A New Statistical Methodology
GNANADESIKAN • Methods for Statistical Data Analysis of Multivariate Observations
GOLDBERGER • Econometric Theory
GOLDSTEIN and DILLON • Discrete Discriminant Analysis
GROSS and CLARK • Survival Distributions: Reliability Applications in the Biomedical Sciences
GROSS and HARRIS • Fundamentals of Queueing Theory
GUPTA and PANCHAPAKESAN • Multiple Decision Procedures: Theory and Methodology of Selecting and Ranking Populations
GUTTMAN, WILKS, and HUNTER • Introductory Engineering Statistics, *Second Edition*
HAHN and SHAPIRO • Statistical Models in Engineering
HALD • Statistical Tables and Formulas
HALD • Statistical Theory with Engineering Applications
HARTIGAN • Clustering Algorithms
HILDEBRAND, LAING, and ROSENTHAL • Prediction Analysis of Cross Classifications
HOEL • Elementary Statistics, *Fourth Edition*
HOLLANDER and WOLFE • Nonparametric Statistical Methods
JAGERS • Branching Processes with Biological Applications
JESSEN • Statistical Survey Techniques
JOHNSON and KOTZ • Distributions in Statistics
Discrete Distributions
Continuous Univariate Distributions—1
Continuous Univariate Distributions—2
Continuous Multivariate Distributions
JOHNSON and KOTZ • Urn Models and Their Application: An Approach to Modern Discrete Probability Theory
JOHNSON and LEONE • Statistics and Experimental Design in Engineering and the Physical Sciences, Volumes I and II, *Second Edition*
JUDGE, GRIFFITHS, HILL and LEE • The Theory and Practice of Econometrics
KALBFLEISCH and PRENTICE • The Statistical Analysis of Failure Time Data
KEENEY and RAIFFA • Decisions with Multiple Objectives
LANCASTER • An Introduction to Medical Statistics
LEAMER • Specification Searches: Ad Hoc Inference with Nonexperimental Data
McNEIL • Interactive Data Analysis
MANN, SCHAFER and SINGPURWALLA • Methods for Statistical Analysis of Reliability and Life Data
MEYER • Data Analysis for Scientists and Engineers
MILLER, EFRON, BROWN, and MOSES • Biostatistics Casebook
OTNES and ENOCHSON • Applied Time Series Analysis: Volume I, Basic Techniques
OTNES and ENOCHSON • Digital Time Series Analysis
POLLOCK • The Algebra of Econometrics
PRENTER • Splines and Variational Methods
RAO and MITRA • Generalized Inverse of Matrices and Its Applications
SCHUSS • Theory and Applications of Stochastic Differential Equations
SEAL • Survival Probabilities: The Goal of Risk Theory
SEARLE • Linear Models
SPRINGER • The Algebra of Random Variables
UPTON • The Analysis of Cross-Tabulated Data
WEISBERG • Applied Linear Regression
WHITTLE • Optimization Under Constraints